ANALYSIS IN EUCLIDEAN SPACE

Kenneth Hoffman
Professor Emeritus
Massachusetts Institute of Technology

DOVER PUBLICATIONS, INC.
Mineola, New York

Bibliographical Note

This Dover edition, first published in 2007, is an unabridged republication of the work originally published by Prentice-Hall, Inc., Englewood Cliffs, N.J., in 1975.

Library of Congress Cataloging-in-Publication Data

Hoffman, Kenneth. $l \infty S_4 \in l_4 2 \tau \tau$
   Analysis in Euclidean space / Kenneth Hoffman — Dover ed.
      p. cm.
   Originally published: Englewood Cliffs, N.J. : Prentice-Hall, 1975.
   Includes bibliographical references and index.
   ISBN 0-486-45804-0 (pbk.)
    1. Mathematical analysis. 2. Algebraic spaces. I. Title.

QA300.H63 2007
515—dc22

2006053462

Manufactured in the United States of America
Dover Publications, Inc., 31 East 2nd Street, Mineola, N.Y. 11501

Contents

To my parents

Preface

This textbook has been developed for use in the two-semester introductory course in mathematical analysis at the Massachusetts Institute of Technology. The aim of the course is to introduce the student to basic concepts, principles, and methods of mathematical analysis.

The presumed mathematical background of the students is a solid calculus course covering one and (some of) several variables, plus (perhaps) elementary differential equations and linear algebra. The linear algebra background is not necessary until the second semester, since it enters the early chapters only through certain examples and exercises which utilize matrices. At M.I.T. the introductory calculus course is condensed into one year, after which the student has available a one-semester course in differential equations and linear algebra. Thus, over half the students in the course are sophomores. Since many students enter M.I.T. having had a serious calculus course in high school, there are quite a few freshmen in the course. The remainder of the students tend to be juniors, seniors, or graduate students in fields such as physics or electrical engineering. Since very little prior experience with rigorous mathematical thought is assumed, it has been our custom to augment the lectures by structured tutorial sessions designed to help the students in learning to deal with precise mathematical definitions and proofs. It is to be expected that at many institutions the text would be suitable for a junior, senior, or graduate course in analysis, since it does assume a considerable technical facility with elementary mathematics as well as an affinity for mathematical thought.

The presentation differs from that found in existing texts in two ways. First, a concerted effort is made to keep the introductions to real and complex analysis close together. These subjects have been separated in the curriculum for a number of years, thus tending to delay the introduction to complex function theory. Second, the generalizations beyond R^n are presented for subsets of normed linear spaces rather than for metric spaces. The pedagogical advantage of this is that the original material can be developed on the familiar terrain of Euclidean space and then simply observed to be largely valid for normed linear spaces, where the symbolism is just like that of R^n. The students are prepared for the generalization in much the same way that high-school algebra prepares one for manipulation in a commutative ring with identity.

The first semester covers the bulk of the first five chapters. It emphasizes the Four C's: completeness, convergence, compactness, and continuity. The basic results are presented for subsets of and functions on Euclidean space of n dimensions. This presentation includes (of course) a rigorous review of the intellectual skeleton of calculus, placing greater emphasis on power series expansions than one normally can in a calculus course. The discussion proceeds (in Chapter 5) into complex power series and an introduction to the theory of complex-analytic functions. The review of linear geometry in Section 1.6 is usually omitted from the formal structure of the first semester. The instructor who is pressed for time or who is predisposed to separate real and complex analysis may also omit all or part of Sections 5.5–5.10 on analytic functions and Fourier series without interrupting the flow of the remainder of the text.

The second semester begins with Chapter 6. It reviews the main results of the first semester, the review being carried out in the context of (subsets of and functions on) normed linear spaces. The author has found that the student is readily able to absorb the fact that many of the arguments he or she has been exposed to are formal and are therefore valid in the more general context. It is then emphasized that two of the most crucial results from the first semester—the completeness of R^n and the Heine-Borel theorem—depend on finite-dimensionality. This leads naturally to a discussion of (i) complete (Banach) spaces, the Baire category theorem and fixed points of contractions, and (ii) compact subsets of various normed linear spaces, in particular, equicontinuity and Ascoli's theorem. From there the course moves to the Lebesgue integral on R^n, which is developed by completing the space of continuous functions of compact support. Most of the basic properties of integral and measure are discussed, and a short presentation of orthogonal expansions (especially Fourier series) is included. The final chapter of the notes deals with differentiable maps on R^n, the implicit and inverse function theorems, and the change of variable theorem. This chapter may be presented earlier if the instructor finds it desirable, since the only dependence on Lebesgue integration is the proof of the change of variable theorem.

A few final remarks. Some mathematicians will look at these notes and say, "How can you teach an introductory course in analysis which never mentions partial differential equations or calculus of variations?" Others will ask, "How can you teach a basic course in analysis which devotes so little attention to applications, either to mathematics or to other fields of science?" The answer is that there is no such thing as *the* introductory course in analysis. The subject is too large and too important to allow for that. The three most viable foci for organization of an introductory course seem to be (i) emphasis on general concepts and principles, (ii) emphasis on hard mathematical analysis (the source of the general ideas), and (iii) emphasis on applications to science and engineering. This text was developed for the first type of course. It can be very valuable for a certain category of students, principally the students going on to graduate school in mathematics, physics, or (abstract) electrical engineering, etc. It is not, and was not intended to be, right for all students who may need some advanced calculus or analysis beyond the elementary level.

Thanks are due to many people who have contributed to the development of this text over the last eight years. Colleagues too numerous to mention used the classroom notes and pointed out errors or suggested improvements. Three must be singled out: Steven Minsker, David Ragozin, and Donald Wilken. Each of them assisted the author in improving the notes and managing the pedagogical affairs of the M.I.T. course. I am especially grateful to David Ragozin, who wrote an intermediate version of the chapter on Lebesgue integration. I am indebted to Mrs. Sophia Koulouras, who typed the original notes, and to Miss Viola Wiley, who typed the revision and the final manuscript. Finally, my thanks to Art Wester and the staff of Prentice-Hall, Inc.

KENNETH HOFFMAN

Preface
to the Student

This textbook will introduce you to many of the general principles of mathematical analysis. It assumes that you have a mathematical background which includes a solid course (at least one year) in the calculus of functions of one and several variables, as well as a short course in differential equations. It would be helpful if you have been exposed to introductory linear algebra since many of the exercises and examples involve matrices. The material necessary for following these exercises and examples is summarized in Section 1.6, but a linear algebra background is not essential for reading the book since it does not enter into the logical development in the text until Chapter 6.

You will meet a large number of concepts which are new to you, and you will be challenged to understand their precise definitions, some of their uses, and their general significance. In order to understand the meaning of this in quantitative terms, thumb through the Index and see how many of the terms listed there you can describe precisely. But it is the qualitative impact of the definitions which will loom largest in your experience with this book. You may find that you are having difficulty following the "proofs" presented in the book or even in understanding what a "proof" is. When this happens, look to the definitions because the chances are that your real difficulty lies in the fact that you have only a hazy understanding of the definitions of basic concepts or are suffering from a lack of familiarity with definitions which mean exactly what they say, nothing less and nothing more.

You will also learn a lot of rich and beautiful mathematics. To make the learning task more manageable, the notes have been provided with supplementary material and mechanisms which you should utilize:

1. *Appendix:* Note that the text proper is followed by an Appendix which discusses sets, functions, and a bit about cardinality (finite, infinite, countable, and uncountable sets). Read the first part on sets and functions and then refer to the remainder when it comes up in the notes.

2. *Bibliography:* There is a short bibliography to which you might turn if you're having trouble or want to go beyond the notes.

3. *List of Symbols:* If a symbol occurs in the notes which you don't recognize, try this list.

4. *Index:* The Index is fairly extensive and can lead you to various places where a given concept or result is discussed.

One last thing. Use the Exercises to test your understanding. Most of them come with specific instructions or questions, "Find this", "Prove that", "True or false?". Occasionally an exercise will come without instructions and will be a simple declarative sentence, "Every differentiable function is continuous". Such statements are to be proved. Their occurrence reflects nothing more than the author's attempt to break the monotony of saying, "Prove that . . ." over and over again. The exercises marked with an asterisk are (usually) extremely difficult. Don't be discouraged if some of the ones without asterisks stump you. A few of them were significant mathematical discoveries not so long ago.

KENNETH HOFFMAN

ANALYSIS IN
EUCLIDEAN SPACE

1. Numbers and Geometry

1.1. The Real Number System

The basic prerequisite for reading this book is a familiarity with the real number system. That familiarity should include a facility both with the elementary algebra of real numbers and with a few inequalities derived from the natural ordering of those numbers. This section is designed to emphasize some properties of the number system which may be less familiar.

The first thing we shall do is to list a few fundamental properties of algebra and order from which all of the properties of the real number system can be deduced. Let R be the set of real numbers.

A. Field Axioms. On the set R there are two operations, as follows. The first operation, called *addition*, associates with each pair of elements x, y in R an element $(x + y)$ in R. The second operation, called *multiplication*, associates with each pair of elements x, y in R an element xy in R. These two operations have the following properties.

1. Addition is commutative,

$$x + y = y + x$$

for all x and y in R.

2. Addition is associative,

$$(x + y) + z = x + (y + z)$$

for all x, y, and z in R.

3. There is a unique element 0 (zero) in R such that $x + 0 = x$ for all x in R.

4. To each x in R there corresponds a unique element $-x$ in R such that $x + (-x) = 0$.

5. Multiplication is commutative,

$$xy = yx$$

for all x and y in R.

6. Multiplication is associative,

$$(xy)z = x(yz)$$

for all x, y, and z in R.

7. There is a unique element 1 (one) in R such that $x1 = x$ for all x in R.

8. To each non-zero x in R there corresponds a unique element x^{-1} (or $1/x$) in R such that $xx^{-1} = 1$.

9. $1 \neq 0$.

10. Multiplication distributes over addition,

$$x(y + z) = xy + xz$$

for all x, y, and z in R.

B. Order Axioms. There is on R a relation $<$, called *less than*, with these properties.

1. If x and y are in R, one and only one of the following holds:

$$x < y; \qquad x = y; \qquad y < x.$$

2. $x < y$ if and only if $0 < y - x$.

3. If $0 < x$ and $0 < y$, then $0 < (x + y)$ and $0 < xy$.

C. Completeness Axiom. If S and T are non-empty subsets of R such that

(i) $R = S \cup T$;

(ii) $s < t$ for every s in S and every t in T,

then either there exists a largest number in the set S or there exists a smallest number in the set T.

These properties are usually summarized by saying that the set of real numbers, with its usual addition, multiplication, and ordering, is (A) a field, which (B) is ordered and which (C) is complete in that ordering. Briefly, the real number system is a complete ordered field.

From the field axioms (A), we could deduce the various algebraic relations which we shall use; however, we shall not do that. We shall use without comment basic identities such as the binomial theorem

$$(x + y)^n = \sum_{k=0}^{n} \binom{n}{k} x^k y^{n-k}$$

or the telescoping property of a geometric series

$$1 - x^{n+1} = (1 - x)(1 + x + x^2 + \cdots + x^n).$$

We could define the set of **positive integers**

$$Z_+ = \{1, 2, 3, \ldots\}$$

(from the axioms) as the set consisting of the numbers $1, 1 + 1, 1 + 1 + 1, \ldots$; and then we could prove the principle of **mathematical induction**: If S is a subset of R such that

(a) $1 \in S$;
(b) if $x \in S$ then $(x + 1) \in S$,

then S contains every positive integer. Then we could define the set of **integers**

$$Z = \{\ldots, -2, -1, 0, 1, 2, \ldots\}$$

and the set of **rational numbers**

$$Q = \left\{ \frac{m}{n}; m \in Z, n \in Z_+ \right\}$$

and diligently verify that

$$\frac{m}{n} + \frac{p}{q} = \frac{mq + np}{nq}$$

and so on. Perhaps (logically) we should carry out those deductions; however, that would be time-consuming and it would be of virtually no help in understanding analysis.

A similar comment is applicable to a few inequalities which can be deduced (easily) from the order axioms (B). If $x < y$, then $x + z < y + z$; if $x < y$ and $0 < c$, then $cx < cy$. We use $x \leq y$ to mean $x < y$ or $x = y$. It is understood that $y > x$ means the same thing as $x < y$. The **absolute value** of a number x is defined by

$$|x| = \begin{cases} x, & \text{if } x \geq 0 \\ -x, & \text{if } x < 0 \end{cases}$$

and absolute value has these properties:

$$|xy| = |x||y|$$
$$|x + y| \leq |x| + |y|$$
$$|x - y| \geq |x| - |y|.$$

These inequalities will be used with little or no comment.

Now, one might reasonably ask this. If we are not going to deduce the various properties of the real number system from (A), (B), and (C), why do we bother to list just those particular properties and to assert that they determine the real number system? There are two principal reasons.

First, analysis is based upon the concept of number, and so we are obligated to state clearly what the real number system is. One way do to that is to state that the system is characterized by two theorems: (i) There exists a complete ordered field. (ii) Any two such fields are isomorphic; that is, there exists a 1:1 correspondence between their members which preserves addition, multiplication, and order. The second reason for listing (A), (B), and (C) is that it will help us understand the completeness property (C). A fair fraction of introductory analysis consists of learning the meaning of the completeness of the real number system and learning to use various reformulations of it.

As we have suggested, we shall not prove here that the real number system exists or that it is unique. What we assume is a familiarity with calculations in an ordered field. The one aspect of the number system with which we do not assume much familiarity is the completeness. In the next two sections we begin to look at some implications of completeness. Right now, let us try to be clear about what it says.

Intuitively, property (C) is intended to say that if one thinks of real numbers as corresponding to points on a line, then the line has no holes in it. How can one subdivide the "line" R into the union of two non-empty sets, S and T, such that every number in S is less than every number in T? The only way to do that is to cut the line at some point, to let S be everything on one side of the cut and to let T be everything on the other side of the cut. Of course, the point where we cut must be put either in S or in T, and it will accordingly be the largest number in S or the smallest number in T.

Precisely, suppose we choose any real number c. From c we obtain two slightly different subdivisions as described in (C):

$$S = \{s \in R; s \leq c\}$$
$$T = \{t \in R; t > c\}$$

or

$$S = \{s \in R; s < c\}$$
$$T = \{t \in R; t \geq c\}.$$

The completeness property states that there are no other examples, the first type being the one in which S has a largest member, the second type being the one in which T has a smallest member.

EXAMPLE 1. Let us look at the rational number system, which consists of Q (the set of rational numbers), together with the addition, multiplication, and ordering inherited from R. Since sums, differences, products, and quotients of rational numbers are rational, we see that if we substitute Q for R in (A), the field axioms are satisfied. Similarly, Q satisfies the order axioms (B). Thus, the rational number system is an ordered field; however, it is *not* complete. Long ago, the Greeks noted that (loosely speaking) the set of rational numbers had holes in it—for instance, at the place where $\sqrt{2}$ ought to be. More precisely, they proved that there is no rational number x such that $x^2 = 2$. That means that we can cut the set of rationals at $\sqrt{2}$ and show that Q is not complete, because the set $S = \{s \in Q; s < \sqrt{2}\}$ has no largest member and the set $T = \{t \in Q; t > \sqrt{2}\}$ has no smallest member. Let us describe the situation without mentioning $\sqrt{2}$.

Suppose we define

(1.1)
$$S = \{s \in Q; \text{ either } s < 0 \text{ or } 0 \leq s \text{ and } s^2 < 2\}$$
$$T = \{t \in Q; 0 < t \text{ and } t^2 > 2\}.$$

Clearly S and T are non-empty subsets of Q and each number in S is less than each number in T. Here is the important point. Since there does not exist any x in Q with $x^2 = 2$, it follows that

$$Q = S \cup T.$$

Does T have a smallest member? If $t \in T$, then $t^2 > 2$. If r is a very small positive rational number, then we shall have $(t - r)^2 > 2$ (as well as $t - r > 0$); i.e., we shall have $(t - r) \in T$. Hence T has no smallest member. By similar reasoning, S has no largest member. Thus, the rational number system does not have the completeness property (C).

Why can't we give the same example in the real number system? Of course, the completeness property says that we cannot. But, let's try it and see exactly what goes wrong. We define sets S and T as in (1.1), but replace Q by R. Again, we conclude that S and T are non-empty and that every number in S is less than every number in T. Again, we can show that S has no largest member and that T has no smallest member. The completeness property (C) leaves us with only one possibility, namely, that $R \neq S \cup T$, i.e., that some real number belongs neither to S nor to T. It is very easy to see that if x is a real number such that $x \notin S$ and $x \notin T$, then $x^2 = 2$. Thus, one of the things which completeness guarantees is that there exists in R a square root for the number 2.

Exercises

In Exercises 1–10, deduce the stated properties of real numbers from the basic properties (A), (B), and (C).

1. If $x < y$ and $z < w$, then $x + z < y + w$.

2. If $x < 0$, then $-x > 0$.

3. If $x + y = x$, then $y = 0$.

4. For each x in R, $x0 = 0$.

5. If $x < y$ and $y < z$, then $x < z$.

6. If $xy = 0$, then either $x = 0$ or $y = 0$.

7. (a) $(-x)y = -(xy)$. *Hint:* $[x + (-x)]y = ?$
(b) $(-x)(-y) = xy$.

8. For each x in R, $x^2 \geq 0$.

9. If $x < y$, then $x < \frac{1}{2}(x + y) < y$.

10. If $x^2 + y^2 = 0$, then $x = y = 0$.

11. Use the completeness of the real number system to prove that each positive real number has a unique positive square root.

12. The set of integers, with the addition, multiplication, and ordering inherited from R, is not a complete ordered field Precisely which of the conditions listed under the headings (A), (B), (C) are not satisfied?

1.2. *Consequences of Completeness*

We shall discuss a few applications of the completeness of the real number system. First, we need some basic terminology.

Definition. Let A *be a set of real numbers, i.e., a subset of* R. *We say that* A *is* **bounded above** *if there exists a number* b $\in$ R *such that*

$$a \leq b, \quad \text{for all a} \in A.$$

Any such b *is called an* **upper bound** *for the set* A. *We say that* A *is* **bounded below** *if there exists a number* c $\in$ R *such that*

$$c \leq a, \quad \text{for all a} \in A.$$

Any such c *is called a* **lower bound** *for the set* A. *We say that* A *is* **bounded** *if* A *is bounded above and bounded below.*

There are various simple observations we should make. The set A is bounded below if and only if the set $-A = \{-x; x \in A\}$ is bounded above. If b is an upper bound for $-A$, then $-b$ is a lower bound for A. Such things are immediate from the fact that the condition $x > y$ is equivalent to $-x < -y$. The set A is bounded if and only if the set $|A| = \{|x|; x \in A\}$ is bounded above. If b is an upper bound for $|A|$, then

$$-b \leq x \leq b, \quad x \in A.$$

On the other hand, if

$$c \leq x \leq d, \quad x \in A$$

then the larger of $|c|$ and $|d|$ is an upper bound for the set $|A|$.

EXAMPLE 2. The set of positive real members

(1.2) $$R_+ = \{x \in R; x > 0\}$$

is an elementary example of a subset of R which is not bounded. It is bounded below; in fact, any $x \leq 0$ is a lower bound for R_+. But it is not bounded above. If b were an upper bound, we could deduce in order: $b \geq 0$, $(b + 1) \in R_+$, $b \geq b + 1$, ??

EXAMPLE 3. The set of positive integers Z_+ is bounded below. It is not bounded above. One of the properties of the real number system with which the reader is supposed to be familiar is the Archimedean ordering property, which states that if $b \in R$, then there is a positive integer greater than b. We shall call that a theorem (Theorem 2), and prove it as an exercise in the use of completeness.

Theorem 1. *Let* A *be a non-empty subset of* R *which is bounded above. Then* A *has a least (smallest) upper bound.*

Proof. Let T be the set of all upper bounds for A:

$$T = \{b \in R; \ x \leq b \text{ for all } x \in A\}.$$

Let S be the complement of T

$$S = \{x \in R; \ x \notin T\}.$$

We can see easily that

(i) $R = S \cup T$;
(ii) if $s \in S$ and $t \in T$, then $s < t$.

We defined S so that (i) would be true. What does (ii) say? It says, if $s \notin T$ and $t \in T$, then $s < t$; or, if $t \in T$ and $s \geq t$, then $s \in T$. The last statement is clearly true. Look at the definition of T.

The hypothesis that A is bounded above is precisely the statement that T is non-empty. The hypothesis that A is non-empty tells us that S is non-empty, as follows. Choose any $x \in A$. Then S contains every number $y < x$, because, if $y < x$ then y is not an upper bound for A.

The completeness condition now tells us that either S has a largest member or T has a smallest member. But S does not have a largest member. Let $s \in S$, that is, let $s \notin T$. Then s is not an upper bound for A. Consequently there exists a number $a \in A$ with $a > s$. The number $d = \frac{1}{2}(a + s)$ satisfies

$$s < d < a.$$

Since $d < a$, we have $d \in S$. Since $s < d$, we see that s is not the largest member of S.

Therefore, T has a smallest (least) number in it. That is a number c such that

(1.3)
    (i) c is an upper bound for the set A;
    (ii) if b is an upper bound for A, then $c \leq b$.

In other words, c is the least upper bound for A.

Evidently, there is a companion result which asserts that, if a subset A of R is non-empty and bounded below, it has a greatest lower bound. That is a number c such that

(1.4)
    (i) c is a lower bound for A;

    (ii) if b is a lower bound for A, then $c \geq b$.

Notation and Terminology. Let A be a non-empty subset of R. If A is bounded above, the least upper bound for A is also called the **supremum** of A and is denoted by

$$\sup A.$$

If A is bounded below, the greatest lower bound for A is also called the **infimum** of A and is denoted by

$$\inf A.$$

One might wonder why we introduce other names for least upper bound and greatest lower bound. One reason is that they occur so often that they must be abbreviated, and "lub" and "glb" leave a little to be desired.

Theorem 1 is a reformulation of the completeness of the real number system. In Section 1.1, if one assumes Theorem 1 instead of property (C), then it is easy to prove (C) as a theorem. The two properties are only slightly different. Let's use Theorem 1 to prove that the set of positive integers is not bounded.

Theorem 2 (*Archimedean Ordering Principle*). *If* x *is a real number, there exists a positive integer* n *such that* x < n.

Proof. Suppose Z_+ is bounded above. Let $c = \sup Z_+$. Since c is the least upper bound for Z_+, $c - 1$ is not an upper bound for Z_+. Therefore, there exists a positive integer n such that $c - 1 < n$. So $c < n + 1$. But that says that c is not an upper bound for Z_+. (?)

Corollary. *If* x > 0, *there exists a positive integer* n *such that* 1/n < x.

Corollary. *If* y − x ≥ 1, *there is an integer* n *such that* x ≤ n ≤ y.

Proof. According to Theorem 2, there exists an integer m such that $x \leq m$. There are at most finitely many integers k such that $x \leq k \leq m$. (That follows from the principle of mathematical induction.) Let n be the least of those integers. It is a simple matter to verify that $x \leq n \leq y$.

Corollary. *If* A *is a bounded set of integers, then* sup A *and* inf A *are integers.*

Corollary. *If* x < y, *there exists a rational number* r *such that* x < r < y.

Proof. Choose a positive integer n such that $n(y - x) > 1$. Then find an integer m such that $nx < m < ny$. Let $r = m/n$.

Theorem 3. *Let* x *be a positive real number and let* n *be a positive integer. There is precisely one positive real number* y *such that* $y^n = x$.

Proof. Let us make a simple basic observation. If $s \geq 0$ and $t \geq 0$, then $t \geq s$ if and only if $t^n \geq s^n$. That follows from the fact that

$$t^n - s^n = (t - s)f(t, s)$$

where $f(t, s) = t^{n-1} + t^{n-2}s + \cdots + ts^{n-2} + s^{n-1}$. Since $f(t, s) > 0$ unless $s = t = 0$, the numbers $t^n - s^n$ and $t - s$ have the same sign.

Obviously (then) we cannot have two distinct positive nth roots. The only problem is to prove that there exists at least one.

Let

$$A = \{y \in R; y > 0 \text{ and } y^n \geq x\}.$$

Then A is bounded below. Furthermore A is non-empty. In case $x \leq 1$, we have $1^n \geq x$ so that $1 \in A$; and, in case $x > 1$ we have

$$x^n - x = x(x^{n-1} - 1)$$
$$\geq 0$$

so that $x \in A$. Let $c = \inf A$. Certainly $c \geq 0$, and the claim is that $c^n = x$.

First, we show that $c^n \leq x$. Suppose $c^n > x$. Then we can find a small positive number r such that $(c - r)^n > x$. (See following lemma.) By the definition of A, $(c - r) \in A$. But, $c - r < c$ and c is a lower bound for A, a contradiction. It must be that $c^n \leq x$.

The fact that no lower bound for A is greater than c will imply that $c^n \geq x$. Suppose $c^n < x$. We can find a small positive number r such that $(c + r)^n < x$. Thus $(c + r)^n < x \leq y^n$ for all $y \in A$, which yields $c + r < y$ for all $y \in A$. So, $c + r$ is a lower bound for A. But $c + r > c$; hence, something is wrong. We conclude that $c^n \geq x$.

Lemma. *Let* n *be a positive integer. Let* a, b, *and* c *be real numbers such that* $a < c^n < b$. *There exists a number* $\delta > 0$ *such that* $a < (c + r)^n < b$ *for every* r *which satisfies* $|r| < \delta$.

Proof. We have

$$t^n - c^n = (t - c)f(t, c)$$

where

$$f(t, c) = t^{n-1} + t^{n-2}c + \cdots + tc^{n-2} + c^{n-1}.$$

If we apply this with $t = c + r$, we obtain

$$(c + r)^n - c^n = rf(c + r, c)$$

and hence

$$|(c + r)^n - c^n| \leq |r||f(c + r, c)|.$$

Now

$$|f(c + r, c)| \leq (|c| + |r|)^{n-1} + (|c| + |r|)^{n-2}|c| + \cdots$$
$$+ (|c| + |r|)|c|^{n-2} + |c|^{n-1}.$$

If $|r| \leq 1$, then

$$|f(c + r, c)| \leq (|c| + 1)^{n-1} + (|c| + 1)^{n-2}|c| + \cdots$$
$$+ (|c| + 1)|c|^{n-2} + |c|^{n-1}$$
$$= f(1 + |c|, |c|).$$

Therefore,

$$|(c + r)^n - c^n| \leq |r| f(1 + |c|, |c|), \quad \text{if } |r| \leq 1.$$

Although it is not necessary, we shall rewrite this inequality in a more concrete form: Since $(1 + |c|) - |c| = 1$, the definition of f tells us that

$$f(1 + |c|, |c|) = (1 + |c|)^n - |c|^n.$$

So our inequality says

$$(1.5) \qquad |(c + r)^n - c^n| \leq |r|[(1 + |c|)^n - |c|^n], \quad \text{if } |r| \leq 1.$$

We are told that $a < c^n < b$ and we want to ensure that $a < (c + r)^n < b$, provided $|r|$ is small. Let s be the smaller of the two numbers $c^n - a$ and $b - c^n$. Then, if

$$|(c + r)^n - c^n| \leq s$$

we shall have $a < (c + r)^n < b$. Define δ by

$$\delta[(1 + |c|)^n - |c|^n] = s.$$

From (1.5) we then have

$$a < (c + r)^n < b, \quad \text{provided } |r| < \delta.$$

The reader may already be familiar with the conclusion of the last lemma—the nth power function is continuous. One should look at the proof anyway, since one cannot have too much experience in handling inequalities.

The unique $y > 0$ such that $y^n = x$ is denoted either $\sqrt[n]{x}$ or $x^{1/n}$. Remember that $x^{1/n} > 0$. If n is even, there is another real number y such that $y^n = x$, namely, $y = -x^{1/n}$. If n is odd, there is no other real nth root.

Exercises

1. Is the set of rational numbers bounded below?

2. Give an example of a bounded set A such that sup A is in A but inf A is not in A.

3. Find all non-empty bounded sets A such that sup $A \leq$ inf A.

4. Is the empty set bounded above? Does it have a least upper bound?

5. Every subset of a bounded set is bounded. Any set which contains an un-bounded set is unbounded. (Unbounded means not bounded.)

6. If A is bounded above and B is bounded below, then the intersection $A \cap B$ is bounded.

7. Prove that, if x is any real number, then

$$x = \sup \{r \in Q; r < x\}.$$

8. If $x < y$, there exists an irrational (not rational) number t such that $x < t < y$.

9. Verify that every non-empty set of positive integers contains its infimum.

10. Let A be a subset of R which has uncountably many points in it. Prove that there exists a non-empty set $B \subset A$ such that sup B is not in B. (Uncountable is defined in the Appendix.)

11. Prove the completeness property (C) from Theorem 1.

12. Prove that, if a subset S of R (with the inherited addition, multiplication, and ordering) is a complete ordered field, then $S = R$.

*13. Let R and S be complete ordered fields. Show that R and S are isomorphic, i.e., show that there is a 1:1 correspondence between the members of R and the members of S which preserves addition, multiplication, and order.

1.3. Intervals and Decimals

This is a short section, in which we shall discuss the decimal represen-tations of real numbers. We shall not use these representations very much. The purpose of the section is twofold. It provides us with some concrete objects to which we can point and say, "There, if you will, are the real numbers." More important, it will make us think about the relation of intervals to the completeness of the real number system.

Definition. *An* **interval** *is a set* $I \subset R$ *such that*

  (i) I *contains at least two points*;
  (ii) *if* $x < t < y$ *and if* $x, y \in I$*, then* $t \in I$.

There are four types of bounded intervals, to which we shall refer repeatedly:

The **open interval** $(a, b) = \{x \in R; a < x < b\}$
The **closed interval** $[a, b] = \{x \in R; a \le x \le b\}$
The **semi-closed interval** $(a, b] = \{x \in R; a < x \le b\}$
The **semi-closed interval** $[a, b) = \{x \in R; a \le x < b\}$.

It is understood that a, b are real numbers with $a < b$.

There are five types of unbounded intervals, to which we shall refer occasionally:

$$(a, \infty) = \{x \in R; a < x\}$$
$$[a, \infty) = \{x \in R; a \leq x\}$$
$$(-\infty, b) = \{x \in R; x < b\}$$
$$(-\infty, b] = \{x \in R; x \leq b\}$$
$$(-\infty, \infty) = R.$$

We have left for the exercises the proof that every interval is of one of the nine types listed. In the notations for unbounded intervals, there occur the symbols "$-\infty$" and "∞". There are no objects $-\infty$ or ∞ in the real number system; indeed, we have assigned no meaning whatever to "$-\infty$" and "∞". The "far left" and the "far right" have their uses, but we'll talk about that later.

The decimal representation of a real number simply locates the number in a nested sequence of intervals, the lengths of which go down by a factor of 10 each time. The Archimedean ordering property and mathematical induction locate each $x \in R$ in the semi-closed interval $[n, n + 1)$, defined by some integer n. That n is the **greatest integer in** x:

(1.6) $$n = \sup \{k \in Z; k \leq x\}.$$

Then $(x - n) \in [0, 1)$, and we shall confine our discussion of decimal representation to numbers in that interval.

Consider a number $x \in [0, 1)$. To any such x will correspond a decimal expansion

$$x \sim .a_1 a_2 a_3 \ldots$$

where the "digits" are integers a_n between 0 and 9. The process by which we arrive at the expansion is assumed to be familiar. We subdivide $[0, 1)$ into 10 intervals I_k of length $\frac{1}{10}$, and we enumerate them by $k = 0, \ldots, 9$:

$$I_k = \left[\frac{k}{10}, \frac{k + 1}{10}\right), k = 0, \ldots, 9.$$

We locate the I_k which contains x, and that k is the digit a_1 (see Figure 1). An alternative way of describing this first digit in the decimal representation of x is

$$a_1 = \sup \left\{k \in Z; \frac{k}{10} \leq x\right\}.$$

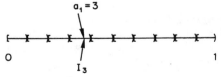

FIGURE 1

Next, we subdivide the interval I_{a_1} into 10 intervals, each of length $1/10^2$. The intervals are

$$[a_1 10^{-1} + k10^{-2}, a_1 10^{-1} + (k + 1)10^{-2}), \qquad k = 0, \ldots, 9.$$

The one of those intervals which contains x determines the digit a_2. In other words,

$$a_2 = \sup \{k \in Z; a_1 10^{-1} + k10^{-2} \leq x\}.$$

We repeat the subdivision process and continue. What do we end up with as a description of the decimal representation? It is a recursive definition of the digits $a_1, a_2, a_3, \ldots$.

(i) a_1 is the largest integer k such that $k10^{-1} \leq x$.

(ii) After $a_1, \ldots, a_{n-1}$ have been determined, a_n is the largest integer k such that

$$a_1 10^{-1} + a_2 10^{-2} + \cdots + a_{n-1} 10^{-(n-1)} + k10^{-n} \leq x.$$

What we have done is to place x successively in the semi-closed intervals $J_1, J_2, J_3, \ldots$ defined by

$$(1.7) \quad \begin{aligned} J_n &= [a_1 10^{-1} + a_2 10^{-2} + \cdots + a_n 10^{-n}, a_1 10^{-1} + a_2 10^{-2} \\ &\quad + \cdots + (a_n + 1)10^{-n}). \end{aligned}$$

The intervals J_n are "nested"

$$J_1 \supset J_2 \supset J_3 \supset \cdots$$

and x belongs to the intersection of all the J_n. In fact

$$(1.8) \qquad \bigcap_n J_n = \{x\};$$

that is, no other number belongs to every J_n. Why? If $a, b \in J_n$, then $|a - b| < 10^{-n}$. If y (as well as x) belongs to every J_n, then

$$|y - x| < 10^{-n}, \qquad n = 1, 2, 3, \ldots.$$

Hence $y - x = 0$.

We know that, by the scheme just described, there is associated with each $x \in [0, 1)$ a sequence of digits $a_1, a_2, a_3, \ldots$; and, we know (1.8) that different x's have different sequences of digits associated with them. So, it is legitimate to employ the shorthand

$$x = .a_1 a_2 a_3 \ldots.$$

What is really being abbreviated is

$$(1.9) \qquad x = \sum_{n=1}^{\infty} a_n 10^{-n},$$

but we'll worry about that later.

What interests us now is this. Is every sequence of digits $a_1, a_2, a_3, \ldots$ the decimal representation of a number $x \in [0, 1)$? Obviously not, if we follow the ground rules thus far. Take $a_n = 9$ for every n. The intersection

$$\bigcap_n [1 - 10^{-n}, 1)$$

is empty so that we cannot have $x \sim .999\ldots$ for any x in $[0, 1]$. We know how to fix that. Had we considered $[0, 1]$ instead of $[0, 1)$, then $.999\ldots$ would have arisen as the decimal representation of the number 1. But there are still other sequences of digits which do not occur in our process. Obviously, in the scheme we described, no decimal representation will arise which is ultimately all 9's:

(1.10) $.a_1 a_2 \ldots a_m 999 \ldots$

But, we know how to fix that also. We allow a very slight ambiguity in the decimal representation by agreeing that (1.10) represents the same number as does

(1.11) $.a_1 a_2 \ldots (a_m + 1)000 \ldots$

provided $a_m \neq 9$.

The fuss about repeating 9's is not, however, at the heart of the question of whether a sequence of digits $a_1, a_2, \ldots$ need represent any real number. The central problem is this. If someone hands us a sequence of digits $.a_1 a_2 a_3 \ldots$, where will we find the real number which it represents? The digits give us a nested sequence of intervals

$$J_1 \supset J_2 \supset J_3 \supset \ldots$$

defined by (1.7). The x we want is supposed to be (in) the intersection of that sequence of intervals. But the intersection may be empty, because of the repeated 9's business. So, we must replace the semi-closed interval

$$J_n = [b_n, c_n)$$

by the closed interval

$$\bar{J}_n = [b_n, c_n].$$

The intersection of the sets $\bar{J}_n$ will catch the right-hand end point if the 9's repeat. What we want to assert is that

$$\bigcap_n \bar{J}_n = \{x\}$$

where $x \in [0, 1]$. Since the length of $\bar{J}_n$ is 10^{-n}, there cannot be more than one point in the intersection. It is the completeness of the real number system which guarantees that there exists at least one x in the intersection.

Theorem 4 (*Nested Intervals Theorem*). *Let*

$$I_1 \supset I_2 \supset I_3 \supset \cdots$$

be a nested sequence of (bounded) closed intervals in R. *Then there exists a real number* x *which belongs to every* I_n.

Proof. Let $I_n = [b_n, c_n]$. Then

$$b_1 \leq b_2 \leq b_3 \leq \cdots \leq c_3 \leq c_2 \leq c_1.$$

Let B be the set which consists of all the numbers b_n, $n = 1, 2, 3, \ldots$. Then B is non-empty, and B is bounded above because any c_n is an upper bound for B. Let

$$x = \sup B.$$

We claim that $x \in I_n$ for every n, i.e., that $b_n \leq x \leq c_n$ for every n. Certainly $b_n \leq x$, since x is an upper bound for B. As we remarked, every c_n is an upper bound for B; hence $x \leq c_n$.

With the nested intervals theorem our discussion of decimals is complete. We have a correspondence between the set of all numbers $x \in [0, 1]$ and the set of all sequences $.a_1 a_2 a_3 \ldots$ of digits ($a_k \in Z$, $0 \leq a_k \leq 9$). The correspondence is $1 : 1$, except that the sequences (1.10) and (1.11) must be identified.

Exercises

1. Prove that every interval in R is of one of the nine types which we listed.

2. Let $\{I_\alpha\}$ be any collection of intervals in R. Prove that the intersection

$$\bigcap_\alpha I_\alpha$$

is one of the following:

    (a) The empty set.
    (b) A set with precisely one member.
    (c) An interval.

3. Let A be a bounded subset of R. What is the intersection of all closed intervals which contain A?

4. Give an example of a *nested* sequence of *open* intervals for which the intersection is empty. Give similar examples for which the intersection is a set with one member, an open interval, a closed interval, a semi-closed interval.

5. What kind of a set can the intersection of a nested sequence of closed intervals be?

6. Suppose you were working in the rational number system. Describe a nested sequence of closed intervals for which the intersection is empty.

7. Use the nested intervals theorem to give a binary representation for each point in $[0, 1]$:

$$x \sim .a_1 a_2 a_3 \ldots$$

where the digits a_n are either 0 or 1.

8. In Section 1.1, assume (A), (B), and the nested intervals theorem, and the Archimedean ordering property. Prove the completeness property (C).

1.4. Euclidean Space

We presume that the reader knows something about Euclidean space of n-dimensions. If n is a positive integer, then

$$R^n = R \times \cdots \times R$$

is the set of all n-tuples of real numbers. The points in R^n will sometimes be called vectors, and our standard notation for those points will be

$$X = (x_1, \ldots, x_n)$$
$$Y = (y_1, \ldots, y_n)$$

and so on. The number x_i is the ith (standard) **coordinate** of X.

There is a natural (vector) addition on R^n, defined by adding the coordinates:

$$X + Y = (x_1 + y_1, \ldots, x_n + y_n).$$

There is a product, called **scalar multiplication**, defined for vectors $X \in R^n$ and numbers $c \in R$ by

$$cX = (cx_1, \ldots, cx_n).$$

With this addition and scalar multiplication, R^n is a vector space. This means that the vector addition satisfies conditions A(1)–A(4) of Section 1.1 and that the scalar multiplication satisfies

$$c(X + Y) = cX + cY$$
$$(b + c)X = bX + cX$$
$$1X = X.$$

The zero vector for addition is the **origin** $0 = (0, \ldots, 0)$.

If X and Y are vectors in R^n, the (standard) **inner product** of X and Y is the number

$$(1.12) \qquad \langle X, Y \rangle = x_1 y_1 + \cdots + x_n y_n.$$

In many books, this is called the dot product and is denoted by $X \cdot Y$. Evidently, the inner product has these properties:

$$
(1.13) \qquad
\begin{aligned}
&\text{(i) } \langle X, Y \rangle = \langle Y, X \rangle; \\
&\text{(ii) } \langle cX + Y, Z \rangle = c\langle X, Z \rangle + \langle Y, Z \rangle; \\
&\text{(iii) } \langle X, X \rangle \geq 0; \text{ if } \langle X, X \rangle = 0 \text{ then } X = 0.
\end{aligned}
$$

If $X \in R^n$, the **length** (norm) of X is

$$|X| = \langle X, X \rangle^{1/2}.$$

The **distance** from X to Y is $|X - Y|$. In order to see that length and distance have their expected properties, it is most convenient to verify **Cauchy's inequality:** If $x_1, \ldots, x_n$ and $y_1, \ldots, y_n$ are real numbers, then

$$(1.14) \quad (x_1 y_1 + \cdots + x_n y_n)^2 \leq (x_1^2 + \cdots + x_n^2)(y_1^2 + \cdots + y_n^2).$$

Lemma (*Cauchy's Inequality*). *If* X *and* Y *are vectors in* $\mathbf{R}^n$, *then*

(1.15) $|\langle \mathbf{X}, \mathbf{Y} \rangle| \leq |\mathbf{X}| \, |\mathbf{Y}|.$

Furthermore, equality holds if and only if one of the two vectors is a scalar multiple of the other.

Proof. If $Y = 0$, the inequality is a trivial equality. If $Y \neq 0$, use the fact that

$$0 \leq \langle X - cY, X - cY \rangle$$
$$= |X|^2 - 2c\langle X, Y \rangle + c^2 |Y|^2$$

and apply it with

$$c = \frac{\langle X, Y \rangle}{|Y|^2}.$$

The result is Cauchy's inequality. If equality holds, then $X = cY$.

Length has these properties:

(i) $|X| \geq 0$; if $|X| = 0$ then $X = 0$;
(ii) $|cX| = |c| \, |X|$;
(iii) $|X + Y| \leq |X| + |Y|$.

The triangle inequality (iii) follows from Cauchy's inequality, because

$$|X + Y|^2 = |X|^2 + 2\langle X, Y \rangle + |Y|^2$$
$$\leq |X|^2 + 2|X| \, |Y| + |Y|^2$$
$$= (|X| + |Y|)^2.$$

Let us say a brief word about geometry. There is a fuller discussion in Section 1.6. Normally, when we discuss a "vector" X in R^n, we are thinking of the line segment from the origin to X, rather than the point X. If we have two vectors, X and Y, and if neither is a scalar multiple of the other, then those two vectors span a plane in R^n. That plane passes through the origin, and it consists of all vectors $aX + bY$ with $a, b \in R$. The vectors X and Y are two of the edges of a parallelogram in that plane. One diagonal of that parallelogram extends from the origin to the point $X + Y = (x_1 + y_1, \ldots, x_n + y_n)$, as in Figure 2. Suppose we let θ, $0 < \theta < \pi$, be the angle between the vectors X and Y. Then θ measures the extent to which Cauchy's inequality fails to be an equality:

(1.16) $\langle X, Y \rangle = |X| \, |Y| \cos \theta.$

It is particularly easy to verify (1.16) after one knows something about orthogonal bases, because the use of such bases shows that (1.16) need only be verified in R^2. See also Exercises 3 and 4 of Section 1.5.

EXAMPLE 4. Let us look at an application of Cauchy's inequality to matrices. For simplicity, we'll talk only about square matrices. A $k \times k$ matrix with real entries is (represented as) a square array of real numbers

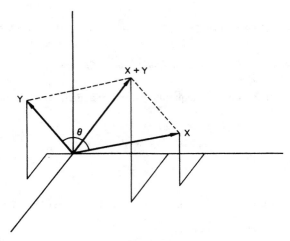

Figure 2

having k rows and k columns:

$$A = [a_{ij}], \quad 1 \le i \le k, \quad 1 \le j \le k.$$

We add matrices by adding the corresponding entries

$$A + B = [a_{ij} + b_{ij}]$$

and similarly

$$cA = [ca_{ij}].$$

Thus, the space of $k \times k$ matrices is just R^{k^2}, with the k^2 coordinates listed in k rows and k columns. The norm of the matrix A is given by

$$|A|^2 = \sum_{i,j} a_{ij}^2.$$

We shall denote the space of $k \times k$ matrices by $R^{k \times k}$, to remind us of the arrangement of the k^2 coordinates into rows and columns. This arrangement is pertinent in problems which involve matrix multiplication. Matrix product is defined by $C = AB$, where

$$c_{ij} = \sum_r a_{ir} b_{rj}.$$

It is associative: $(AB)C = A(BC)$, and it distributes over addition: $A(B+C) = AB + AC, (A + B)C = AC + BC$. It is not commutative unless $k = 1$, that is, generally $AB \ne BA$. What really interests us is the relation of norm to matrix product, $|AB| \le |A||B|$.

Theorem 5. *If* A *and* B *are* k $\times$ k *matrices, then*

$$|AB| \le |A||B|.$$

Proof.

$$|AB|^2 = \sum_{i,j} \left[\sum_r a_{ir} b_{rj} \right]^2$$

$$\leq \sum_{i,j} \sum_r a_{ir}^2 \sum_s b_{sj}^2 \qquad \text{(Cauchy)}.$$

Now

$$|A|^2 = \sum_{i,r} a_{ir}^2$$

$$|B|^2 = \sum_{s,j} b_{sj}^2$$

and so it is apparent that

$$|AB|^2 \leq |A|^2 |B|^2.$$

We might remark that matrix multiplication can be used to express inner products on the space of matrices in the following way

(1.17) $\langle A, B \rangle = \text{trace } (AB^t)$.

The **trace** of a matrix is the sum of its diagonal entries. The matrix B^t is the **transpose** of B. Its i, j entry is b_{ji}. In the case $B = A$, (1.17) says

$$|A|^2 = \text{trace } (AA^t).$$

1.5. Complex Numbers

The complex number system is (essentially) obtained by adjoining to the real number system a square root for the number -1. The enlarged "system" has (in one sense) less structure, because the ordering of the real numbers does not extend to an ordering of the complex numbers. But, the complex system is richer in ways which make it indispensable for understanding parts of mathematics. For instance, we obtain complex numbers by introducing a zero for the polynomial $x^2 + 1$; but it turns out that every non-constant polynomial with real (or even complex) coefficients has a zero in the set of complex numbers. That is the so-called fundamental theorem of algebra, which we shall prove later.

We mildly caution the reader on two points: (1) This section is brief; however, the importance of complex numbers in this book should not be underestimated. (2) Mathematicians have retained the mystical terminology of "complex" and "real" and "imaginary" numbers; however, the terms are not now intended to suggest anything about reality or the absence thereof.

We list some properties which characterize the complex number system. Let C be the set of complex numbers.

1. C is a field; i.e., there is an addition and a multiplication on C which satisfy the field axioms (A) of Section 1.1.

2. C contains R as a subfield.

3. There is an element $i \in C$ such that $i^2 = -1$.
4. If a subfield of C contains R and i, then that subfield is (all of) C.

The subset S is a **subfield** if S contains 0 and 1 and is closed under the formation of sums, differences, products, and quotients. Thus, (2) says that R sits in C and that when we restrict to R the addition and multiplication of C, they become the addition and multiplication of R. If S is a subfield which contains R and i, then S contains every element of C of the form $x + iy$, with x and y in R. On the other hand, one can show that the set of those numbers is a subfield; hence, by (4) it exhausts C.

Therefore C consists of the numbers

(1.18) $$z = x + iy, \qquad x, y \in R$$

added and multiplied according to the usual rules of algebra (the field axioms), with $i^2 = -1$. The representation (1.18) of the number z is unique. We call x the **real part** of z:

$$x = \text{Re} \, (z)$$

and we call y the **imaginary part** of z:

$$y = \text{Im} \, (z).$$

Notice that the imaginary part is real (a real number). The (complex) **conjugate** of x is the number

(1.19) $$z^* = x - iy.$$

Note that $(z + w)^* = z^* + w^*$ and $(zw)^* = z^*w^*$. Since

(1.20) $$zz^* = x^2 + y^2 \geq 0$$

it makes sense to define the **absolute value** of z

(1.21) $$|z| = (zz^*)^{1/2}.$$

In connection with (1.20), we might remark that if $w \in C$ and we write $w \geq 0$, that is understood to mean that w is real and $w \geq 0$.

It is a straightforward matter to verify that

(1.22)
$$|z + w| \leq |z| + |w|$$
$$|zw| = |z| \, |w|.$$

Since absolute value preserves products, the number $z/|z|$ has absolute value 1 (if $z \neq 0$). Hence each non-zero complex number z is uniquely expressible in the form

(1.23) $$z = rw, \qquad r > 0, \quad |w| = 1.$$

There is a 1 : 1 correspondence between complex numbers and points in R^2 which is so immediate that we often identify C and R^2 as sets. The number $z = x + iy$ corresponds to the point (x, y) in R^2. Addition of complex numbers corresponds to the vector addition in R^2; briefly, we add complex numbers by adding their real and imaginary parts:

$$\operatorname{Re}(z + w) = \operatorname{Re}(z) + \operatorname{Re}(w)$$
$$\operatorname{Im}(z + w) = \operatorname{Im}(z) + \operatorname{Im}(w).$$

Thus, we have the parallelogram picture of addition in C. Absolute value corresponds to length in R^2:

$$|z| = (x^2 + y^2)^{1/2}.$$

Conjugation corresponds to reflection about the real line.

The geometric interpretation of multiplication involves angles. Let us look at complex numbers w of absolute value 1:

$$w = u + iv$$
$$u^2 + v^2 = 1.$$

These points (u, v) comprise the circle of radius 1 centered at the origin—usually called the **unit circle**. Each w on the unit circle is uniquely located by the angle θ from the vector 1 to the vector w. (See Figure 3.) Furthermore,

$$u = \cos \theta$$
$$v = \sin \theta$$

because that is the usual definition of $\cos \theta$ and $\sin \theta$. Thus

(1.24) $$w = \cos \theta + i \sin \theta.$$

If angles are measured by numbers, i.e., if θ "is" a number in our discussion, then (1.24) determines a unique θ, $0 \leq \theta < 2\pi$. Any number $\theta + 2k\pi$, $k \in Z$, would then serve as well.

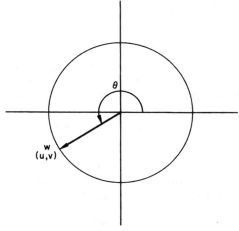

FIGURE 3

Now
$$\cos\theta + i\sin\theta = e^{i\theta}.$$

(We'll discuss that carefully later.) It is then clear that each non-zero $z \in C$ can be expressed

(1.25) $z = re^{i\theta}, \qquad r > 0, \quad \theta \in R.$

Of course $r = |z|$. There are many such θ's, all differing by integer multiples of 2π. Each θ is an **argument** of z. If we also have
$$w = \rho e^{it}, \qquad \rho > 0, \quad t \in R$$
then
$$zw = (r\rho)e^{i(t+\theta)}.$$

In other words, multiplication of complex numbers multiplies the absolute values and adds the arguments. If the same fact is verified without using the exponential function, it amounts to the verification of the trigonometric identities
$$(\cos\theta + i\sin\theta)(\cos t + i\sin t) = \cos(\theta + t) + i\sin(\theta + t)$$
that is,
$$\cos(\theta + t) = \cos\theta\cos t - \sin\theta\sin t$$
$$\sin(\theta + t) = \sin\theta\cos t + \cos\theta\sin t.$$

Each non-zero complex number z has n distinct nth roots. Write
$$z = |z|e^{i\theta}, \qquad 0 < \theta \le 2\pi$$
and let $\alpha = \theta/n$. Then the numbers
$$z_k = |z|^{1/n}e^{i\theta_k}$$
$$\theta_k = \alpha + k\frac{2\pi}{n}, \qquad k = 0,\dots,n-1$$
are the nth roots:
$$z_k^n = z, \qquad k = 0,\dots,n-1.$$

On a few occasions, we shall want to discuss **complex n-space**, C^n. This is the set of all n-tuples of complex numbers $(z_1,\dots,z_n)$. Vector addition and scalar multiplication are defined formally as in R^n, replacing R by C. The standard inner product on C^n is
$$\langle Z, W \rangle = z_1w_1^* + \cdots + z_nw_n^*.$$
It has the properties (1.13), except that
$$\langle W, Z \rangle = \langle Z, W \rangle^*.$$

Thus, Cauchy's inequality relating inner product and length is valid, with essentially the same proof. Apply the inequality $0 \le \langle Z + cW, Z + cW \rangle$ with
$$c = -\frac{\langle Z, W \rangle}{\langle W, W \rangle}.$$

In terms of coordinates, Cauchy's inequality states that

$$|z_1w_1^* + \cdots + z_nw_n^*|^2 \leq (|z_1|^2 + \cdots + |z_n|^2)(|w_1|^2 + \cdots + |w_n|^2).$$

There is a (real vector space) isomorphism between C^n and R^{2n}. If $z_j = x_j + iy_j$, then

$$(z_1, \ldots, z_n) \longrightarrow (x_1, y_1, x_2, y_2, \ldots, x_n, y_n)$$

maps C^n onto R^{2n}. It preserves sums, multiplication by real scalars, and length:

$$|z_1|^2 + \cdots + |z_n|^2 = x_1^2 + y_1^2 + \cdots + x_n^2 + y_n^2.$$

EXAMPLE 5. The discussion of matrices in Example 4 can be extended immediately to the complex case. The space of $k \times k$ matrices with complex entries behaves like complex k^2-space. In this case, the inner product is described this way:

$$\langle A, B \rangle = \text{trace} (AB^*)$$

where B^* is the **conjugate transpose** of B. Its i, j entry is b_{ji}^*. The verification that

$$|AB| \leq |A||B|$$

is as in Theorem 5, since we have Cauchy's inequality for complex numbers. We shall (of course) denote the space of complex $k \times k$ matrices by $C^{k \times k}$.

Exercises

1. If z is a complex number and $|1 - z| = |1 + z| = 1$, then $z = 0$.

2. If $|z| < 1$ and $|w| = 1$, then

$$\left| \frac{w + z}{1 + z^*w} \right| = 1.$$

3. If we identify C with R^2, then the inner product of z and w is

$$\langle z, w \rangle = \text{Re} (zw^*).$$

4. If we identify C with R^2,

$$\langle z, w \rangle = |z||w| \cos \alpha$$

where α is the angle from z to w.

5. Let $0 < r < 1$. Then

$$\sup \left\{ \left| \frac{w + r}{w - r} \right|; \ w \in C, |w| = 1 \right\} = \frac{1 + r}{1 - r}.$$

6. Each complex $k \times k$ matrix A is uniquely expressible in the form $A = A_1 + iA_2$ where the matrices A_j have real entries. Is it true that

$$|A|^2 = |A_1|^2 + |A_2|^2 ?$$

7. Each complex $k \times k$ matrix A is uniquely expressible in the form $A = A_1 + iA_2$ where $A_j = A_j^*$ (the conjugate transpose). Is it true that

$$|A|^2 = |A_1|^2 + |A_2|^2 ?$$

1.6. Linear Geometry

We shall review a few basic facts from linear analytic geometry. We hope that the reader is acquainted with this material. It is not essential for reading this book; however, we shall refer to this material often in the examples and exercises.

Suppose that X and Y are distinct points in R^n. The (straight) **line** through X and Y is

$$\{tX + (1 - t)Y; t \in R\}.$$

A subset S of R^n is called **flat** if it has this property: If X and Y are in S, then the line through X and Y is contained in S. A (linear) **subspace** of R^n is a flat subset which contains the origin.

A subspace is more commonly defined as a non-empty subset S such that

(i) if X and Y are in S, then $(X + Y)$ is in S;
(ii) if X is in S and c is any real number, then cX is in S.

In this formulation, one may replace (i) and (ii) by the single condition: If X and Y are in S, then $cX + Y$ is in S for all $c \in R$. This is equivalent to the flat subset characterization which we used as the definition. The essential point is that if $X_1, \ldots, X_k$ are vectors in the subspace S, then every **linear combination**

$$c_1 X_1 + \cdots + c_k X_k$$

of those vectors is in S. If we start with any vectors $X_1, \ldots, X_k$ in R^n, the set of all linear combinations of those vectors is a subspace, called the **subspace spanned by** $X_1, \ldots, X_k$.

Note that a line is a flat subset and consequently is a subspace if and only if it passes through the origin. Planes through the origin will be (defined as) the 2-dimensional subspaces. Let us say something about dimension.

The vectors $V_1, \ldots, V_k$ are **linearly dependent** if it is possible to express the 0 vector as a linear combination of them in some non-trivial way

$$c_1 V_1 + \cdots + c_k V_k = 0, \qquad c_j \neq 0 \text{ for at least one } j.$$

Linearly independent means not linearly dependent. If $V_1, \ldots, V_k$ are linearly independent, then $V_i \neq 0$ and also $V_i \neq V_j$ when $i \neq j$.

The most basic fact about linear dependence is this. *If $V_1, \ldots, V_k$ are vectors in R^n and if $k < n$, then $V_1, \ldots, V_k$ are linearly dependent.* In short, any $n + 1$ vectors in R^n are linearly dependent. That is a reformulation of the basic theorem on systems of linear equations. For, suppose we wish to find numbers $c_1, \ldots, c_k$ such that $c_1 V_1 + \cdots + c_k V_k = 0$. If

$$V_i = (v_{i1}, \ldots, v_{in}), \qquad 1 \le i \le k$$

then the condition on the numbers c_i is

$$c_1 v_{11} + \cdots + c_k v_{k1} = 0$$
$$c_1 v_{12} + \cdots + c_k v_{k2} = 0$$
$$\vdots \qquad \qquad \vdots$$
$$c_1 v_{1n} + \cdots + c_k v_{kn} = 0.$$

If $k > n$, this homogeneous system of n linear equations in k unknowns has a solution for $(c_1, \ldots, c_k)$ which is non-trivial (not every $c_i = 0$).

If S is a subspace of R^n, the **dimension** of S is the maximum number of linearly independent vectors which can be found in S. More precisely, dim S is the largest non-negative integer k such that some k-tuple of vectors in S is linearly independent. Of course dim $S \le n$. It is not difficult to see that dim $S < n$, except for the one subspace $S = R^n$.

An (ordered) **basis** for a subspace S is a k-tuple of vectors $V_1, \ldots, V_k$ such that

(i) $V_1, \ldots, V_k$ are linearly independent;
(ii) S is the subspace spanned by $V_1, \ldots, V_k$.

The simplest example of a basis is the **standard basis** for R^n

$$E_1 = (1, 0, 0, \ldots, 0)$$
$$E_2 = (0, 1, 0, \ldots, 0)$$
(1.26)
$$\vdots \qquad \qquad \vdots$$
$$E_n = (0, 0, 0, \ldots, 1).$$

The unique expression for $X = (x_1, \ldots, x_n)$ as a linear combination of those vectors is

$$X = x_1 E_1 + \cdots + x_n E_n.$$

Theorem 6. *If* S *is a subspace of* R^n, *then* S *has a basis, and every basis for* S *consists of precisely* dim S *vectors.*

Proof. If $d = $ dim S, then S has a basis consisting of d vectors: We can find vectors $Y_1, \ldots, Y_d$ in S which are independent. By the definition of d, we know that, for any $X \in S$, the vectors $Y_1, \ldots, Y_d, X$ are dependent. But that means that X is a linear combination of the vectors Y_j.

Suppose $V_1, \ldots, V_k$ is any (ordered) basis for the subspace S. Each X in S can be expressed as a linear combination

(1.27) $$X = c_1 V_1 + \cdots + c_k V_k.$$

Furthermore, the numbers $c_1, \ldots, c_k$ are uniquely determined by X (and the basis). If we had another expression for X as a linear combination, we could subtract and contradict the independence of $V_1, \ldots, V_k$. The k-tuple $(c_1, \ldots, c_k)$ is the k-tuple of **coordinates of X relative to the ordered basis** $V_1, \ldots, V_k$. It is clear that (1.27) defines a $1:1$ correspondence between the set of all vectors X in S and the set of all k-tuples $(c_1, \ldots, c_k)$ in R^k. If we add two vectors X and Y in S, the corresponding coordinates add; and, multiplication of X by the number t multiplies each coordinate c_j by t. Thus, as far as linear operations are concerned, S behaves just like R^k (S is isomorphic to R^k). In particular, any $k + 1$ vectors in S are linearly dependent. If $d = \dim S$, there exist d independent vectors in S. Thus $d \leq k$. But $k \leq d$ by the definition of d.

Suppose $V_1, \ldots, V_n$ is a basis for R^n. Then we can describe each X in R^n by its coordinates relative to that basis, as well as by its standard coordinates. The standard coordinates are the coordinates relative to the basis $E_1, \ldots, E_n$ (1.26). How do we get from one set of coordinates to the other? If

$$(1.28) \qquad X = (x_1, \ldots, x_n) = c_1 V_1 + \cdots + c_n V_n$$

then

$$x_j = c_1 v_{1j} + \cdots + c_n v_{nj}, \qquad 1 \leq j \leq n.$$

That says that

$$(1.29) \qquad\qquad X = CQ$$

where $X = (x_1, \ldots, x_n)$ and $C = (c_1, \ldots, c_n)$ are $1 \times n$ matrices and where Q is the $n \times n$ matrix

$$Q = \begin{bmatrix} v_{11} & \cdots & v_{1n} \\ \cdot & & \cdot \\ \cdot & & \cdot \\ \cdot & & \cdot \\ v_{n1} & \cdots & v_{nn} \end{bmatrix}.$$

Since $V_1, \ldots, V_n$ is a basis, each vector E_i is a linear combination of $V_1, \ldots, V_n$:

$$E_i = p_{i1} V_1 + \cdots + p_{in} V_n, \qquad 1 \leq i \leq n.$$

The $n \times n$ matrix $P = [P_{ij}]$ then has the property that

$$(1.30) \qquad\qquad C = XP$$

for each X in R^n. The matrices P and Q are (easily seen to be) related by

$$(1.31) \qquad\qquad PQ = QP = I$$

where I is the $n \times n$ identity matrix: $I_{ij} = \delta_{ij}$. Thus, the matrix Q is **invertible** and $P = Q^{-1}$ (similarly $Q = P^{-1}$).

If we begin with n vectors $V_1, \ldots, V_n$ in R^n, they form a basis for R^n

if and only if the matrix $Q = [v_{ij}]$ is invertible. In turn, that happens if and only if the determinant of Q is different from zero.

The simplest bases are orthonormal bases, and we should say something about them. The vectors X, Y in R^n are called **orthogonal** (perpendicular) if $\langle X, Y \rangle = 0$. The Pythagorean property is immediate from the definition: If X and Y are orthogonal, then

$$|X + Y|^2 = |X|^2 + |Y|^2.$$

The k-tuple of vectors $V_1, \ldots, V_k$ is called **orthogonal** if V_i is orthogonal to V_j for $i \neq j$. The k-tuple is called **orthonormal** if it is orthogonal and each V_i has length 1. Thus, orthonormality means

(1.32) $$\langle V_i, V_j \rangle = \delta_{ij}.$$

Suppose we have an orthonormal k-tuple $V_1, \ldots, V_k$. If X is a linear combination of those vectors

$$X = c_1 V_1 + \cdots + c_k V_k$$

it is easy to compute from (1.32) that

(1.33) $$c_i = \langle X, V_i \rangle, \qquad 1 \leq i \leq k.$$

In particular, if $X = 0$, then $c_i = 0$ for all i. Therefore, orthonormality implies linear independence.

Suppose X is any vector in R^n. Let

(1.34)
$$Y = \langle X, V_1 \rangle V_1 + \cdots + \langle X, V_k \rangle V_k$$
$$Z = X - Y.$$

By the remark in the last paragraph

$$\langle X, V_i \rangle = \langle Y, V_i \rangle.$$

Hence Z is orthogonal to every V_i. Of course, it may be that $Z = 0$; but, clearly $Z = 0$ if and only if X is in the subspace spanned by $V_1, \ldots, V_k$.

Theorem 7. *Every subspace of* R^n *has an orthonormal basis.*

Proof. If $V_1, \ldots, V_k$ is an orthonormal k-tuple of vectors in S, how large can k be? Certainly $k \leq \dim S$, since orthonormality implies independence. Look at the largest possible k. Then every vector X in S must be a linear combination of $V_1, \ldots, V_k$. Otherwise, (1.34) would yield a vector Z in S so that the $(k + 1)$-tuple $V_1, \ldots, V_k, Z/|Z|$ was orthonormal.

Suppose $V_1, \ldots, V_n$ is an orthonormal n-tuple of vectors in R^n. Then they constitute an (orthonormal) basis for R^n. The coordinates of a vector X relative to the basis $V_1, \ldots, V_n$ are easy to compute as in (1.33):

(1.35)
$$X = c_1 V_1 + \cdots + c_n V_n$$
$$c_i = \langle X, V_i \rangle.$$

This is a great simplification over the situation for a general basis, where one needs to invert the matrix $Q = [v_{ij}]$ before many coordinates can be computed. In the orthonormal case, the inverse of Q is the transpose matrix $Q^t = [v_{ji}]$. The conditions $\langle V_i, V_j \rangle = \delta_{ij}$ say precisely that

(1.36) $$QQ^t = I$$

Such a matrix Q is called an **orthogonal matrix.**

The vectors $V_1, \ldots, V_n$:

$$V_i = (v_{i1}, \ldots, v_{in})$$

form an orthonormal basis for R^n if and only if the matrix $Q = [v_{ij}]$ is an orthogonal matrix. If Q is orthogonal, then the coordinates $C = (c_1, \ldots, c_n)$ of the vector X relative to the orthonormal basis $V_1, \ldots, V_n$ are given by

$$C = XQ^t$$

or

$$c_i = \langle X, V_i \rangle.$$

If one wishes to study a subspace S, it is often most convenient to use the **orthogonal complement** of S:

(1.37) $$S^\perp = \{X \in R^n; \langle X, Y \rangle = 0 \text{ all } Y \in S\}.$$

Theorem 8. *Let* S *be a subspace of* R^n. *Each vector* X *in* R^n *is uniquely expressible in the form*

(1.38) $$X = Y + Z, \quad Y \in S, \quad Z \in S^\perp.$$

Proof. Suppose we have X decomposed as in (1.38). Let $V_1, \ldots, V_k$ be any orthonormal basis for S. Since Y is in S

$$Y = \langle Y, V_1 \rangle V_1 + \cdots + \langle Y, V_k \rangle V_k.$$

Since Z is orthogonal to each V_i, we have $\langle Y, V_i \rangle = \langle X, V_i \rangle$. Hence,

(1.39)
$$Y = \langle X, V_1 \rangle V_1 + \cdots + \langle X, V_k \rangle V_k$$
$$Z = X - Y.$$

That determines Y and Z uniquely. On the other hand, given X, we can define Y and Z by those formulas and clearly Y is in S and Z is in $S^\perp$.

If X is in R^n, the vector Y in Theorem 8 is called the **orthogonal projection** of X on S. Even in the proof of the theorem we used a particular orthonormal basis to define Y (1.39); however, as the proof shows, Y is independent of the basis. Notice how norms and inner products behave relative to orthogonal projections. Suppose Y_1 is the orthogonal projection of X_1 on S and Y_2 is the orthogonal projection of X_2:

$$X_1 = Y_1 + Z_1$$
$$X_2 = Y_2 + Z_2.$$

Then

$$\langle X_1, X_2 \rangle = \langle Y_1, Y_2 \rangle + \langle Z_1, Z_2 \rangle$$
$$|X_j|^2 = |Y_j|^2 + |Z_j|^2.$$

We have described subspaces by bases. There is a dual point of view in which we describe a subspace by giving a set of equations for it. The equations are linear; that is, they are of the form $f(X) = 0$ where f is a **linear function**

$$R^n \xrightarrow{f} R$$
$$f(cX + Y) = cf(X) + f(Y).$$

If f is such a (real-valued) linear function on R^n, there are numbers $a_1, \ldots,$ a_n such that f has the form

$$f(X) = a_1 x_1 + \cdots + a_n x_n.$$

Suppose $V_1, \ldots, V_n$ is an ordered basis for R^n. There is an associated k-tuple of **coordinate functions** $f_1, \ldots, f_n$. The function f_i assigns to the vector X its ith coordinate relative to the basis $V_1, \ldots, V_n$. In short

$$X = f_1(X)V_1 + \cdots + f_n(X)V_n.$$

Certainly $f_1, \ldots, f_n$ are linear functions. Their specific form is

$$f_j(X) = p_{1j}x_1 + \cdots + p_{nj}x_n$$

where $P = [p_{ij}]$ is the matrix of (1.30):

$$E_i = p_{i1}V_1 + \cdots + p_{in}V_n.$$

The functions $f_1, \ldots, f_n$ are uniquely determined by the fact that they are linear and that

$$f_i(V_j) = \delta_{ij}.$$

Theorem 9. *If* S *is a* k*-dimensional subspace of* R^n *then* S *is the solution space for a system of* $(n - k)$ *homogeneous linear equations.*

Proof. Let $V_1, \ldots, V_k$ be a basis for S. We can find vectors $V_{k+1}, \ldots,$ V_n such that $V_1, \ldots, V_n$ is a basis for R^n. Consider the corresponding coordinate functions $f_1, \ldots, f_n$. Obviously S consists of all vectors X in R^n such that

$$f_i(X) = 0, \qquad i = k + 1, \ldots, n.$$

Notice that in the proof we could use orthonormal bases. The conclusion is that there are $(n - k)$ vectors V_{k+1}, V_n in R^n such that

$$S = \{X; \langle X, V_i \rangle = 0, i = k + 1, \ldots, n\}.$$

A **hyperplane** in R^n is a level set of a non-zero linear function f:

$$H = \{X; f(X) = c\}$$

A hyperplane is a flat subset of dimension $(n - 1)$. For, if we are given such an H and if we choose any X_0 such that $f(X_0) = c$, then $f(X - X_0) = 0$ for every X in H. Thus X is in H if and only if

$$X = X_0 + Y, \qquad f(Y) = 0.$$

The set of vectors which f sends into 0 is an $(n - 1)$-dimensional subspace. Thus, every hyperplane is a subspace of dimension $(n - 1)$, translated by a fixed vector in R^n. Similarly, we see that each non-empty flat subset of R^n is a subspace, translated by a fixed vector. Theorem 9 says that every k-dimensional subspace of R^n is the intersection of $n - k$ hyperplanes (through the origin). Thus, we see that every non-empty flat subset F is an intersection of hyperplanes; in fact, F can be described by specifying (not more than n) linear conditions on the coordinates of the vectors in F.

2. Convergence and Compactness

2.1. Convergent Sequences

Analysis is founded upon the concept of limit; indeed, the central role played by limiting operations is essentially what defines the branch of mathematics known as analysis. We shall first take up the idea of the limit of a sequence. It is the simplest type of limit, and yet a thorough comprehension of it enables one to understand other limits rather easily.

We shall be working in R^m, Euclidean space of dimension m. The first thing we need is some language for describing when points are near to one another.

Definition. *If* $X_0 \in R^m$ *and* $r > 0$, *the* **open ball** *of radius* r *about the point* X_0 *is*

(2.1) $$B(X_0; r) = \{X; |X - X_0| < r\}.$$

The **closed ball** *of radius* r *about the point* X_0 *is*

(2.2) $$\bar{B}(X_0; r) = \{X; |X - X_0| \leq r\}.$$

A **neighborhood** *of* X_0 *is a subset (of* R^m*) which contains an open ball about* X_0.

A neighborhood of X contains every point which is (sufficiently) close to X. The most important neighborhoods of X are the open balls about X. Notice that the concept of open [closed] ball reverts to open [closed] disk in R^2 and (symmetric) open [closed] interval in R^1.

Definition. *The sequence* $\{X_n\}$ **converges to the point** X *if every neighborhood of* X *contains* X_n, *except for a finite number of values of* n. *If* $\{X_n\}$ *converges to* X, *then we write*

(2.3) $$X = \lim_n X_n.$$

Suppose that $X = \lim X_n$. If $r > 0$, the open ball $B(X; r)$ contains X_n, except for certain integers $n_1, \ldots, n_k$, which presumably depend on r. Let N_r be the maximum of those numbers, and then N_r is a positive integer such that

$$|X - X_n| < r, \qquad n > N_r.$$

Since every neighborhood of X contains some $B(X; r)$, we see that X_n converges to X if and only if, for each positive number r, there exists a positive integer N_r such that

(2.4) $$|X - X_n| < r, \qquad n > N_r.$$

The terminology concerning convergence is frequently bent in various ways. Often, we omit the braces and say "X_n converges to X". We say that the sequence $\{X_n\}$ **converges** (or, **is convergent**; or, **has a limit**) if there exists an X such that $\{X_n\}$ converges to X. If no such X exists, we say that the sequence **diverges**. If $\{X_n\}$ converges to X, we sometimes call X the **limit** of the sequence and we sometimes write

$$X = \lim X_n$$

or

$$X = \lim_{n \to \infty} X_n$$

instead of (2.3).

Be careful about trying to bend the wording of the definition of convergent sequence. There is a distinct difference between "contains X_n, except for a finite number of n's" and "contains all except a finite number of X_n's". In R^1, the sequence $X_n = (-1)^n$ would converge to every $x \in R$, if we used the second wording. Every open interval contains all except a finite number of x_n's, because there are only two x_n's. On the other hand, no interval of length less than 2 has the property that it contains x_n, except for a finite number of values of n. Hence, this sequence $\{x_n\}$ does not converge.

Lemma. *If*

$$X = \lim_n X_n \quad and \quad Y = \lim_n Y_n$$

then

(i) *for every real number* c,

$$cX + Y = \lim_n (cX_n + Y_n)$$

(ii) $$\langle X, Y \rangle = \lim_n \langle X_n, Y_n \rangle.$$

Proof. (*i*) Part (i) is trivial when $c = 0$. Suppose $c \neq 0$, and let r be a positive number. Since

$$|(cX + Y) - (cX_n + Y_n)| \leq |c||X - X_n| + |Y - Y_n|$$

the distance from $cX_n + Y_n$ to $cX + Y$ will be less than r provided

$$|Y - Y_n| < \frac{r}{2}$$

$$|X - X_n| < \frac{r}{2|c|}.$$

Since $\{X_n\}$ converges to X, there is a positive integer M such that

$$|X - X_n| < \frac{r}{2|c|}, \qquad n \geq M.$$

Similarly, since $\{Y_n\}$ converges to Y, there is a positive integer N such that

$$|Y - Y_n| < \frac{r}{2}, \qquad n \geq N.$$

Let $K = \max(M, N)$. Then we shall have

$$|(cX + Y) - (cX_n + Y_n)| < r, \qquad n \geq K.$$

(ii) There is a standard sort of manipulation for this type of argument:

$$\langle X, Y \rangle - \langle X_n, Y_n \rangle = \langle X, Y - Y_n \rangle + \langle X - X_n, Y_n \rangle.$$

Thus by Cauchy's inequality

$$|\langle X, Y \rangle - \langle X_n, Y_n \rangle| \leq |X||Y - Y_n| + |Y_n||X - X_n|.$$

Since $Y = \lim_n Y_n$, we have

$$|Y - Y_n| < 1$$

and thus

$$|Y_n| < 1 + |Y|$$

except for a finite number of n's. If b is the larger of $|X|$ and $1 + |Y|$, we have

$$|\langle X, Y \rangle - \langle X_n, Y_n \rangle| \leq b[|X - X_n| + |Y - Y_n|]$$

except for finitely many n. The remainder of the proof is like the proof of part (i).

Three special cases of this lemma should be noticed. If

$$X = \lim_n X_n$$

$$Y = \lim_n X_n$$

then

$$Y - X = \lim_n (X_n - X_n) = 0.$$

Hence, *a sequence has at most one limit.* In part (ii), if $X_n = Y_n$, we have

$$|X|^2 = \lim_n |X_n|^2.$$

It is then easy to show that we can remove the exponents 2; but, note that it is trivial by another means to see that

$$|X| = \lim_n |X_n|$$

if X_n converges to X, because

$$||X| - |X_n|| \leq |X - X_n|.$$

In R^1, part (ii) states that $x_n y_n$ converges to xy if x_n converges to x and y_n converges to y.

Theorem 1. *The sequence* $\{X_n\}$ *in* $\mathbf{R}^m$:

$$X_n = (x_{n1}, \ldots, x_{nm})$$

converges if and only if each of the m *coordinate sequences* $\{x_{nj}\}$ *converges. If*

$$x_j = \lim_n x_{nj}, \qquad 1 \leq j \leq m$$

then X_n *converges to* $X = (x_1, \ldots, x_n)$.

Proof. Suppose that $\{X_n\}$ converges to a vector $X = (x_1, \ldots, x_m)$. Let j be an index, $1 \leq j \leq m$. Since

$$|x_{nj} - x_j| \leq |X_n - X|$$

and $\lim |X_n - X| = 0$, the sequence $\{x_{nj}\}$ converges to x_j.

Now, suppose that for each j, $1 \leq j \leq m$, the coordinate sequence $\{x_{nj}\}$ is known to converge. Let $x_j = \lim_n x_{nj}$, and define $X = (x_1, \ldots, x_n)$. Observe that

$$|X - X_n| \leq |x_1 - x_{n1}| + |x_2 - x_{n2}| + \cdots + |x_n - x_{nm}|.$$

Each sequence $\{|x_j - x_{nj}|\}$ converges to 0; hence, the previous lemma tells us that their sum converges to 0 also. We conclude that $\lim |X - X_n| = 0$, i.e., $X = \lim X_n$.

EXAMPLE 1. Convergence of sequences $\{z_n\}$ of complex numbers is related to Theorem 1. Of course, we say that z_n converges to z if $|z_n - z|$ converges to 0. There is a natural correspondence between C and R^2, whereby $z = x + iy$ corresponds to the point (x, y). Theorem 1 tells us that $z = \lim_n z_n$ if and only if $Re\,(z_n)$ converges to $Re\,(z)$ and $Im\,(z_n)$ converges to $Im\,(z)$. Analogous remarks apply to the space C^k.

EXAMPLE 2. We shall refer from time to time to convergence of sequences of matrices. Suppose we talk about convergence of sequences of $k \times k$ matrices with real [complex] entries. We are then operating in the space $R^{k \times k}$ [$C^{k \times k}$], Euclidean space with the k^2 coordinates arranged in k rows and k columns. By Theorem 1, that is equivalent to convergence of the corresponding entries. Note that, if A_n converges to A and B_n converges to B, then the matrix product $A_n B_n$ converges to AB. The proof is as in part (ii) of the last lemma:

$$|AB - A_n B_n| \leq |A||B - B_n| + |A - A_n||B_n|.$$

We have used here the complex version of Theorem 5 of Chapter 1:

$$|ST| \leq |S||T|.$$

EXAMPLE 3. In many problems, we are given a sequence $\{X_n\}$ and we are interested in the convergence of the successive sums

(2.5) $$S_n = X_1 + \cdots + X_n.$$

We then speak of the **infinite series**

(2.6) $$\sum_n X_n$$

and we call $S_1, S_2, \ldots$ (2.5) the **partial sums** of the series. We say that the series **converges** if the sequence $\{S_n\}$ converges; and if

$$S = \lim_n S_n$$

we call S the **sum** of the series, and we write

(2.7) $$S = \sum_n X_n \quad \text{or} \quad S = \sum_{n=1}^{\infty} X_n.$$

The symbolism (2.6) for the series is a little sloppy, but very convenient. When we say that $\sum_n X_n$ "is" an infinite series, we mean that $\{X_n\}$ is a sequence and what interests us is the question of the convergence of the partial sums S_n. But, if S is a vector and we write (2.7), that means that the series converges and S is the sum.

In Section 1.3, we associated with each number $x \in [0, 1]$ a decimal representation

$$x \sim .a_1 a_2 a_3. \ldots .$$

In series notation that is now seen to be:

$$x = \sum_n a_n 10^{-n}.$$

If z is a complex number and $|z| < 1$, then the series

$$\sum_{n=0}^{\infty} z^n$$

converges, and its sum is $(1 - z)^{-1}$. For this series,

$$S_n = 1 + z + \cdots + z^n = (1 - z^{n+1})(1 - z)^{-1}.$$

Since $|z| < 1$, $\lim_k z^k = 0$. Thus $1 - z^{n+1}$ converges to 1 and S_n converges to $(1 - z)^{-1}$. Symbolically,

$$\frac{1}{1-z} = \sum_{n=0}^{\infty} z^n, \qquad |z| < 1.$$

Exercises

1. Let X be a point of R^m. What is the union of all neighborhoods of X? What is the intersection of all neighborhoods of X? Is that intersection a neighborhood of X?

2. Let S be a bounded non-empty set of real numbers. Prove that there exists a sequence of points $x_n \in S$ such that

$$\lim_n x_n = \sup S.$$

If $\{X_n\}$ is any sequence of points in S which converges, then

$$\inf S \le \lim_n x_n \le \sup S.$$

3. True or false? If $\{z_n\}$ is a sequence of complex numbers which converges, then the sequence $\{z_n^{-1}\}$ converges.

4. True or false? If x_n converges to x, then the greatest integer in x_n converges to the greatest integer in x.

5. True or false? If X_n converges to X, then the angle between (the vectors) X_n and X converges to 0.

6. If

$$x_n = 1 + 2^{-1} + \cdots + 2^{-n}$$

and x_n converges to x, find the smallest N such that

$$|x - x_n| < 10^{-2}, \qquad n \ge N.$$

7. If $x_n \ge 0$ and x_n converges to x, then $\sqrt{x_n}$ converges to $\sqrt{x}$.

8. Let A be a square matrix. Look at the sequence of its powers. Show that if A^n converges to B, then $AB = B$. Give an example where the sequence $\{A^n\}$ does not converge, yet $|A^n|$ remains bounded.

9. Let z be a complex number. Prove that the sequence

$$\frac{z^n}{n!}$$

is bounded. From the fact that it is bounded, show that it converges to 0. From this fact prove that, if $\epsilon > 0$, there is a constant K such that

$$\frac{|z^n|}{n!} \le K\epsilon^n$$

for all except a finite number of n's.

10. Let S be a (linear) subspace of R^m. If X is a vector in R^m, let $P(X)$ be the orthogonal projection of X onto the subspace S. Show that if X_n converges to X, then $P(X_n)$ converges to $P(X)$.

11. Let $\{A_n\}$ be a sequence of invertible $k \times k$ matrices. Suppose A_n converges to A but A is not invertible. Show that

$$\text{``}\lim_n |A_n^{-1}| = \infty\text{''}.$$

12. Prove that if $x_n \in R$ and x_n converges to x, then the sequence of arithmetic means

$$s_n = \frac{1}{n}(x_1 + \cdots + x_n)$$

also converges to x.

2.2. Convergence Criteria

It is extremely important to develop criteria which will ensure that a sequence converges, in spite of the fact that we do not know the limit explicitly. How can we tell if a sequence converges? One crude test is the **boundedness** of the sequence

$$|X_n| \leq M, \qquad n = 1, 2, 3, \ldots$$

Every convergent sequence is bounded; hence, boundedness is a necessary condition for convergence. But it is a very long way from being sufficient. Many bounded sequences fail to converge. The interesting sufficient conditions are derived from the completeness of the real number system.

Theorem 2. *Let*

$$x_1 \leq x_2 \leq x_3 \leq \cdots$$

be a monotone-increasing sequence of real numbers. The sequence converges if and only if it is bounded.

Proof. The content here is in the "if" half of the theorem. The hypothesis is then that

$$x_1 \leq x_2 \leq x_3 \leq \cdots \leq b$$

where $b \in R$. Thus b is an upper bound for the set of values of the sequence. Let x be the least upper bound for that set:

$$x = \sup \{x_n; n \in Z_+\}.$$

Let $r > 0$. Then $x - r$ is not an upper bound for the set of x_n's. Thus, there exists a positive integer N_r such that

$$x_{N_r} > x - r.$$

Since $x_n \leq x_{n+1} \leq x$, we have

$$x - r < x_n \leq x < x + r, \qquad n > N_r.$$

Theorem 2 is another in the list of reformulations of the completeness of the real number system. It is just the sequential form of the existence of least upper bounds. Evidently, there is a companion result about monotone decreasing sequences.

Let $x_1, x_2, x_3, \ldots$ be any bounded sequence of real numbers. For each n, define

(2.8)
$$a_n = \inf \{x_k; k \geq n\}$$
$$b_n = \sup \{x_k; k \geq n\}.$$

Then we have two monotone sequences, one increasing and the other decreasing:

$$a_1 \leq a_2 \leq a_3 \leq \cdots \leq b_3 \leq b_2 \leq b_1.$$

The **limit inferior** and **limit superior** of the sequence $\{x_n\}$ are then defined by

(2.9)
$$\liminf x_n = \lim_n a_n$$
$$\limsup x_n = \lim_n b_n.$$

Obviously,

$$\liminf x_n \leq \limsup x_n.$$

Equality holds if and only if $\{X_n\}$ converges. If it converges, it converges to the common value of the lim inf and lim sup.

Theorem 3. *Let $\{x_n\}$ be a bounded sequence of real numbers. The sequence converges if and only if*

$$\liminf x_n = \limsup x_n.$$

If it converges, it converges to the common value of the limit inferior and limit superior.

Proof. Define a_n and b_n by (2.8). First, suppose the sequence $\{x_n\}$ converges to the number x. Given $\epsilon > 0$, we can find an N such that

$$x - \epsilon < x_n < x + \epsilon, \qquad n \geq N.$$

Therefore,

$$x - \epsilon \leq a_n \leq x + \epsilon$$
$$x - \epsilon \leq b_n \leq x + \epsilon, \qquad n \geq N.$$

It should now be clear that

$$x = \lim_n a_n = \lim_n b_n.$$

Now, suppose that we are given a sequence $\{x_n\}$ with

$$\liminf x_n = \limsup x_n.$$

Let $x = \lim a_n = \lim b_n$ and let us show that x_n converges to x. Let $\epsilon > 0$. There exist positive integers M, N such that

$$x - \epsilon < a_n < x + \epsilon, \qquad n \geq M$$
$$x - \epsilon < b_n < x + \epsilon, \qquad n \geq N.$$

From the definitions of a_n and b_n, we then have

$$x - \epsilon < x_n, \qquad n \geq M$$
$$x_n < x + \epsilon, \qquad n \geq N.$$

Thus,

$$|x - x_n| < \epsilon, \qquad n \geq \max(M, N).$$

One frequently sees the notations $\overline{\lim}\, x_n$ and $\underline{\lim}\, x_n$ employed for lim sup x_n and lim inf x_n, respectively. In attempting to picture the geometrical relationship of the superior and inferior limits to the sequence $\{x_n\}$, the following observation (which we state only for lim sup) is sometimes useful. If $s = \lim \sup x_n$ and $\epsilon > 0$, then $x_n \geq s + \epsilon$ holds for only a finite number of values of n and $x_n > s - \epsilon$ holds for infinitely many values of n.

The completeness of the real number system will (evidently) be reflected in some type of completeness of Euclidean space R^m. The sequential form of that completeness is conveniently phrased in terms of the Cauchy convergence criterion.

Start with a sequence $\{X_n\}$ in R^m. *If* it converges to some point X, then the various X_n's with large subscripts must be close to one another, because they are all close to the limit point X.

Definition. *A* **Cauchy sequence** *is a sequence* $\{X_n\}$ *with this property. For each* $\epsilon > 0$, *there exists a positive integer* N_ϵ *such that*

(2.10) $$|X_k - X_n| < \epsilon, \qquad k > N_\epsilon, \quad n > N_\epsilon.$$

We have just commented that *every convergent sequence is a Cauchy sequence.* In that case, a positive integer N_ϵ (2.10) can be determined precisely this way. If X_n converges to X, choose N_ϵ so that

$$|X - X_n| < \frac{\epsilon}{2}, \qquad n > N_\epsilon.$$

Then (2.10) is satisfied.

Theorem 4 (*Completeness of* R^m). *Every Cauchy sequence in* R^m *converges.*

Proof. Suppose

$$Y_n = (x_{n1}, \ldots, x_{nm}), \qquad n = 1, 2, 3, \ldots$$

is a sequence of points in R^m. For each coordinate index j

$$|x_{kj} - x_{nj}| \leq |X_k - X_n|.$$

Therefore, if $\{X_n\}$ is a Cauchy sequence, each of the m sequences $\{x_{nj}\}$ is a Cauchy sequence in R. If we show that each of those sequences converges, we will know that the sequence $\{X_n\}$ converges (Theorem 1). In other words, we need only prove the theorem in R^1.

Let $\{x_n\}$ be a Cauchy sequence of real numbers. How will we find a number $x \in R$ to which it converges? First, let us note that the sequence is bounded. By the Cauchy condition, there is a positive integer N such that

$$|x_k - x_n| < 5, \qquad k \geq N, \quad n \geq N.$$

In particular,

$$|x_N - x_n| < 5, \qquad n \geq N.$$

Let M be the largest of the numbers

$$|x_1|, \ldots, |x_{N-1}|, 5 + |x_N|$$

and plainly $|x_n| \leq M$ for all n.

Now, let

$$x = \lim \inf x_n.$$

We remind the reader what that means:

$$a_n = \inf\{x_k; k \geq n\}$$
$$a_1 \leq a_2 \leq a_3 \leq \cdots$$
$$x = \lim_n a_n.$$

We have just used the completeness of the real numbers, in the form of Theorem 2, to tell us that x exists. We claim that x_n converges to x.

Let $\epsilon > 0$. We wish to show that x_n is in the open interval $(x - \epsilon, x + \epsilon)$, except for a finite number of n's. Why is any x_n in that interval? Look at the definition of x. We can find N so that

$$(2.11) \qquad x - \epsilon < a_N \leq x.$$

Now, look at the definition of a_N. Together with (2.11), it tells us two things:

(i) $x_n > x - \epsilon$ for all $n \geq N$.

(ii) There exists some $k \geq N$ such that $x_k < x + (\epsilon/2)$.

(In (ii), we could replace $x + (\epsilon/2)$ by any number greater than x.)

Now apply the Cauchy condition. There is a positive integer P such that

$$(2.12) \qquad |x_k - x_n| < \frac{\epsilon}{2}, \qquad k \geq P, \quad n \geq P.$$

We may assume that the N in (2.11) is greater than P, because, if (2.11) holds for a particular N, it holds for every larger N. If $N \geq P$, then (i), (ii), and (2.12) tell us that

$$x - \epsilon < x_n < x + \epsilon, \qquad n \geq N.$$

EXAMPLE 4. One of the most useful special cases of Theorem 3 is the following. Suppose $\{X_n\}$ is a sequence and

(2.13) $|X_n - X_{n+1}| < 2^{-n}, \quad n = 1, 2, 3, \ldots$

Then $\{X_n\}$ is a Cauchy sequence. Why? Suppose $k < n$. Then

$$|X_k - X_n| \leq |X_k - X_{k+1}| + |X_{k+1} - X_{k+2}| + \cdots + |X_{n-1} - X_n|$$
$$\leq 2^{-k} + 2^{-(k+1)} + \cdots + 2^{-(n-1)}$$
$$= 2(2^{-k} - 2^{-n})$$
$$< 2^{-(k-1)}.$$

Consequently, any sequence in R^m which satisfies (2.13) is convergent.

EXAMPLE 5. The Cauchy condition on a particular sequence frequently arises in this way. In addition to the sequence of points X_n, we have a sequence of sets $S_1, S_2, S_3, \ldots$ with these properties:

(i) $S_1 \supset S_2 \supset S_3 \supset \cdots$;
(ii) $X_n \in S_n$;
(iii) $\lim_n \text{diam}(S_n) = 0$.

Here diam (S) is the **diameter** of the set S:

$$\text{diam}(S) = \sup\{|X - Y|; X \in S, Y \in S\}.$$

This is finite if the set S is bounded, i.e., if S is contained in some ball about the origin. Clearly (i), (ii), and (iii) imply that we have a Cauchy sequence:

$$|X_k - X_n| \leq \text{diam}(S_N), \quad k \geq N, \quad n \geq N.$$

In fact, the existence of such a sequence of sets is just a reformulation of the Cauchy property.

Exercises

1. True or false? If $|X_1| \geq |X_2| \geq \cdots$, then the sequence $\{X_n\}$ converges.

2. Give an example of a sequence for which

$$\lim_n |X_n - X_{n+1}| = 0,$$

but which is not a Cauchy sequence.

3. Describe a Cauchy sequence of rational numbers which does not converge *in the rational number system.*

4. If z is a complex number, what is

$$\lim \inf |z^n|?$$

5. True or false? If the sequence $\{X_n\}$ converges, then the set of norms

$$\{|X_n|; n \in Z_+\}$$

has a largest member.

6. True or false? If the infinite series $\sum X_n$ converges, then the set of norms

$$\{|X_n|; n \in Z_+\}$$

has a largest number in it.

7. Let $\{x_n\}$ be a bounded sequence of real numbers. Let A be the set of real numbers t such that $x_n < t$ for only a finite number of n's. Show that

$$\lim \inf x_n = \sup A.$$

8. True or false? If z is a complex number and $|z| \geq 1$, then $\sum z^n$ diverges.

9. True or false? If A is a $k \times k$ matrix and $|A| \geq 1$, then $\sum A^n$ diverges.

10. Let

$$x_n = (1 - 2^{-1})(1 - 2^{-2}) \cdots (1 - 2^{-n}).$$

Prove that the sequence converges and $\lim x_n \neq 0$.

11. Let $\{x_n\}$ and $\{y_n\}$ be (bounded) sequences of real numbers. Prove that

$$\lim \sup (x_n + y_n) \leq \lim \sup x_n + \lim \sup y_n.$$

12. Prove this generalization of Example 4. If $\{X_n\}$ is a sequence in R^m such that

$$\sum_{n=1}^{\infty} |X_n - X_{n+1}| < \infty$$

then $\{X_n\}$ is a Cauchy sequence.

13. True or false? In R^1, every convergent sequence is the sum of an increasing sequence and a decreasing sequence.

14. Assume the monotone convergence theorem. Prove that every non-empty set of real numbers which is bounded above has a least upper bound.

15. If you knew that every Cauchy sequence in R converged, how would you prove the monotone convergence theorem? (You will need to use the Archimedean ordering property.)

2.3. Infinite Series

Now we shall see what the convergence criterion of the last section tells us about infinite series. We shall concentrate on the two most important classes of infinite series, namely, series with positive terms and absolutely convergent series.

We shall make frequent use of the series analogue of sums and scalar multiples of convergent sequences: If $\sum X_n = S$ and $\sum Y_n = T$, then $\sum (cX_n + Y_n) = cS + T$.

Suppose that we have an infinite series in which the terms are non-negative real numbers:

$$\sum_n x_n$$

$$x_n \geq 0.$$

Then the partial sums constitute an increasing sequence:

$$s_n = \sum_{k=1}^{n} x_k$$

$$s_1 \leq s_2 \leq s_3 \leq \cdots.$$

If that sequence is bounded, the monotone convergence theorem tells us that it converges and hence (by definition) the infinite series converges. If the sequence $\{s_n\}$ is not bounded, it does not converge and so the infinite series diverges.

The two possibilities for a series with non-negative terms are conveniently described by the notation

$$\sum_n x_n < \infty \qquad \text{(convergence)}$$

$$\sum_n x_n = \infty \qquad \text{(divergence)}.$$

EXAMPLE 6. Two simple examples with which the reader should be familiar are

$$\sum_n \frac{1}{n}, \qquad \sum_n \frac{1}{n^2}.$$

The first series diverges, because the grouping

$$(2.14) \qquad 1 + \tfrac{1}{2} + (\tfrac{1}{3} + \tfrac{1}{4}) + (\tfrac{1}{5} + \tfrac{1}{6} + \tfrac{1}{7} + \tfrac{1}{8}) + \cdots$$

shows that the $2n$th partial sum exceeds $(n+1)\tfrac{1}{2}$. There are several ways to show that the second series converges. One way is to verify by mathematical induction that

$$(2.15) \qquad 1 + \frac{1}{2} + \frac{1}{9} + \cdots + \frac{1}{n^2} \leq 2 - \frac{1}{n}.$$

It then follows that

$$\sum_n \frac{1}{n^2} \leq 2.$$

Notice that, once we have verified (2.15), the monotone convergence theorem guarantees that the series converges, but it does not tell us what the sum of the series is. As a matter of fact,

$$\sum_n \frac{1}{n^2} = \frac{\pi^2}{6}.$$

but that is hardly obvious.

Given the non-negative series $\sum x_n$, it is not always easy to determine whether or not the partial sums are bounded. Often, a certain amount of cleverness or experience is needed. A beginner would probably have to fiddle with the harmonic series $\sum 1/n$ for quite a while to guess whether or not it converged; and, even if he guessed that it diverged, it might be another while before he thought of proving it by grouping the terms as in (2.14). We showed that

$$\sum_n \frac{1}{n^2} < \infty$$

by verifying the inequality (2.15), which is probably not the first thing one would think of, upon being confronted with that infinite series.

As one builds a stockpile of specific series which are known to converge or diverge, it becomes possible to test new series for convergence by comparing them with the known series. If

$$\sum_n y_n < \infty$$

and $0 \leq x_n \leq y_n$, obviously

$$\sum_n x_n < \infty.$$

In fact, it is enough to know that $x_n \leq y_n$ for all sufficiently large n:

$$x_n \leq y_n, \qquad n \geq N.$$

Similarly, if

$$\sum_n y_n = \infty$$

and $x_n \geq y_n \geq 0$ for all sufficiently large n, then $\sum x_n$ diverges.

EXAMPLE 7. In Section 2.1, we verified that the series $\sum z^n$ converges for all complex numbers of absolute value less than 1:

$$\frac{1}{1-z} = \sum_{n=0}^{\infty} z^n, \qquad |z| < 1.$$

In particular,

(2.16) $$\sum_n x^n < \infty, \qquad 0 \leq x < 1.$$

A number of different series can be seen to converge by comparing their terms with the terms of a geometrical series (2.16). For instance,

(2.17) $$\sum_n \frac{n^2}{2^n} < \infty$$

because

(2.18) $$\frac{n^2}{2^n} < \left(\frac{2}{3}\right)^n, \qquad \text{for large } n.$$

Why? The inequality (2.18) says that

$$(\tfrac{3}{4})^n n^2 < 1, \qquad \text{for large } n$$

which is true, because

$$\lim_n n^2 t^n = 0$$

for any fixed t such that $0 \leq t < 1$. This last assertion can be proved as follows. Since

$$\lim_n \frac{n}{n+1} = 1$$

and $t < 1$, there exists an N such that

$$t < \left(\frac{n}{n+1}\right)^2, \qquad n \geq N.$$

This inequality may be rewritten

$$(n+1)^2 t^{n+1} < n^2 t^n, \qquad n \geq N.$$

Thus the sequence $\{n^2 t^n\}$ is monotone decreasing for large n and, accordingly, it converges. If the limit were not 0, we could divide and obtain the contradiction

$$1 = \frac{\lim n^2 t^n}{\lim (n+1)^2 t^{n+1}}$$

$$= \lim \left(\frac{n}{n+1}\right)^2 t^{-1}$$

$$= t^{-1}.$$

The reader should be aware that the series

$$\sum_n \frac{x^n}{n!}$$

converges for every real number x. For the moment, let's worry only about non-negative numbers x. If $0 \leq x < 1$, the series obviously converges. The interesting point is that it converges for large x. Fix such an x. What happens to

$$\frac{x^n}{n!}$$

as n gets large? We pass from the nth term to the $(n+1)$th term by multiplying by $x/(n+1)$. Once n is large enough so that $n + 1 > x$, we are multiplying by a number less than 1. It should be clear now that the series converges, and converges faster than a geometric series. If we want to be precise, we can phrase it this way. Given x, choose a positive integer N such that

$$\frac{x}{N+1} = t < 1.$$

Then

$$\sum_n \frac{x^n}{n!} \leq \sum_{n=1}^{N} \frac{x^n}{n!} + \sum_{n \geq N+1} \frac{x^n}{n!}$$

$$\leq \sum_1^N \frac{x^n}{n!} + \frac{x^N}{N!} \sum_{k=1}^{\infty} t^k$$

$$< \infty.$$

The series which we have just been discussing is the power series for the exponential function and we shall say more about it shortly. The reader should be familiar with the case $x = 1$ which defines the number e:

$$e = \sum_{n=0}^{\infty} \frac{1}{n!} = 1 + 1 + \frac{1}{2!} + \frac{1}{3!} + \cdots.$$

Now that we have indicated what the monotone convergence theorem tells us about non-negative series, let us see what the Cauchy convergence criterion tells us about infinite series. Suppose that we have a (now vector-valued) series with terms X_n and partial sums S_n:

$$S_n = \sum_{k=1}^{n} X_k.$$

By definition, the series converges if the sequence $\{S_n\}$ converges. According to Theorem 4, the series converges if and only if $\{S_n\}$ is a Cauchy sequence. The crudest possible estimate for the distance from S_k to S_n (with $n > k$) is

$$|S_k - S_n| = |X_{k+1} + \cdots + X_n|$$
$$\leq |X_{k+1}| + \cdots + |X_n|.$$

How can we guarantee that the last sum is small, provided that k and n are both large? Well

$$|X_{k+1}| + \cdots + |X_n| < \epsilon, \qquad N \leq k < n$$

means that

$$(2.19) \qquad \sum_{n=N+1}^{\infty} |X_n| \leq \epsilon.$$

What does it mean to say that, for each $\epsilon > 0$, there exists an N so that (2.19) holds? It means that

$$(2.20) \qquad \sum_{n} |X_n| < \infty$$

i.e., that the series of numbers (2.20) converges. An infinite series such that (2.20) holds is said to **converge absolutely**. What we just observed was this. *If an infinite series (in R^m) converges absolutely, then it converges.* The converse is by no means true. Many series converge without converging absolutely. In a sense, absolutely convergent series are the ones which "really" converge, because for such series we can claim that the sum of the series is just the sum of all the vectors X_n. We may commute and associate the vectors in any way before we sum. Let us see why.

Suppose that we have an absolutely convergent series $\sum X_n$. The sum S satisfies

$$|S - \sum_{1}^{n} X_k| \leq \sum_{n+1}^{\infty} |X_k|.$$

In fact, if A is any subset of the positive integers, then

$$(2.21) \qquad |S - \sum_{k \in A} X_k| \leq \sum_{k \notin A} |X_k|.$$

It is this fact which makes all rearrangements possible. Let $A_1, A_2, \ldots$ be a sequence of subsets of the positive integers such that each n is in precisely one A_k:

$$Z_+ = \bigcup_n A_n$$

$$A_n \cap A_k = \varnothing, \qquad n \neq k.$$

For each n, the series

$$\sum_{k \in A_n} X_k$$

converges, because it converges absolutely:

$$\sum_{k \in A_n} |X_k| \leq \sum_k |X_k| < \infty,$$

and we assert that

$$\sum_k X_k = \sum_n \left(\sum_{k \in A_n} X_k \right).$$

In other words, if

$$Y_n = \sum_{k \in A_n} X_k$$

then the series $\sum_n Y_n$ converges and

$$\sum_n Y_n = S = \sum_k X_k.$$

That conclusion can be obtained directly from (2.21):

$$|S - Y_1| \leq \sum_{k \notin A_1} |X_k|$$

$$|S - (Y_1 + Y_2)| \leq \sum_{k \notin (A_1 \cup A_2)} |X_k|, \qquad \text{etc.}$$

EXAMPLE 8. If a series converges but is not absolutely convergent, one must be very careful about rearranging the order of the terms or grouping them in various ways in order to find the sum of the series. For instance, the alternating harmonic series

$$\sum_n (-1)^n \frac{1}{n}$$

converges (see Exercise 4); however, we cannot compute the sum as

$$\sum_{n \text{ even}} x_n + \sum_{n \text{ odd}} x_n$$

because neither of those series converges. A well-known fact, which we shall not stop to prove, is this. If $\sum_n x_n$ is an infinite series of real numbers which converges but does not converge absolutely, then not only are there rearrangements of the order of the terms which yield divergent series, in fact, if s is any real number, the order of the terms can be so rearranged as to yield a series which converges to s. Evidently, unless a series is absolutely convergent, one should not try to think of the sum of the series as being "the sum of all the terms".

EXAMPLE 9. Let A be a $k \times k$ matrix with complex entries. Suppose $|A| < 1$. Then the series

$$\sum_{n=0}^{\infty} A^n, \quad (A^0 = I)$$

converges absolutely because $|A^n| \leq |A|^n$, $n \neq 0$. It should converge to

$$\frac{1}{1 - A}$$

but that doesn't exactly make sense. What does make sense is the inverse of the matrix $I - A$. Let

$$(2.22) \qquad\qquad B = \sum_{0}^{\infty} A^n, \quad |A| < 1.$$

It is easy to verify that

$$(2.23) \qquad\qquad B(I - A) = (I - A)B = I$$

i.e., $I - A$ is invertible and $B = (I - A)^{-1}$.

Let A be any $k \times k$ matrix with complex entries. We define the **exponential** of A to be

$$\exp(A) = e^A = I + A + \frac{1}{2!}A^2 + \cdots$$

$$(2.24) \qquad\qquad = \sum_{0}^{\infty} \frac{1}{n!} A^n.$$

Again, the series converges absolutely because (see Example 7)

$$\sum_{1}^{\infty} \left| \frac{1}{n!} A^n \right| \leq \sum_{1}^{\infty} \frac{1}{n!} |A|^n = e^{|A|} - 1.$$

If B is a matrix which commutes with A, $AB = BA$, then we can show that

$$e^{(A+B)} = e^A e^B = e^B e^A.$$

For any fixed matrix M

$$\sum_{0}^{\infty} \frac{1}{n!} A^n M = e^A M.$$

Thus,

$$e^A e^B = \sum_{0}^{\infty} \frac{1}{n!} A^n e^B$$

$$= \sum_{0}^{\infty} \frac{1}{n!} A^n \left(\sum_{k=0}^{\infty} \frac{1}{k!} B^k \right)$$

Since

$$\sum_{n} \sum_{k} \frac{1}{n!} \frac{1}{k!} |A^n B^k| < \infty$$

we may regroup to obtain

$$e^A e^B = \sum_{N=0}^{\infty} \sum_{k+n=N} \frac{1}{k! \, n!} A^n B^k.$$

Since $AB = BA$, that says

$$e^A e^B = \sum_N \frac{1}{N!}(A + B)^N = e^{(A+B)}.$$

Exercises

1. Let x and t be any positive real numbers. Show that

$$\frac{x^n}{n!} < t^n$$

for all sufficiently large n.

2. True or false? If $0 \le t < 1$, then

$$\sum_n n!\, t^{n^2} < \infty.$$

3. In view of Exercises 1 and 2, what can you say about the behavior of the nth root of $n!$ as n gets large?

4. Let $\{x_n\}$ be a sequence of non-negative real numbers which converges monotonely to 0. Prove that the infinite series

$$\sum_n (-1)^n x_n$$

converges.

5. Let $\sum X_n$ be an absolutely convergent series of vectors in R^k, and let $\{Y_n\}$ be a bounded sequence of vectors in R^k. Show that the series of numbers

$$\sum_n \langle X_n, Y_n \rangle$$

is absolutely convergent.

6. Prove the inequality (2.21), meaning: If $A \subset Z_+$ then $\sum_{k \in A} X_k$ converges and (2.21) holds.

7. Let $\{X_n\}$ be a sequence of vectors in R^k such that

$$\sum_n |X_n - X_{n+1}| < \infty$$

Prove that the sequence $\{X_n\}$ converges. (Such a sequence is called a *fast Cauchy sequence*.)

8. By comparison with a geometric series, prove the following about a (bounded) sequence of complex numbers $\{x_n\}$.

   (a) (Ratio test) Let

$$s = \lim \sup \left| \frac{x_{n+1}}{x_n} \right|.$$

The series $\sum_n x_n$ converges absolutely if $s < 1$.

   (b) (Root test) Let

$$t = \lim \sup \sqrt[n]{|x_n|}.$$

The series $\sum_n x_n$ converges absolutely if $t < 1$ and diverges if $t > 1$.

9. Show that

$$\sum_n n^{-(3/2)} < \infty$$

by using induction to verify that

$$s_n \leq 3 - 2n^{-(1/2)}.$$

10. Let $\{x_n\}$ be a monotone decreasing sequence of positive numbers. Prove that $\sum_n x_n$ converges if and only if

$$\sum_k 2^k x_{2^k}$$

converges.

11. Use the result of Exercise 10 to investigate the convergence of

$$\sum_n n^{-q}, \qquad q > 0.$$

12. Let $x_1, x_2, \ldots$ be any sequence of non-negative numbers such that

$$\sum_n x_n < \infty.$$

Prove that there exist positive integers $N_1, N_2, \ldots$ with these properties:

(i) $N_1 \leq N_2 \leq N_3 \leq \cdots$;
(ii) $\lim_k N_k = \infty$, that is, $\{N_k\}$ is unbounded;
(iii) $\sum_k N_k x_k < \infty$.

13. Prove that

$$e^x \geq 1 + x$$

for all real numbers x. One procedure might be to verify in order:

$$e^x \geq 1 + x, \qquad x \geq 0$$

$$e^x \leq \frac{1}{1-x}, \qquad 0 \leq x < 1$$

$$e^x \geq 1 + x, \qquad x \geq -1$$

$$e^x \geq 1 + x, \qquad x \in R.$$

14. Use the result of Exercise 13 to show that

$$\exp\left(\frac{|z|}{|z|-1}\right) \leq |1 + z| \leq \exp(|z|)$$

for all complex numbers z with $|z| < 1$.

15. If A is a $k \times k$ matrix, show that matrix e^A is invertible.

16. Let A and B be $k \times k$ matrices of norm less than 1. Discover a simple relationship between $(I - AB)^{-1}$ and $(I - BA)^{-1}$. In what sense does that relationship hold for all A and B?

17. Let $\{w_n\}$ be a sequence of complex numbers. The infinite product

$$\prod_n w_n$$

is said to converge if the sequence of partial products

$$p_n = \prod_{k=1}^{n} w_k = w_1 w_2 \cdots w_n$$

converges to a *non-zero* number w. Prove the following, by using the results of Exercises 7 and 13: If $\{z_n\}$ is a sequence of complex numbers $(z_n \neq -1)$ such that

$$\sum_n |z_n| < \infty$$

then the infinite product

$$\prod_n (1 + z_n)$$

converges.

18. Let $\sum x_n$ be a series of real numbers which converges but does not converge absolutely. Show that, if t is a real number, the order of the terms in the series can be rearranged so that the rearranged series converges to t. *Hint*: Take enough positive terms so that their sum just exceeds t, then add enough negative terms so the sum is just under t, etc.

*19. Let $\sum X_n$ be a series of vectors in R^m which converges but does not converge absolutely. Let S be the set of all vectors which are sums of rearrangements of the series. What kind of a set can S be?

*20. Prove that

$$\det e^A = e^{\operatorname{tr}\,(A)}$$

for all $k \times k$ complex matrices A ($\det$ = determinant function; tr = trace function).

2.4. Sequential Compactness

The sequence $\{X_n\}$ converges to the point X if X_n is near X for all sufficiently large n. There are many situations in which we are given a sequence and we don't need to know that it converges. What we need to know is that there is a point X at which the sequence accumulates, i.e., X_n is near X for infinitely many values of n.

Definition. *The point* X *is a* **point of accumulation (accumulation point)** *of the sequence* $\{X_n\}$ *if every neighborhood of* X *contains* X_n *for infinitely many values of* n.

We can say it another way: X is an accumulation point of $\{X_n\}$ if, for each $\epsilon > 0$ and each positive integer n, there exists $k > n$ such that

$$|X - X_k| < \epsilon.$$

If $\{X_n\}$ converges to X, then clearly X is the unique point of accumulation of the sequence.

EXAMPLE 10. Let $r_1, r_2, r_3, \ldots$ be the sequence which consists of the positive rational numbers, enumerated according to the scheme

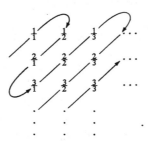

Then, every non-negative real number is an accumulation point of the sequence $\{r_n\}$.

EXAMPLE 11. Beware of working with coordinates when discussing accumulation points. Consider in R^2

$$X_n = (0, 1), \qquad n \text{ odd}$$
$$X_n = (1, 0), \qquad n \text{ even.}$$

The sequence of first coordinates is 0, 1, 0, 1, ... which has two accumulation points in R, 0 and 1. The sequence of second coordinates is 1, 0, 1, 0, ... and it has the same accumulation points. In particular, 0 is an accumulation point for the first coordinates and for the second coordinates. We cannot conclude that (0, 0) is a point of accumulation of the sequence in R^2.

Definition. *The sequence* $Y_1, Y_2, Y_3, \ldots$ *is a* **subsequence** *of the sequence* $X_1, X_2, X_3, \ldots$ *if there exist positive integers* $n_1, n_2, n_3, \ldots$ *such that*

(i) $n_1 < n_2 < n_3 < \cdots$;
(ii) $Y_k = X_{n_k}$.

In other words, a subsequence of $\{X_n\}$ is any sequence $X_{n_1}, X_{n_2}, \ldots$ with $n_1 < n_2 < \cdots$. We shall usually describe this by saying that $\{X_{n_k}\}$ is a subsequence of $\{X_n\}$. The point of introducing subsequences at this stage could hardly be missed.

Lemma. *The point* X *is a point of accumulation of the sequence* $\{X_n\}$ *if and only if some subsequence of* $\{X_n\}$ *converges to* X.

Proof. We merely remark that, if X is an accumulation point, then

$$|X - X_{n_1}| < 1$$

for some n_1. We have

$$|X - X_{n_2}| < \tfrac{1}{2}$$

for infinitely many n_2's. Choose one so that $n_2 > n_1$. Continue in that way and obtain

$$|X - X_{n_k}| < \frac{1}{k}$$

$$n_1 < n_2 < n_3 < \cdots.$$

The completeness of the real number system guarantees that bounded sequences in R^m have accumulation points. A sequence can wander aimlessly; however, if it stays in a bounded part of R^m, it must accumulate somewhere. This property is usually called the "sequential compactness" of bounded parts of R^m.

Theorem 5 (Bolzano-Weierstrass). *Every bounded sequence in* R^m *has a point of accumulation. Equivalently, every bounded sequence in* R^m *has a convergent subsequence.*

Proof. We shall work with the coordinates, and, as we noted in Example 11, we must exercise some care. Let the sequence be

$$X_n = (x_{n_1}, \ldots, x_{nm})$$

as usual. Since $|x_{nj}| \leq |X_n|$, each of the m coordinate sequences is bounded. Suppose we have proved the theorem for bounded sequences in R^1. The proof for R^m could then be given this way. Pick a subsequence $X_{n_1}, X_{n_2}, \ldots$ for which the first coordinates converge in R^1. That subsequence is bounded. Hence, it has a subsequence for which the second coordinates converge. This new sequence has the property that the first coordinates of the vectors converge *and* the second coordinates of the vectors converge. In a finite number of steps we shall arrive at a subsequence for which each of the coordinate sequences converges.

So, our problem is to prove the theorem for a bounded sequence $\{x_n\}$ in R^1. That is now easy, because the limit inferior of the sequence is a point of accumulation of the sequence. If we don't wish to refer to that concept, we can define directly

$$x = \sup \{t; x_n \geq t \text{ for infinitely many } n\}$$

and verify that x is an accumulation point of the sequence.

Corollary. *A bounded sequence in* R^m *converges if and only if it has precisely one point of accumulation.*

Let us outline another proof of the Bolzano-Weierstrass theorem which has a slightly different intuitive basis. We can take the bounded sequence, multiply it by a non-zero scalar and then translate it so that the new sequence is in the box

$$B = \{X; 0 \leq x_j \leq 1, j = 1, \ldots, m\}.$$

Therefore, we may as well assume that the given sequence is in B. This box is the Cartesian product of the closed unit interval with itself m times:

$$B = I \times \cdots \times I$$
$$= \{X; x_j \in I, 1 \leq j \leq m\}.$$

Now cut I in half on each coordinate axis, and see how that subdivides B. It exhibits B as the union of 2^m boxes, each of which is the Cartesian product

$$J_1 \times \cdots \times J_m = \{X; x_j \in J_j, 1 \leq j \leq m\}$$

of m closed intervals. Each J_j is either $[0, \frac{1}{2}]$ or $[\frac{1}{2}, 0]$. These 2^m boxes overlap a little, but that is irrelevant. What matters is this. One of those boxes must contain X_n for infinitely many values of n. Pick one such box and call it B_1. Now subdivide B_1 into 2^m boxes, as we did with B, and let B_2 be one which contains X_n for infinitely many values of n. (See Figure 4). If we continue this subdivision process, we obtain a nested sequence of boxes

$$B_1 \supset B_2 \supset B_3 \supset \cdots$$

each of which contains X_n for infinitely many n; and the edges of B_n have length 2^{-n}. In particular,

$$\lim_n \operatorname{diam}(B_n) = 0.$$

Clearly we can choose $n_1 < n_2 < \cdots$ such that

$$X_{n_k} \in B_k.$$

This (sub) sequence is therefore a Cauchy sequence and converges (Example 5).

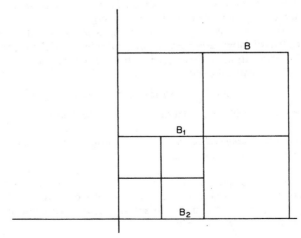

FIGURE 4

After we had chosen the boxes B_n, we did not need to use the Cauchy criterion. We might, for example, have applied the nested intervals theorem in each coordinate.

Exercises

1. Prove this. In R^1, the bounded sequence $\{x_n\}$ has a smallest and a largest accumulation point. They are (respectively) lim inf x_n and lim sup x_n.

2. Let z be a complex number of absolute value 1:

$$z = e^{i\theta}, \qquad 0 \le \theta < 2\pi$$

What are the accumulation points of the sequence $\{z^n\}$? Distinguish between the case where θ is a rational multiple of 2π and the case where it is not.

3. Let $\{X_n\}$ be a sequence. Suppose that $Y_1, Y_2, \ldots$ is a sequence of accumulation points of $\{X_n\}$. Show that any accumulation point of the sequence $\{Y_n\}$ is an accumulation point of the sequence $\{X_n\}$.

4. If $\{X_n\}$ is a Cauchy sequence and if some subsequence converges to X, then $X = \lim X_n$.

5. Let $\{x_n\}$ be a bounded sequence of real numbers such that $|x_n - x_{n+1}| = 1$ for each n. Show that the sequence has only a finite number of accumulation points.

6. If $\{x_n\}$ is a bounded sequence of real numbers such that $|x_n - x_{n+1}| \ge 1$ for each n, can it have an infinite number of accumulation points?

7. Suppose that $\{x_n\}$ is a bounded sequence of real numbers such that

$$\limsup_n x_n < \sup_n x_n.$$

Show that, among the numbers x_n, there is a largest one.

8. Find a sequence $\{z_n\}$ of complex numbers with $|z_n| < 1$ such that

(a) no point z with $|z| < 1$ is a point of accumulation of the sequence;
(b) every point z with $|z| = 1$ is an accumulation point of the sequence.

*9. True or false? Let A be a $k \times k$ matrix with complex entries. The set of all accumulation points of the sequence $\{A^n\}$ is bounded.

2.5. Open and Closed Sets

The title of this section mentions two very special classes of sets. One can do analysis in Euclidean space without either concept; however, quite a bit is lost if one does. The concepts of open set and closed set shed light on questions of convergence and on the geometry of mappings.

Definition. *The set* U *is* **open** *(in* R^m*) if it is a neighborhood of each of its points.*

Thus, U is open if and only if the following condition is satisfied: For each point X in U, there exists a positive number r such that the open ball $B(X; r)$ is contained (entirely) in U.

Theorem 6. *The union of any collection of open sets is open. The intersection of any finite collection of open sets is open.*

Proof. Suppose $\{U_\alpha\}$ is a family of open sets. Let

$$U = \bigcup_\alpha U_\alpha.$$

If $X \in U$, then $X \in U_\alpha$, for some α. Therefore (that) U_α is a neighborhood of X. Since U contains U_α, U is a neighborhood of X.

Suppose $U_1, \ldots, U_n$ are open sets. Let

$$U = \bigcap_k U_k.$$

Let $X \in U$. For each k, $1 \leq k \leq n$, there is a number $r_k > 0$ such that

$$B(X; r_k) \subset U_k.$$

Let r be the least of the numbers $r_1, \ldots, r_n$. Then

$$B(X; r) \subset U.$$

Theorem 6 is virtually trivial. It is called a theorem, because it states properties of open sets which are used so often.

EXAMPLE 12. Every open ball $B(X; r)$ is an open set. If $Y \in B(X; r)$ then $B(Y; t) \subset B(X; r)$ where

$$t = r - |X - Y|.$$

Thus, the union of any collection of open balls is open:

$$\bigcup_\alpha B(X_\alpha; r_\alpha).$$

Furthermore, every open set is of the last type. It is the union of those open balls which it contains.

EXAMPLE 13. Let us look at open sets in R^1. Each open interval (a, b) is an open set in R^1. On the other hand, an interval $(a, b]$ is not open in R^1, because $b \in (a, b]$ but no open interval about b is contained in $(a, b]$. The unbounded interval (a, ∞) is open in R^1.

Every open set in R^1 is a union of open intervals (a, b). In this 1-dimensional case, the open set U can be expressed as a union of intervals in a very special way. That is because open intervals in R^1 have this special property: If several open intervals have a point in common, their union is an open interval. (See Exercise 11.) Take any x in the open set U. Let I_x be the union of all those open intervals which contain the point x and are contained in the set U. Then I_x is an open interval (possibly unbounded).

Therefore, for each $x \in U$ there is a largest open interval I_x which contains x and is contained in U. If $y \in I_x$, then plainly $I_y = I_x$. In other words, all the points in I_x belong to the same largest interval in U, namely I_x. Consequently, if $x, y \in U$ then either $I_x = I_y$ or the intersection of I_x and I_y is empty. How many different intervals I_x are there? Only a countable number. (See Exercise 12 and Appendix.) Thus every open set in R^1 is uniquely expressible as the union of a countable collection of open intervals which are pairwise disjoint.

EXAMPLE 14. Let's look at the space of $k \times k$ matrices (real or complex entries). Let U be the set of invertible matrices, i.e., matrices A such that there exists A^{-1} with $AA^{-1} = A^{-1}A = I$. Is U an open set? What would that say? It would say that, if the matrix A is invertible then every matrix (sufficiently) near A is also invertible. We showed earlier (Example 9) that every matrix near the identity matrix is invertible: If $|T| < 1$, then

$$(I - T)^{-1} = \sum_{n=0}^{\infty} T^n;$$

or if $|I - S| < 1$, then

$$S^{-1} = \sum_{n=0}^{\infty} (I - S)^n.$$

So maybe U is open. If A is invertible, how close must B stay to A to guarantee that B is invertible? Now

$$A - B = A(I - A^{-1}B),$$

or

$$I - A^{-1}B = A^{-1}(A - B).$$

Thus,

$$|I - A^{-1}B| \leq |A^{-1}||A - B|.$$

Suppose

$$|A - B| < \frac{1}{|A^{-1}|}.$$

Then

$$|I - A^{-1}B| < 1;$$

hence, $A^{-1}B$ is invertible, and since A^{-1} is invertible, that makes B invertible. To summarize, the set U of invertible matrices is open because, if $A \in U$, then U contains the open ball of radius $|A^{-1}|^{-1}$ about the point A.

Definition. *The point* X *is a* **cluster point** *of the set* S *if every neighborhood of* X *contains a point of* S *which is different from* X.

Lemma. *Let* S *be a subset of* R^m *and let* $X \in R^m$. *The following are equivalent* (*all true or all false*).

(i) X *is a cluster point of the set* S.

(ii) *Every neighborhood of* X *contains infinitely many points of* S.

(iii) *There exists a sequence* $\{X_n\}$ *in* S *such that* $X_n \neq X$ *and* $X = \lim_n X_n$.

Proof. Exercise.

The reader may have noticed the similarity of the concepts of "cluster point of a set" and "accumulation point of a sequence". It is important to be clear about the relationship between the two ideas. If $\{X_n\}$ is a sequence in R^m, then its image

$$S = \{X_n; n \in Z_+\}$$

is a subset of R^m. It is the set of all distinct vectors which occur as one (or more) of the terms of the sequence. A point X is a cluster point of S if each neighborhood of X contains infinitely many points of S. Such an X is surely an accumulation point of the sequence, because each neighborhood, containing infinitely many points of S, must contain X_n for infinitely many values of n. On the other hand, if Y is a point of accumulation of the sequence it need not be a cluster point of the set S. A simple example should make this clear. The sequence of real numbers

$$0, 1, 0, 1, 0, 1, \ldots$$

has two points of accumulation, 0 and 1. The image of the sequence is $S = \{0, 1\}$, and it has no cluster points at all. If all of the terms of the sequence $\{X_n\}$ are distinct, then every accumulation point of the sequence is a cluster point of S.

The terms "cluster point", "limit point", and "accumulation point" are used interchangeably in most parts of mathematics. We have elected to apply "accumulation" to sequences and "cluster" to sets, as a reminder of the distinction discussed in the last paragraph.

Definition. *The set* K *is* **closed** *if every cluster point of* K *is in* K.

A closed set is one which is closed under (the process of taking) limits. From the last lemma, it should be clear that these conditions on a set K are equivalent:

(i) K is closed.

(ii) If $\{X_n\}$ is a sequence of points in K and if the sequence converges, then the limit of the sequence is in K.

Theorem 7. *A set* S *is open if and only if its complement* (*complementary set*) *is closed.*

Proof. Let T be the complement of S:

$$T = \{X \in R^m; X \notin S\}.$$

To say that T is closed is to say that, if $X \notin T$, then X is not a cluster point of T. Think about that.

Corollary. *The intersection of any collection of closed sets is closed. The union of any finite collection of closed sets is closed.*

Proof. This follows from Theorems 6 and 7. We describe the proof of the first statement. Let $\{K_\alpha\}$ be a collection of closed sets. For each α, let U_α be the complement of K_α. Then each U_α is an open set and, therefore, the union

$$U = \bigcup_\alpha U_\alpha$$

is an open set. This union is the complement of

$$K = \bigcap_\alpha K_\alpha,$$

the intersection of the complements of the sets U_α. Since U is open, K is closed.

We should remark that closed balls and closed intervals are closed sets. In view of Theorem 7, there is no need for a separate list of examples of closed sets. Every example of an open set provides an example of a closed set (and vice versa). But, the human mind being what it is, it doesn't follow that just because we know about open sets we'll recognize a closed set when we bump into it.

EXAMPLE 15. Let's look at a famous closed set—the **Cantor set.** We shall refer to it often. Start with the closed interval $[0, 1]$ in R^1. Remove the open middle one-third, i.e., the open interval $(\frac{1}{3}, \frac{2}{3})$. What remains is the union of two disjoint closed intervals. Remove the open middle one-third of each of those intervals. (See Figure 5.) Now, remove the open middle one-third of each of the four remaining closed intervals. Continue ad infinitum. What remains is the Cantor set K.

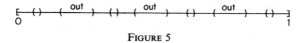

FIGURE 5

We obtained K by removing from $[0, 1]$ an open set U. That set is the union of the sequence of open intervals: $(\frac{1}{3}, \frac{2}{3}), (\frac{1}{9}, \frac{2}{9}), (\frac{7}{9}, \frac{8}{9}), \ldots$. Since U is open and $[0, 1]$ is closed,

$$K = \{x \in [0, 1]; x \notin U\}$$

is a closed set. The set K is very thin. The lengths of the open intervals in U add up to

$$\frac{1}{3} + 2 \cdot \frac{1}{9} + 4 \cdot \frac{1}{27} + \cdots = \sum_{n=1}^{\infty} \frac{2^{n-1}}{3^n}$$

$$= \tfrac{1}{2} \sum_{n=1}^{\infty} (\tfrac{2}{3})^n$$

$$= \tfrac{1}{2}[(1 - \tfrac{2}{3})^{-1} - 1]$$

$$= 1.$$

But, there are a lot of points in K—uncountably many. In particular, K contains many more points than the end points of the deleted intervals. (Those are the obvious points in K.)

We can describe the Cantor set very nicely, if we use the ternary rather than the decimal expansion of points in the interval $[0, 1]$. The ternary expansion represents each $x \in [0, 1]$ as the sum of a series

$$x = \sum_{n=1}^{\infty} a_n 3^{-n}$$

where the "digits" a_n are 0, 1, or 2. The digits are defined by locating the point x in a sequence of intervals, the lengths of which go down by a factor of 3 each time:

$$a_1 = \sup \left\{ k \in Z; \frac{k}{3} \leq x \right\}$$

$$a_2 = \sup \{ k \in Z; a_1 3^{-1} + k 3^{-2} \leq x \}$$

$$\vdots$$

$$a_n = \sup \{ k \in Z; a_1 3^{-1} + \cdots + a_{n-1} 3^{-(n-1)} + k 3^{-n} \leq x \}$$

$$\vdots$$

The Cantor set contains all those points x which have a ternary representation in which the digits are either 0 or 2 (no 1's). So clearly there are just as many points in K as there are in the interval $[0, 1]$: Under the mapping (function)

$$\sum_{n=1}^{\infty} a_n 3^{-n} \longrightarrow \sum_{n=1}^{\infty} \frac{a_n}{2} 2^{-n} \qquad (a_n = 0 \text{ or } 2)$$

the image of K covers $[0, 1]$.

The "sequential compactness" of R^m can be reformulated in the language of this section. For emphasis, we shall record two reformulations.

Theorem 8 (*Bolzano-Weierstrass*). *Every bounded and infinite subset of* R^m *has a cluster point.*

Proof. If S is an infinite set, we can select a sequence of points X_n in S such that $X_k \neq X_n$ whenever $k \neq n$ (a sequence of distinct points). For such a sequence, "contains infinitely many X_n's" is the same as "contains X_n for infinitely many n's".

Theorem 9. *Let*

$$K_1 \supset K_2 \supset K_3 \supset \cdots$$

be a nested sequence of bounded closed sets in R^m. If each K_n is non-empty, then the intersection

$$\bigcap_n K_n$$

is non-empty.

Proof. For each n, there exists a point X_n in K_n. The sequence $\{X_n\}$ is bounded and, accordingly, it has a point of accumulation X. Since $X_k \in K_n$ for $k \geq n$ and since K_n is closed, X must be in K_n.

Here is an amusing application of the weaker result.

EXAMPLE 16. An analyst (of the mathematical variety) might proceed this way to show that the medians of a triangle meet in a point. Let the triangle be ABC and let X, Y, Z be the midpoints of the sides. By comparison of similar triangles, it is clear that (1) the median lines of ABC are also the median lines of XYZ, (2) diam $XYZ = \frac{1}{2}$ diam ABC. (See Figure 6.) Replace ABC by XYZ and repeat. Then repeat again, etc. Do you see where Theorem 9 comes in?

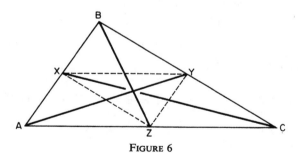

FIGURE 6

One might think of Theorem 9 as a slightly more geometrical way of stating the Bolzano-Weierstrass theorem. If (in Theorem 9) one knows that diam (K_n) converges to 0, then the intersection of all the K_n's will consist of precisely one point. That result is weaker than Theorem 9. It is (essentially) a reformulation of the fact that each Cauchy sequence in R^m converges.

Exercises

1. Show that there are (at least) two subsets of R^m which are both open and closed—R^m and the empty set.

2. If S is a subset of R^m and if S is both open and closed, then either $S = R^m$ or S is empty. (You might prove it first in R^1.)

3. If S is a subset of R^m, what is the union of all closed subsets of S?

4. Which of the following sets of complex numbers are closed? Which are open?

 (a) all z such that $z = z^*$;
 (b) all z such that $zz^* > 2$;
 (c) all z such that $|z| \leq 1$ and $z \neq 0$;
 (d) all z such that $|z|$ is rational.

5. True or false? If S is closed, then S contains a cluster point of S.

6. True or false? If S is a bounded infinite subset of R, then among the cluster points of S there is a largest one.

7. True or false? If every subset of S is closed, then S contains only a finite number of points.

8. True or false? If every subset of S is open, then S contains only a finite number of points.

9. If S is a set, the X_0-**translate** of S is $X_0 + S = \{X_0 + Y;\ Y \in S\}$. Show that each translate of an open set is open and each translate of a closed set is closed.

10. What can you say about scalar multiples of open [closed] sets?

11. If $\{I_\alpha;\ \alpha \in A\}$ is a family of open intervals on the real line and if the intersection

$$\bigcap_\alpha I_\alpha$$

is non-empty, then the union is an open interval. (Remember the definition of interval.)

12. Let $\{I_\alpha;\ \alpha \in A\}$ be a family of open intervals on the real line which are pairwise disjoint:

$$I_\alpha \cap I_\beta = \varnothing, \qquad \alpha \neq \beta.$$

Prove that A is a countable set.

13. Let A be a subset of R^m and let B be a subset of R^n. The Cartesian product $A \times B$ is a subset of R^{m+n}. Prove that

 (a) if A and B are open, then $A \times B$ is open;
 (b) if A and B are closed, then $A \times B$ is closed.

14. Is the set of orthogonal $k \times k$ matrices open?

15. Let M be a linear subspace of R^m. Let K be a closed subset of R^m. Project the set K orthogonally onto M. Do you end up with a closed set?

16. Every (linear) subspace of R^m is closed.

17. A set K is **perfect** if K is closed and every point of K is a cluster point of K. Show that the Cantor set is a perfect set.

*18. Every non-empty perfect set is uncountable.

*19. Every closed set is the union of a perfect set and a countable set.

2.6. *Closure and Interior*

Let S be a subset of R^m. The **closure** of S is the intersection of all closed sets which contain S. The **interior** of S is the union of all open sets which are contained in S. The **boundary** of S is the intersection of the closure of S with the closure of the complement of S.

The closure of S will be denoted $\bar{S}$. Evidently $\bar{S}$ is a closed set. It is the smallest closed set which contains S. If X is a point in R^m, these conditions are equivalent:

   (i) $X \in \bar{S}$.
   (ii) Either $X \in S$ or X is a cluster point of S.
   (iii) There exists a sequence of points in S which converges to X.

The interior of S will be denoted S^o (although we shall not use the notation very much). Evidently S^o is an open set—the largest open subset of S. The following conditions on the point X are equivalent:

   (i) $X \in S^o$.
   (ii) Some open ball $B(X; r)$ is contained in S.
   (iii) There does not exist any sequence of points in the complement of S which converges to X.

The following conditions on X are equivalent:

   (i) X is in the boundary of S.
   (ii) Every neighborhood of X contains a point in S and a point in the complement of S.

EXAMPLE 17. Consider the open ball $B(X_0; r)$. It is an open set and therefore is equal to its interior. The closure of B is the closed ball $\bar{B}(X_0; r)$. The boundary of B is the **sphere** of radius r about X_0:

$$S(X_0; r) = \{X; |X - X_0| = r\}.$$

In R^2, the "sphere" is a circle. In R^1, the "sphere" consists of the two endpoints of the interval B.

EXAMPLE 18. Let's look at some subsets of R^2. In order to understand closure, interior, and boundary, one usually begins by drawing a set S as in Figure 7. Indicated are a point X in S^o, a point Y on the boundary of

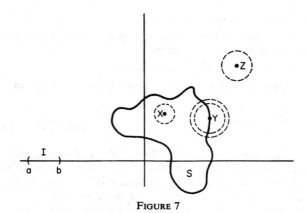

FIGURE 7

S, and a point Z which is *not* in $\bar{S}$. The boundary of (this particular) S is a curve, and it does not matter whether S is the open region bounded by the curve or the closed region bounded by the curve. The interior, the closure and the boundary are the same for those two regions. A region of either type is a *very* special set. From the point of view of things which we are now discussing, the nicest set is one which is closed, bounded and equal to the closure of its interior (or, the interior of such a set).

In the same figure, let T be the set obtained by deleting from S that part of the real axis which lies in S. The points on the deleted line segment are in the boundary of T. Yet, somehow they seem "interior" to T in a weak sense. They are not in the interior of T, because they are not even in T. But those points are in the interior of $\bar{T}$.

Let's look at a set which is less nice. Let V be the set of all points (x, y) in R^2 such that both x and y are rational numbers. Now V is (to say the least) scattered all over the place. Every point of R^2 is a cluster point of V; hence $\bar{V} = R^2$. Evidently, the interior of V is the empty set, because every point of R^2 is also a cluster point of the complement of V. What is the boundary of V? It's also all of R^2.

One final set which we discuss is the interval I on the real axis:

$$I = \{(x, 0\}; x \in (a, b)\}.$$

It should be clear that $\bar{I}$ is the corresponding closed interval on the real axis:

$$\bar{I} = \{(x, 0); x \in [a, b]\}.$$

The interior of I (in R^2) is empty, because each point of I is in the boundary of I. The boundary of I is $\bar{I}$. If we were discussing the interval (a, b) in $^-R^1$, the interior and the boundary would be quite different. Hence, always bear in mind that such sets are not intrinsically attached to the geometric object which we envision. Closure, boundary, and interior are defined relative to the ambient space.

Definition. *Let* S *be a subset of* T. *We say that* S *is* **dense in** T *if every point of* T *is in the closure of* S.

Density has to do with approximation. To say S is dense in T means that every point of T can be approximated (as closely as one might wish) by points in S, i.e., every point in T is the limit of a sequence of points from S. The set of rational numbers is dense in R^1. More generally, the set of points $X = (x_1, \ldots, x_m)$ in R^m such that every x_j is rational is dense in R^m. Every set is dense in its closure. If K is a closed set, a subset S is dense in K if and only if $\bar{S} = K$.

EXAMPLE 19. Once again, let us consider the space of $k \times k$ matrices (real or complex entries). Let S be the set of singular matrices, i.e., matrices which are not invertible. Then S is a closed set, because its complement is the set of invertible matrices, which is an open set (Example 14, p. 57). The interior of S is the empty set. What does that say? It says that, if A is a singular matrix, then every neighborhood of A contains an invertible matrix. In other words, if A is singular, we can perturb A just a little and obtain an invertible matrix. How? Look at $A + cI$. For a $k \times k$ matrix A, there are at most k numbers c for which $A + cI$ is singular, because when $A + cI$ is singular

$$\det (A + cI) = 0$$

and the determinant of $A + cI$ is a polynomial in c of degree k. The zeros of that polynomial are the **characteristic values** (eigenvalues) of A. One of those characteristic values is 0, because A is singular. Let the non-zero characteristic values be $c_1, \ldots, c_r$. Let δ be the minimum of $|c_1|, \ldots, |c_r|$. Then $A + cI$ is invertible for every c such that $0 < |c| < \delta$. In particular, $A + \epsilon I$ is invertible for all small positive ϵ. Thus, the interior of the set of singular matrices is empty. We can phrase that in terms of the complement. The set of invertible $k \times k$ matrices is a dense open subset of Euclidean space of dimension k^2.

Exercises

1. The interior of any set S is the complement of the closure of the complement of S.

2. The union of the interior of S and the interior of $R^m - S$ is the complement of the boundary of S.

3. Find all subsets of R^m for which the boundary is empty.

4. True or false? If S and S^o have the same boundary, then S^o is dense in $\bar{S}$.

5. True or false? If S and $\bar{S}$ have the same boundary, then S is bounded.

6. True or false? If S^o is bounded and the boundary of S is bounded, then S is bounded.

7. Show that the set S in R^m is open if and only if $S \cap \bar{A} \subset \overline{S \cap A}$ for all $A \subset R^m$.

8. Let S be a linear subspace of R^m. Show that, if the interior of S is non-empty, then $S = R^m$.

9. Let f be a real-valued function on the real line

$$R \xrightarrow{\ f\ } R.$$

Show that the graph of f has empty interior (in R^2).

10. Let G be an additive subgroup of R, i.e., a non-empty set of real numbers such that $(x - y) \in G$ whenever $x \in G$ and $y \in G$. Show that either G consists of all integer multiples of some fixed real number or G is dense in R. (*Hint:* Is there a smallest positive number in G?)

11. Let

$$p(x, y) = \sum_{k,n=0}^{N} a_{kn} x^k y^n$$

be a (real) polynomial in two variables, $p \neq 0$. Let K be the zero set of $p : K = \{(x, y); p(x, y) = 0\}$. Show that the interior of K is empty. Generalize to n variables. (The determinant of a matrix is a polynomial function of the entries, so Example 19 is a special case of this exercise.)

12. A set S in R^m is convex if, whenever X and Y are in S, the line segment between X and Y is in S. That is, S is convex if whenever X and Y are in S, $tX + (1 - t)Y$ is in S for all $0 \le t \le 1$.

    (a) The closure of a convex set is convex.
    (b) The interior of a convex set is convex.
    (c) Let S be a set such that $\frac{1}{2}(X + Y)$ is in S, whenever X and Y are in S. If S is closed, then S is convex.

*13. Show that the set of diagonalizable $k \times k$ matrices is a dense subset of the space of $k \times k$ complex matrices. (*Hint:* The matrices which have k distinct characteristic values are diagonalizable.)

*14. Use the result of Exercise 13 to prove the Cayley-Hamilton theorem: Every $k \times k$ matrix satisfies its characteristic equation.

2.7. Compact Sets

Compactness is an important geometrical property enjoyed by every closed bounded subset of R_m. The concept of compactness is concerned with covering a set by open sets.

Definition. *Let* S *be a subset of* R^m. *An (indexed)* **cover** *(or* **covering**) *of* S *is a family of sets* $\{U_\alpha ; \alpha \in A\}$ *such that*

$$S \subset \bigcup_\alpha U_\alpha.$$

A cover of S *is called*

    (a) *an* **open cover** *of* S *if each* U_α *is an open set;*
    (b) *a* **finite cover** *of* S *if the index set* A *is finite;*
    (c) *a* **countable cover** *of* S *if the index set* A *is countable.*

We shall deal almost exclusively with open covers. If $\{U_\alpha\}$ is a cover of S, we also say that the family $\{U_\alpha\}$ **covers** S. A countable cover is a sequence of sets. Thus, a countable open cover of S is a sequence of open sets $\{U_n\}$ such that each point of S is in (at least) one of the sets U_n.

EXAMPLE 20.   Let S be any subset of R^m. Then S has myriad different open covers. For instance, the family consisting of the single set R^m is a finite open cover of S. Or, let $\epsilon > 0$ and let

$$U_X = B(X; \epsilon).$$

Then $\{U_X; X \in S\}$ is an open cover of S, which happens also to be indexed by the set S. Note that this particular cover $\{U_X; X \in S\}$ is also an open cover of the closure $\bar{S}$: Every point in $\bar{S}$ is within distance ϵ of some point of S. Suppose we define

$$V_X = B(X; |X|).$$

Then $\{V_X; X \in S\}$ is an open cover of S if $0 \notin S$. Let

$$B_n = \{X; |X| < n\}.$$

Then $\{B_n\}$ is a countable open cover of S, because it is a countable open cover of R^m.

Suppose that $\{U_\alpha; \alpha \in A\}$ is a cover of the set S. Let B be a subset of the index set A. The family of sets $\{U_\alpha; \alpha \in B\}$ may or may not cover S, i.e., it may or may not be true that

(2.25) $$S \subset \bigcup_{\alpha \in B} U_\alpha.$$

If (2.25) is satisfied, then we call $\{U_\alpha; \alpha \in B\}$ a **cover of** S **subordinate to** $\{U_\alpha; \alpha \in A\}$. It is a common convention in mathematics to shorten this terminology and say that $\{U_\alpha; \alpha \in B\}$ is a **subcover** of the cover $\{U_\alpha; \alpha \in A\}$ provided (2.25) is satisfied. We shall follow this practice but the reader should note how important it is to keep in mind which set is being covered.

Definition. *The set* K *is* **compact** *if every open cover of* K *has a finite subcover.*

In order that K should be compact, the following must be the case. If we are given any family of open sets $\{U_\alpha\}$ which covers K, we can extract from that family a finite number of sets $U_{\alpha_1}, \ldots, U_{\alpha_n}$ which (jointly) cover K. The proof that a given set K is compact must give a prescription for performing that extraction on each of the multitude of different open covers of K. The following theorem certainly provides us with many examples of sets which are *not* compact.

Theorem 10. *Every compact set is closed and bounded.*

Proof. Let K be compact. First, we show that K is bounded. One of the open covers of K is given by the sequence

$$B_n = B(0; n), \qquad n = 1, 2, 3, \ldots.$$

Since K is compact, some finite number of those sets cover K. Hence one of them covers K, because $B_1 \subset B_2 \subset \cdots$.

How do we show that K is closed? Let Y be a point which is not in K. If $X \in K$, let

$$U_X = B(X; \tfrac{1}{2}|X - Y|)$$

Then $\{U_X; X \in K\}$ is an open cover of K. Since K is compact, there exist $X_1, \ldots, X_n$ in K such that

$$K \subset \bigcup_1^n U_{X_k}.$$

Let r be the smallest of the numbers $\tfrac{1}{2}|Y - X_k|$. Then $B(Y; r)$ is a neighborhood of Y which does not contain any point of K. We have proved that, if Y is not in K, then Y is not in the closure of K. Thus, the set K is closed.

EXAMPLE 21. There is one type of (infinite) set which easily can be seen to be compact. Suppose $\{X_n\}$ is a convergent sequence and

$$X = \lim_n X_n.$$

Let S be the image of the sequence together with the limit point X:

$$S = \{Y; Y = X \text{ or } Y = X_n \text{ for some } n\}.$$

Then S is compact. We see this as follows. Let $\{U_\alpha\}$ be any open cover of S. Then one of the sets U_α must contain X. Let's say $X \in U_{\alpha_1}$. Since X_n converges to X, the set U_{α_1} contains X_n except for a finite number of n's. Choose $U_{\alpha_1}, \ldots, U_{\alpha_k}$ so that they cover the X_n's outside U_{α_1} and $\{U_{\alpha_1}, \ldots, U_{\alpha_k}\}$ is a finite subcover.

Theorem 11. *Let* S *be any subset of* R^m. *Every open cover of* S *has a countable subcover.*

Proof. Consider the set of all points in R^m which have rational coordinates. That is a countable set. For each point P with rational coordinates, consider the set of open balls about P which have a rational radius. The collection of all those open balls, as P ranges over all points with rational coordinates, is a countable set. Why? Because we just described it as a countable union of countable sets. (See Appendix.) Thus, there is a sequence of open balls $B_1, B_2, B_3, \ldots$ such that every open ball with rational center and rational radius is one of the sets B_n.

Now, let $\{U_\alpha; \alpha \in A\}$ be an open cover of a set S. For each positive integer n, ask whether or not the ball B_n is contained in one of the sets U_α. If it is, choose one index α_n so that

(2.26) $B_n \subset U_{\alpha_n}.$

The claim is that the sequence of sets U_{α_n} covers S. (Note that α_n is not defined for every n. If that bothers you, define α_n any way you please for the n's where we did not define it.) Let $X \in S$. There exists an α such that $X \in U_\alpha$. Choose an open ball $B(X; r) \subset U_\alpha$. Let P be a point with rational coordinates such that

$$|P - X| < \tfrac{1}{2}r.$$

Choose a rational number t such that

(2.27) $|P - X| < t < \tfrac{1}{2}r.$

From (2.27) it follows that

$$X \in B(P; t) \subset B(X; r).$$

Now $B(P; t) = B_n$ for some n. Therefore that B_n contains X and is inside the set U_α. Since B_n is contained in some U_α, we have the associated set U_{α_n} which contains B_n (2.26). Hence

$$X \in U_{\alpha_n}.$$

Every point of S is in one of the sets U_{α_n}.

The significance of Theorem 11 should be reasonably evident. If we wish to prove that a set is compact, we need only prove that it is "countably compact", i.e., that each countable open cover of it has a finite subcover. We shall use this shortly to prove that each closed and bounded subset of R^m is compact. We first digress a bit to describe compactness in terms of closed sets.

Theorem 12. *Let* K *be a compact set. If* $\{K_\alpha; \alpha \in A\}$ *is a family of closed subsets of* K *and if, for each finite set of indices* $\alpha_1, \ldots, \alpha_n$, *the intersection* $K_{\alpha_1} \cap \cdots \cap K_{\alpha_n}$ *is non-empty, then the intersection*

$$\bigcap_{\alpha \in A} K_\alpha$$

is non-empty.

Proof. Let $\{K_\alpha\}$ be any collection of closed subsets of K. For each index α, let U_α be the (open) complement of K_α. Suppose that the intersection

$$\bigcap_\alpha K_\alpha$$

is empty. This means that its complement

$$\bigcup_\alpha U_\alpha$$

contains K, i.e., that $\{U_\alpha\}$ is an open cover of K. Since K is compact, there are indices $\alpha_1, \ldots, \alpha_n$ such that

$$K \subset \bigcup_{j=1}^n U_{\alpha_j}.$$

But then

$$\bigcap_{j=1}^n K_{\alpha_j}.$$

is empty. Therefore, if the family $\{K_\alpha\}$ is such that this last phenomenon cannot occur, the intersection of all the closed sets K_α must be non-empty.

In analysis, it is important to understand compactness via closed sets. That is why Theorem 12 is called a theorem, in spite of the fact that it is just the rewording of the defining property of compact sets obtained by using complements of open sets rather than the open sets themselves. The usefulness of the reformulation stems from the fact that compactness becomes a tool for proving that things exist: "There exists a point which is in every K_α." We shall repeat part of the reasoning as we prove the most important theorem about compactness.

Theorem 13 (*Heine-Borel*). *In* $\mathbf{R}^m$, *every closed and bounded set is compact.*

Proof. This is the union of Theorem 11 and the Bolzano-Weierstrass theorem. Suppose K is closed and bounded. Theorem 11 tells us that, in order to show that K is compact, we must show that a countable open cover $\{U_n\}$ has a finite subcover. Given such a cover, let

$$K_n = \{X \in K; X \notin U_k, 1 \leq k \leq n\}.$$

The sets K_n are nested

$$K_1 \supset K_2 \supset K_3 \supset \cdots.$$

Since K is closed, each K_n is closed. Since K is bounded, each K_n (i.e., K_1) is bounded. If each K_n were non-empty, Bolzano-Weierstrass (Theorem 9) would tell us that the intersection of all the K_n's was non-empty. But, that would mean that the sequence $\{U_n\}$ did not cover K. Therefore, some K_n must be empty. If K_N is empty, then $U_1, \ldots, U_N$ cover K.

Exercises

1. Which of the following sets of complex numbers are compact?

   (a) all z such that $|z| \geq 1$;
   (b) all z such that $zz^* = 2$;

(c) all z such that $|z|$ is rational and $|z| \leq 1$;

(d) all z such that $e^z = 1$.

2. What is the least number of open intervals of length ϵ which will cover the interval $[0, 1]$?

3. True or false? If K is a set of complex numbers for which the orthogonal projections onto the real axis and the imaginary axis are both compact, then K is compact.

4. True or false? If S is a bounded set, there is a smallest compact set which contains S.

5. Every bounded subset of R^m can be covered by a finite number of open balls of radius ϵ.

6. True or false? If A is closed and B is bounded, then $A \cap B$ is compact.

7. If A and B are compact, then the Cartesian product $A \times B$ is compact.

8. True or false? If the set S has a largest compact subset, then S is compact.

9. True or false? If the boundary of S is compact and if the interior of S is compact, then S is compact.

10. Give an example of a countable compact set which has a countably infinite set of cluster points.

11. If a compact set has only a countable number of cluster points, it is a countable set.

12. Which of the following sets of $k \times k$ (real) matrices are compact?

(a) all A such that $A = A^t$;

(b) all A such that A is orthogonal;

(c) all A such that $A^2 = 0$;

(d) all A such that $A^2 = A$.

13. If $\{I_n\}$ is a sequence of open intervals which covers $[0, 1]$, then

$$\sum_n \text{length}(I_n) > 1.$$

14. If A and B are subsets of R^m, define

$$A + B = \{X + Y; \ X \in A \text{ and } Y \in B\}.$$

(a) If A is any set and U is open, then $A + U$ is open.

(b) If A is closed and B is compact, then $A + B$ is closed.

(c) Give an example of closed sets A and B such that $A + B$ is not closed.

2.8. Relative Topology

In many situations, the space in which we operate will be a subset of R^m, rather than the entire Euclidean space. For instance, we may want to discuss functions which are defined only on an interval $[a, b]$ of the real line. In that case, the interval will become a temporary subuniverse, in which we talk about points being near to one another, convergence, open sets, closed sets, etc. Such concepts in the "sub-space" will be defined by simply disregarding the remainder of R^m.

Fix a set S in R^m. If X is a point of S, a **neighborhood of** X **relative to** S is a set T such that

(i) $T \subset S$;
(ii) T contains $B(X; r) \cap S$ for some $r > 0$.

In other words, a neighborhood of X relative to S is a subset of S which contains every point in S which is sufficiently close to X.

Let V be a subset of S. We say that V is **open relative to** S if V is a relative neighborhood of each of its points. Thus, V is open relative to S if

$$V = S \cap U$$

where U is an open set in R^m.

Let F be a subset of S. We say that F is **closed relative to** S if F contains every cluster point of F which is in the set S. Evidently, the following conditions on a set $F \subset S$ are equivalent:

(i) F is closed relative to S.
(ii) $F = S \cap K$ where K is closed in R^m.
(iii) If $\{X_n\}$ is a sequence in F which converges to a point X *in the set S*, then X is in F.
(iv) The complement of F relative to S, $S - F$, is open relative to S.

There is now a string of obvious comments. The family of sets which are open relative to S is closed under the formation of arbitrary unions and finite intersections. The family of sets which are closed relative to S has the complementary property: arbitrary intersections, finite unions. The set S is both open and closed relative to S. If $\{X_n\}$ is a sequence in S, then $\{X_n\}$ converges to the point $X \in S$ if and only if each neighborhood of X relative to S contains X_n except for finitely many n's.

We could now define interior relative to S, closure relative to S, etc. We shall not make much use of those terms. One further relativization deserves comment. A set is (of course) **compact relative** to S if each cover of it by relatively open sets has a finite subcover. *If K is a subset of S, then K is compact relative to S if and only if K is compact.* That is trivial to verify; however, it tells us something. Vaguely, it says that compactness is an intrinsic property of (compact) geometric objects.

EXAMPLE 22. Most often, we shall deal with relatively open, etc., when the set S is either an open set or a closed set. Suppose that S is an open subset of R^m. Then, "open relative to S" merely means "open set contained in S". "Closed relative to S" is more interesting. For instance, the interval $(a, c]$ is closed relative to the interval (a, b), $a < c < b$.

If S is a closed subset of R^m, then "closed relative to S" becomes "closed subset of S", whereas, "open relative to S" is more interesting. The interval $(\frac{1}{2}, 1]$ is open relative to the interval $[0, 1]$. The set $\{0\}$ is open relative to the set $S = \{0, 1\}$ which consists of the numbers 0 and 1.

EXAMPLE 23. Suppose that S is a linear subspace of R^m. Let us say $k = \dim(S)$. Then S is a closed subset of R^m. Also, S intrinsically looks like R^k. We can decompose each X in R^m into a sum

$$X = Y + Z$$

where Y is in S and Z is orthogonal to (every vector in) S. The vector Y is the orthogonal projection of X on S:

$$R^n \xrightarrow{P} S$$

$$Y = P(X).$$

One can see easily that the set $V \subset S$ is open [closed] relative to S if and only if

$$P^{-1}(V) = \{X; P(X) \in V\}$$

is open [closed] in R^m. This is illustrated in Figure 8.

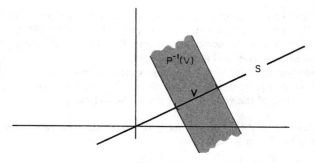

FIGURE 8

One geometric property which is conveniently defined via relatively open (or relatively closed) sets is connectedness. If S is a subset of R^m, we want to say what it means for S to be connected. In Figure 9, there is a set S which probably we would all describe as disconnected. The figure eight, called F, does not appear disconnected, so probably we would call

FIGURE 9

it connected. But suppose we let G be the figure eight with its middle point removed. Is that a disconnected set? Perhaps it ought to be so described, since it falls apart so naturally into two pieces. But, how is that different from decomposing F into two pieces, one being the left and the other being the right half plus the midpoint? Here is the way in which we shall distinguish.

Definition. *The set* S *is* **disconnected** *if there exist two subsets* S_1 *and* S_2 *with these properties:*

(i) S_1 *and* S_2 *are non-empty;* $S_1 \cap S_2 = \varnothing$;
(ii) $S = S_1 \cup S_2$;
(iii) S_1 *and* S_2 *are closed relative to* S.

If S *is not disconnected, then* S *is* **connected.**

Observe that the following conditions on a set S are equivalent:

(i) S is connected.
(ii) If $S = U_1 \cup U_2$ where U_1 and U_2 are disjoint relatively open subsets of S, then either U_1 or U_2 is empty.
(iii) If T is a subset of S which is both open and closed relative to S, then either $T = S$ or T is empty.

In particular, an open set S in R^m is disconnected if and only if $S = U_1 \cup U_2$ where U_1, U_2 are disjoint nonempty open sets in R^m. Similarly, a closed set S is disconnected if and only if $S = K_1 \cup K_2$, where K_1, K_2 are disjoint non-empty closed sets in R^m.

We shall not expend much energy worrying about the connectedness of sets which are neither open nor closed; however, it may increase our understanding of the concept if we prove the following.

Theorem 14. *Let* S *be a subset of* R^m. *If* S *is not connected, there exist open sets* U_1, U_2 *in* R^m *such that*

(a) $S \subset U_1 \cup U_2$;
(b) $S \cap U_1 \neq empty \neq S \cap U_2$;
(c) U_1 *and* U_2 *are disjoint.*

Proof. The fact that S is not connected provides us with open sets U_1, U_2 which satisfy (a), (b) and

(c)$'$ $S \cap U_1 \cap U_2$ is empty.

But U_1 and U_2 may intersect at points outside of S. We must show that we can shrink U_1 and U_2 a bit, so that the new open sets become disjoint but (a) and (b) are not disturbed. As a matter of fact, the shrinking can be done so that neither $U_1 \cap S$ nor $U_2 \cap S$ is changed, and the remainder of the proof has nothing whatever to do with S.

Let

$$T_1 = U_1 - U_2$$
$$T_2 = U_2 - U_1.$$

If $X \in T_1$, then $X \in U_1$ and so there is a positive number $r_1(X)$ such that

$$B(X; r_1(X)) \subset U_1.$$

Similarly, if $Y \in T_2$ then $Y \in U_2$ and

$$B(Y; r_2(Y)) \subset U_2$$

for some $r_2(Y) > 0$. Let

$$V_1 = \bigcup_{X \in T_1} B(X; \tfrac{1}{2}r_1(X))$$
$$V_2 = \bigcup_{Y \in T_2} B(Y; \tfrac{1}{2}r_2(Y)).$$

Then V_j is an open set which contains T_j. (Therefore, $V_j \cap S = U_j \cap S$.) Furthermore, V_1 and V_2 are disjoint. If

$$Z \in V_1 \cap V_2$$

then, for some $X \in T_1$ and $Y \in T_2$

$$|Z - X| < \tfrac{1}{2}r_1(X)$$
$$|Z - Y| < \tfrac{1}{2}r_2(Y).$$

One of the numbers $r_1(X)$, $r_2(Y)$ is at least as large as the other, say, $r_1(X) \geq r_2(Y)$. Then

$$|X - Y| \leq |Z - X| + |Y - Z|$$
$$< \tfrac{1}{2}r_1(X) + \tfrac{1}{2}r_2(Y)$$
$$\leq r_1(X).$$

But that means that Y is in U_1, whereas, $Y \in (U_2 - U_1)$. So V_1 and V_2 are disjoint.

Corollary. *If* K_1 *and* K_2 *are disjoint closed sets in* R^m, *there exist open sets* V_1, V_2 *such that* $K_j \subset V_j$ *and* V_1 *and* V_2 *are disjoint.*

Proof. This is the case of the argument above in which $S = K_1 \cup K_2$, U_1 is the complement of K_2, and U_2 is the complement of K_1.

Exercises

1. In the set of integers, every subset is relatively open (and relatively closed).

2. Every interval in R is connected.

3. If S is a connected subset of R, then S is empty, S contains precisely one point, or S is an interval.

4. Every linear subspace of R^m is connected.

5. If K is an infinite compact set, then K has a subset which is not closed relative to K.

6. Let S be a bounded open set which is connected. Show that any two points in S can be joined by a polygonal path which lies in S, i.e., if $A, B \in S$, there are points $A = X_0, X_1, \ldots, X_n = B$ such that (for each k) the line segment from X_{k-1} to X_k lies in S. (*Hint*: Fix A, and consider the set of points B which can be so joined to A. Show that the set is both open and closed relative to S.)

7. Every convex set is connected. (A set is convex if it contains the line segment joining each pair of its points.)

8. If S is connected and if $S \subset T \subset \bar{S}$, then T is connected. In particular, the closure of a connected set is connected.

9. True or false? The interior of a connected set is connected.

10. True or false? The boundary of a connected compact set is connected.

11. Let K be the Cantor set (Example 15). Show that no connected subset of K contains more than one point.

12. From R^m, remove a hyperplane (level set of a linear functional). The remaining set is disconnected.

13. From C^m, remove a hyperplane (level set of a complex-linear functional.) The remaining set is connected.

14. The set of invertible $k \times k$ matrices with real entries is not a connected subset of $R^{k \times k}$.

15. The set of invertible $k \times k$ matrices with complex entries is a connected subset of $C^{k \times k}$.

16. In R^m, $m \geq 2$, spheres are connected.

17. Consider a polynomial $p(x, y)$ in two (real) variables, $p \neq 0$. Let $K = \{(x, y); p(x, y) = 0\}$.

    (a) Give an example where K is empty.

    (b) Give an example where K contains just one point.

    (c) Give an example where K is not compact.

    (d) If K contains more than one point, can the complement of K be connected?

18. Let S be a subset of R^m. If $X \in S$, let S_X be the union of all connected subsets of S which contain X. Then S_X is (non-empty and) connected. If S_X and S_Y have any point in common, then $S_X = S_Y$. The distinct sets S_X are called the **connected components** of S.

*19. If K is a compact set and $X \in K$, the connected component of K which contains X is the intersection of all relatively open-closed sets which contain X.

3. Continuity

3.1. Continuous Functions

The concept of continuity permeates mathematics. It can be formulated in great generality; however, for the time being, we shall deal with it in the context of maps from one Euclidean space into another. The reader may wish to refer to the Appendix, to review the basic notation and terminology concerning functions.

We consider a function F

$$(3.1) \qquad D \xrightarrow{\;\;F\;\;} R^m, \qquad D \subset R^k$$

which is defined on a subset of R^k and which maps that set into R^m. Roughly speaking, we say that F is continuous if F preserves proximity: If X_1 is near X_2, then $F(X_1)$ is near $F(X_2)$. There is a variety of ways to make this precise. We shall look at several of these ways and show that they are equivalent. In the one which we choose as the definition, we shall describe proximity by open sets, because this lends itself readily to subsequent generalization.

Definition. The function F (3.1) *is* **continuous** *if, for every open set* V *in* R^m, *the inverse image* F^{-1}(V) *is open relative to the domain of definition* D.

Does this say that a continuous function sends points which are close together into points which are close together? In a sense, it does. Let us see why. Suppose F is a continuous function, and let X_0 be a point in the domain D. Let's look at the set of all points which are near the image point $F(X_0)$, i.e., a neighborhood of $F(X_0)$. What we have in mind is a "small" neighborhood; so, let us consider an open ball

$$B_\epsilon = B(F(X_0); \epsilon).$$

We assert that every point of D which is sufficiently near to X_0 is mapped by F into this neighborhood B_ϵ. Why? The definition of continuity tells us that

$$F^{-1}(B_\epsilon) = \{X \in D; F(X) \in B_\epsilon\}$$

is open relative to D. Therefore, this set contains the intersection of D with some open ball about X_0:

(3.2) $$D \cap B(X_0; \delta) \subset F^{-1}(B_\epsilon).$$

In other words, (3.2) says

$$|F(X) - F(X_0)| < \epsilon, \qquad X \in D, \quad |X - X_0| < \delta.$$

Continuity has told us that, if we are given a point $X_0 \in D$ and a number $\epsilon > 0$, there exists some number $\delta > 0$ such that every point of the domain which is within distance δ of X_0 is mapped by F into a point within distance ϵ of $F(X_0)$. This property characterizes continuous functions, as we shall state formally in a moment.

We might note that most of the preceding discussion dealt only with how F behaved near a single point X_0.

Definition. *The function* F *is* **continuous at the point** X_0 *in* D *provided this condition is satisfied: If* V *is a neighborhood of* $F(X_0)$ *in* R^m, *then the inverse image* $F^{-1}(V)$ *is a neighborhood of* X_0 *relative to* D.

It is easy to see that F *is continuous if and only if* F *is continuous at each point of its domain* D. Let us record what we observed about ϵ's and δ's and add a bit more information.

Theorem 1. *Let* X_0 *be a point in* D, *the domain of the function* F. *The following are equivalent:*

(i) F *is continuous at the point* X_0.
(ii) *For each* $\epsilon > 0$, *there exists a number* $\delta > 0$ *such that*

$$|F(X) - F(X_0)| < \epsilon$$

for every point $X \in D$ *with* $|X - X_0| < \delta$.
(iii) *If* $\{X_n\}$ *is a sequence in* D *which converges to* X_0, *then the sequence* $\{F(X_n)\}$ *converges to* $F(X_0)$.

Proof. We noted that (i) implies (ii). Suppose (ii) is satisfied. We shall verify (iii). Let $X_n \in D$ and $X_0 = \lim_n X_n$. Consider an open ball $B(F(X_0); \epsilon)$. From (ii) we obtain $\delta > 0$ such that

(3.3) $$|F(X) - F(X_0)| < \epsilon, \qquad X \in D, \quad |X - X_0| < \delta.$$

Since X_n converges to X_0, we have $|X_n - X_0| < \delta$ except for a finite number of n's. Thus,

$$|F(X_n) - F(X_0)| < \epsilon$$

except for a finite number of n's. Since this is true for every $\epsilon > 0$, the sequence $\{F(X_n)\}$ converges to $F(X_0)$.

Now, suppose condition (iii) is satisfied. We shall verify (i). Let V be a neighborhood of $F(X_0)$. Is $F^{-1}(V)$ a neighborhood of X_0 relative to D? If it is not, then for each n we can select $X_n \in D$ with $|X_0 - X_n| < 1/n$ but $X_n \notin F^{-1}(V)$. Then $\{X_n\}$ converges to X_0. By (iii), $\{F(X_n)\}$ converges to $F(X_0)$. Thus, since V is a neighborhood of X_0, some $F(X_n)$ is in V. But $X_n \notin F^{-1}(V)$, so we cannot have that situation. *Conclusion:* $F^{-1}(V)$ must be a neighborhood of X_0 relative to D.

Since we have shown that

$$(i)\longrightarrow(ii)$$
$$\nwarrow \swarrow$$
$$(iii)$$

it is clear that if F possesses any of the three properties, it possesses the other two as well.

In order to deal successfully with continuity, there are two things which we must acquire rapidly:

1. a basic list of continuous functions;
2. methods for combining continuous functions to produce new continuous functions.

We also need to look at some discontinuous functions, in order to sharpen our understanding of the definition of continuity. Before we turn to examples, let us take note of the most fundamental method for combining continuous functions.

Theorem 2. *Let* F *be a function from (a subset of)* R^k *into* R^m *and let* G *be a function from (a subset of)* R^m *into* R^n *such that the domain of* G *contains the image of* F. *If* F *is continuous at the point* X_0 *and if* G *is continuous at the point* $F(X_0)$, *then the composition* G ∘ F *is continuous at* X_0.

Proof. The statement of this theorem is as long as its proof. Let W be a neighborhood of $(G \circ F)(X_0) = G(F(X_0))$. (See Figure 10.) Then $G^{-1}(W)$ is a neighborhood of $F(X_0)$ relative to D_G, the domain of G. Therefore $G^{-1}(W) = V \cap D_G$, where V is a neighborhood of $F(X_0)$ in R^m. Since D_G contains the image of F, $F^{-1}(V \cap D_G) = F^{-1}(V)$; hence, $F^{-1}(G^{-1}(W))$ is a neighborhood of X_0 relative to D_F. But $F^{-1}(G^{-1}(W)) = (G \circ F)^{-1}(W)$. (See Figure 10.)

The reader should carry out the proof of Theorem 2 using the reformulations of continuity contained in Theorem 1. For example, if $\{X_n\}$ is a

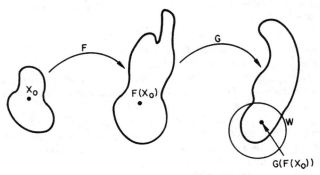

FIGURE 10

sequence of points in the domain of F which converges to $X_0 \in D_F$, then $\{F(X_n)\}$ converges to $F(X_0)$ because F is continuous. Accordingly $G(F(X_n))$ converges to $G(F(X_0))$ because G is continuous.

A function

$$D \xrightarrow{F} R^m$$

may be regarded as an m-tuple of real-valued functions:

$$F = (f_1, \ldots, f_m)$$
$$Y = F(X)$$
$$y_1 = f_1(x_1, \ldots, x_k)$$
$$\cdot \qquad \cdot$$
$$\cdot \qquad \cdot$$
$$\cdot \qquad \cdot$$
$$y_m = f_m(x_1, \ldots, x_k)$$

The function f_j is the composition of F with the jth standard coordinate function on R^m, that is, $f_j(X)$ is the jth coordinate of the point $F(X)$. *The function* $F = (f_1, \ldots, f_m)$ *is continuous if and only if each* f_j *is continuous.* We leave the proof as an exercise.

EXAMPLE 1. Vector addition is a continuous function. Of course, vector addition in R^m is not a function on R^m; it is a function on R^{2m}. Think of R^{2m} as $R^m \times R^m$, each $2m$-tuple being a pair (X, Y) where X is the m-tuple of the first m coordinates and Y is the m-tuple of the last m coordinates:

$$(X, Y) = (x_1, \ldots, x_m, y_1, \ldots, y_m).$$

Addition is the function

$$R^{2m} \xrightarrow{A} R^m$$
$$A(X, Y) = X + Y.$$

We verified the continuity in sequential form in the lemma of Section 2.1. The reader should also ask himself what the largest δ is for which the ball of radius δ about (X, Y) is contained in the inverse image of $B(X + Y; \epsilon)$.

Now, it is easy to see that the *sum of two continuous functions is continuous.* Suppose

$$D \xrightarrow{\;F\;} R^m$$
$$D \xrightarrow{\;G\;} R^m, \qquad D \subset R^k.$$

Then $(F + G)(X) = F(X) + G(X)$. The fact that $F + G$ is continuous is clear, by the same argument which shows that addition is continuous. In fact, the continuity of $F + G$ can be deduced from the continuity of addition, as follows. Let $F \times G$ be the function

$$D \xrightarrow{\;F \times G\;} R^{2m}$$
$$(F \times G)(X) = (F(X), G(X)).$$

If F, G are continuous, obviously $F \times G$ is. Observe that $F + G = A \circ (F \times G)$, where A is addition. By Theorem 2, $F + G$ is continuous.

EXAMPLE 2. Matrix multiplication is continuous. Multiplication of real $k \times k$ matrices is a function from Euclidean space of dimension $2k^2$ into Euclidean space of dimension k^2:

$$R^{k \times 2k} \xrightarrow{\;M\;} R^{k \times k}$$
$$M(A, B) = AB.$$

We verified this continuity in sequential form in Example 2 of Chapter 2. Similarly, the product of two continuous matrix-valued functions is continuous. The matrices may have complex entries as well as real ones. In particular, *the product of two continuous real-valued (complex-valued) functions is continuous.*

EXAMPLE 3. Every polynomial function on R^n is continuous. Such a function has the form

$$(3.4) \qquad f(x_1, \ldots, x_n) = \sum_{k_1, \ldots, k_n = 0}^{N} a_{k_1, \ldots, k_n} x_1^{k_1} \cdots x_n^{k_n}$$

where the $a_{k_1, \ldots, k_n}$ are fixed real numbers. The standard coordinate functions

$$f_j(X) = x_j, \qquad 1 \leq j \leq n$$

are continuous:

$$|f_j(X) - f_j(X_0)| \leq |X - X_0|.$$

Of course, constant functions are continuous. Consequently, by repeated application of the fact that sums and products of continuous functions are continuous, we see that every polynomial function is continuous.

EXAMPLE 4. Every polynomial function on C^n:

(3.5) $$f(z_1, \ldots, z_n) = \sum a_{k_1, \ldots, k_n} z_1^{k_1} \cdots z_n^{k_n}$$

is continuous, by the same reasoning.

EXAMPLE 5. Inversion is continuous, on the set of invertible matrices. Refer to Example 14 of Chapter 2, if necessary. Let A be a fixed $k \times k$ matrix which is invertible. If $|A - B| < |A^{-1}|^{-1}$, then $|I - A^{-1}B| < 1$, so that

$$B^{-1}A = [I - (I - A^{-1}B)]^{-1} = \sum_{n=0}^{\infty} (I - A^{-1}B)^n$$

$$B^{-1} - A^{-1} = \sum_{n=1}^{\infty} (I - A^{-1}B)^n A^{-1}$$

$$|B^{-1} - A^{-1}| \le |A^{-1}| \cdot \sum_{n=1}^{\infty} |I - A^{-1}B|^n.$$

Now

$$|I - A^{-1}B| \le |A^{-1}||A - B|$$

so that

$$|B^{-1} - A^{-1}| \le |A^{-1}| \sum_{n=1}^{\infty} |A^{-1}|^n |A - B|^n$$

that is,

(3.6) $$|B^{-1} - A^{-1}| \le |A^{-1}| \cdot \frac{|A^{-1}||A - B|}{1 - |A^{-1}||A - B|}.$$

It is then clear that $|B^{-1} - A^{-1}|$ is small when $|A - B|$ is small.

The reader to whom the continuity is not clear from (3.6) may wish to carry the detailed reasoning a few steps further. Given an invertible matrix A and a number $\epsilon > 0$, we wish to find $\delta > 0$ such that $|B^{-1} - A^{-1}| < \epsilon$ for all matrices B which satisfy $|B - A| < \delta$. According to the inequalities which we just derived, this will be so provided

$$\delta < |A^{-1}|^{-1}$$

$$|A^{-1}| \cdot \frac{|A^{-1}|\delta}{1 - |A^{-1}|\delta} \le \epsilon.$$

The largest δ which has these properties is the one for which equality holds in the second condition:

$$\delta = \frac{1}{|A^{-1}|} \cdot \frac{\epsilon}{\epsilon + |A^{-1}|}.$$

As a special case of the continuity of inversion, *if f is a real or complex-valued function, then* 1/f *is continuous at* X *if* f *is continuous at* X *and* f(X) $\neq$ 0. This tells us that rational functions (quotients of polynomials) are continuous off the zero sets of their denominators.

If one knows that inversion is continuous on non-zero numbers (1×1 matrices), it follows that rational functions are continuous where defined. One can conclude from that fact that inversion of matrices is continuous: The entries of A^{-1} are rational functions of the entries of A, because the

i, j entry of A^{-1} is

$$(-1)^{i+j}\frac{\det A(j\,|\,i)}{\det A}$$

where $A(j\,|\,i)$ is the $(k-1) \times (k-1)$ matrix obtained by deleting the jth row and ith column of A. But, a careful proof that inversion is continuous on the set of non-zero numbers is not appreciably simpler than the argument we gave for matrices.

EXAMPLE 6. Let D be the set of non-negative real numbers. Let n be a positive integer. The nth root function

$$f(x) = x^{1/n}$$

is continuous. Fix a positive number t. We want to show that $x^{1/n} - t^{1/n}$ is small provided $x - t$ is small. In order to compare the sizes of these two numbers we let $a = t^{1/n}$ and $b = x^{1/n}$ and compare $b^n - a^n$ to $b - a$. Now

$$b^n - a^n = (b - a)(b^{n-1} + b^{n-2}a + \cdots + a^{n-1}).$$

Thus,

$$|b^n - a^n| \geq |b - a|\,nc^{n-1}$$

where c is the smaller of a and b. If we keep x sufficiently near t so that $x > t/2$ then we shall have

$$|x - t| \geq |x^{1/n} - t^{1/n}|\cdot n\Big(\frac{t}{2}\Big)^{(n-1)/n}$$

or

$$|x^{1/n} - t^{1/n}| \leq \frac{1}{n}\Big(\frac{2}{t}\Big)^{(n-1)/n}|x - t|, \qquad x > \frac{t}{2}.$$

From this inequality it should be apparent that $f(x) = x^{1/n}$ is continuous at $x = t$.

EXAMPLE 7. Every linear transformation from R^k into R^m is continuous. A linear transformation is a function

$$R^k \xrightarrow{\;\;T\;\;} R^m$$

such that $T(cX_1 + X_2) = cT(X_1) + T(X_2)$. The defining property extends immediately to the fact that a linear transformation "preserves" linear combinations

$$(3.7) \qquad T(c_1X_1 + \cdots + c_nX_n) = c_1T(X_1) + \cdots + c_nT(X_n).$$

Such transformations are exceedingly special functions. We can describe them all immediately. Let $E_1, \ldots, E_k$ be the standard basis vectors:

$$E_1 = (1, 0, \ldots, 0)$$
$$E_2 = (0, 1, \ldots, 0)$$
$$\vdots \qquad \vdots$$
$$E_k = (0, 0, \ldots, 1).$$

Then $X = x_1 E_1 + \cdots + x_k E_k$, so, if T is linear we have

$$T(X) = x_1 T(E_1) + \cdots + x_k T(E_k).$$

How do we describe all linear transformations from R^k into R^m? Pick any k vectors $Y_1, \ldots, Y_k$ in R^m and define

$$T(X) = x_1 Y_1 + \cdots + x_k Y_k.$$

Then T is linear and $Y_j = T(E_j)$. Furthermore, every linear transformation has that form.

From the specific form of a linear transformation its continuity is apparent. If $Y_j = T(E_j)$, then

$$|T(X)| \le |x_1| |Y_1| + \cdots + |x_k| |Y_k|.$$

By Cauchy's inequality

$$|T(X)| \le M |X|$$

where

$$M = (|Y_1|^2 + \cdots + |Y_k|^2)^{1/2}.$$

Since T is linear,

$$|T(X) - T(X_0)| = |T(X - X_0)|$$
$$\le M |X - X_0|.$$

Exercises

1. The function f, $f(x) = x^2$, is continuous from R into R. Find an explicit formula in terms of t and ϵ for the largest δ such that

$$|f(x) - f(t)| < \epsilon, \qquad |x - t| < \delta.$$

2. Let g be the greatest integer function, i.e., $g(x)$ is the largest integer $\le x$, $x \in R$. At which points is g continuous?

3. Let f be the real-valued function on R^2,

$$f(x, y) = \frac{xy}{x^2 + y^2}, \quad x^2 + y^2 \ne 0$$

$$f(0, 0) = 0.$$

Show that f is not continuous at the origin. On which lines in R^2 is f continuous?

4. Let f be the function of Exercise 3. Describe explicitly all open sets $V \subset R$ for which $f^{-1}(V)$ is open in R^2.

5. Let

$$R^k \xrightarrow{f} R$$

be a continuous function. Suppose that f assumes only rational values. Show that f is constant.

6. Let f be the real-valued function on R^k

$$f(X) = \max_j x_j,$$

i.e., $f(X)$ is the largest coordinate of X. Is f continuous?

7. Length is a continuous function on R^k.

8. If F is continuous, then $|F|$ is continuous.

9. If f and g are continuous real-valued functions on D then the minimum of f and g is continuous:

$$h(X) = \min \{f(X), g(X)\}.$$

10. Conjugation is continuous on the complex numbers.

11. The function

$$D \xrightarrow{f} R$$

is continuous if and only if the sets $\{X; f(X) > t\}$ and $\{X; f(X) < t\}$ are open relative to D (for every $t \in R$).

12. The function f:

$$f(z) = \frac{1 + z}{1 - z}$$

is continuous on the set of complex numbers $z \neq 1$.

   (a) Now $f(0) = 1$. What is the largest δ such that

$$|1 - f(z)| < \tfrac{1}{2}, \qquad |z| < \delta?$$

   (b) If $V = \{z \in C; \text{Re}\,(z) > 0\}$, what is $f^{-1}(V)$?

13. If D is an interval on the real line, a **convex function** on D is a real-valued function f such that

$$f(tx + (1 - t)y) \leq tf(x) + (1 - t)f(y)$$

for all $x, y \in D$ and all $t \in [0, 1]$.

   (a) Interpret the convexity of f in terms of line segments and the graph of f.

   (b) Give an example of a discontinuous convex function.

   (c) Prove that every convex function on an open interval is continuous.

14. Let D be a convex subset of R^k, i.e. a subset such that the line segment from X to Y is in D whenever X and Y are in D.

   (a) Define convex function on D.

   (b) Prove that, if D is open, every convex function on D is continuous.

15. Let T be a function from R^k into R^m which is additive: $T(X_1 + X_2) = T(X_1) + T(X_2)$. If T is continuous, then T is a linear transformation.

3.2. Continuity and Closed Sets

The relationship between continuous functions and closed sets deserves special comment. This little section contains assorted examples and remarks which are elementary but which may increase our understanding of continuity.

Suppose that we have any function

$$D \xrightarrow{F} R^m, \qquad D \subset R^k.$$

If V_1 and V_2 are subsets of R^m, then

$$F^{-1}(V_1 \cup V_2) = F^{-1}(V_1) \cup F^{-1}(V_2)$$
$$F^{-1}(V_1 \cap V_2) = F^{-1}(V_1) \cap F^{-1}(V_2).$$

These two properties of inverse images are basic, in spite of the fact that they are trivial to verify. The second property should not be confused with $F(U_1 \cap U_2) = F(U_1) \cap F(U_2)$, which is generally false. From the two properties, it should be apparent that F *is continuous if and only if* $F^{-1}(K)$ *is closed relative to* D, *for every closed set* K *in* R^m. In particular, if F is continuous and K consists of the single point Y_0 in R^m, then $F^{-1}(K) = \{X \in D; F(X) = Y_0\}$ is closed relative to D; that is, *each level set of a continuous function is closed* (relative to the domain of the function).

If f is a continuous real-valued function on R^k, then $\{X; f(X) \geq 0\}$, $\{X; -1 \leq f(X) \leq 3\}$ and $\{X; f(X) = 0\}$ are closed subsets of R^k, while $\{X; f(X) > 0\}$ is an open set. These are things which a student of analysis must recognize without hesitation.

We have not cited many specific examples of continuous functions. We do know that polynomial functions and a few other functions are continuous, and they provide us with some interesting examples of closed sets.

EXAMPLE 8. Each linear function on R^k is continuous. Therefore, every hyperplane (level set of a non-zero linear function) is a closed set. Each linear subspace of R^k is an intersection of hyperplanes (through the origin), and so, each subspace is a closed set.

The determinant function is continuous on the space of $k \times k$ matrices (real or complex entries), because it is a polynomial function of the entries. Thus, the set of singular matrices (those with det $A = 0$) is a closed set, and the set of invertible matrices is open. In the real case, the set of invertible matrices is disconnected by the determinant function, since it is the union of $\{A; \det A > 0\}$ and $\{A; \det A < 0\}$.

A $k \times k$ matrix is called orthogonal if $AA^t = I$, where A^t denotes the transpose of A (obtained by interchanging the rows and columns of A). The set of orthogonal matrices is a closed subset of $R^{k \times k}$ or $C^{k \times k}$, because it is a level set of the continuous function

$$F(A) = AA^t.$$

In the real case, the set of orthogonal matrices is compact, since

$$|A|^2 = k$$

for a real orthogonal $k \times k$ matrix, and this bounds the set.

At times, it is useful to know that every non-empty closed set in R^k is a level set of a continuous function. Let E be a non-empty closed set in R^k. For each X, consider the **distance from** X **to** E:

$$d(X, E) = \inf \{|X - Y|; Y \in E\}.$$

This can be defined for any non-empty set E, but evidently $d(X, \bar{E}) = d(X, E)$ so we restrict ourselves to closed sets E. We note two things.

(i) $d(X, E) = 0$ if and only if $X \in E$.

(ii) There exists (at least one) $Y \in E$ such that

$$|X - Y| = d(X, E).$$

Property (i) is essentially the definition of closed set. Property (ii) follows from this observation. Given X, choose any point Z in E. Evidently, the points of E which are nearest X are found in the closed ball $\bar{B} = \bar{B}(X; |X - Z|)$. Therefore,

$$d(X, E) = \inf \{|X - Y|; Y \in \bar{B} \cap E\}.$$

The function $f(Y) = |X - Y|$ is continuous on the compact set $\bar{B} \cap E$. Thus its infimum is attained at some point Y. (There may be more than one such Y.)

The function "distance to E" is continuous because

$$|d(X_1, E) - d(X_2, E)| \leq |X_1 - X_2|.$$

Theorem 3. *Let* E *and* K *be disjoint closed subsets of* $\mathbf{R}^k$. *There exists on* $\mathbf{R}^k$ *a real-valued continuous function* f *such that*

(i) $0 \leq f \leq 1$;

(ii) $K = \{X; f(X) = 1\}$;

(iii) $E = \{X; f(X) = 0\}$.

Proof. If K and E are empty, there is nothing to prove. If K is empty but E is not, take

$$f(X) = \frac{d(X, E)}{1 + d(X, E)}$$

and do a similar sort of thing if E is empty while K is not. If they are both non-empty, take

$$f(X) = \frac{d(X, E)}{d(X, K) + d(X, E)}.$$

This function has properties (i), (ii), and (iii).

Theorem 3 is frequently used in the following way. We are given a closed set K and an open set U which contains K. We want to find a continuous function f such that $0 \leq f \leq 1$, $f = 1$ on K and $f = 0$ outside U. By taking $E = R^k - U$ in Theorem 3, we see that there is such a function.

The distance function $d(X, E)$ measures the extent to which a point X can be approximated by points of E. The fact that there exists a $Y \in E$ such that $|X - Y| = d(X, Y)$ says that there exists a point of E which is a best approximation of X by points of E. There is no reason to think that Y is unique. Many points of E might provide a best approximation. There is an important class of sets, closed convex sets, for which there is a unique best approximation. These sets occur in various applications, so let us say a little bit about them.

The set K is **convex** if, for each pair of distinct points X, Y in K, the line segment between X and Y is in K. This means that, if X, Y are in K, then

$$(tX + (1 - t)Y) \in K, \qquad 0 \leq t \leq 1.$$

Closed balls and linear subspaces are examples of closed convex sets. Half-spaces are the basic closed convex sets. A (closed) **half-space** in R^k is a set H of the form

$$H = \{X; f(X) \leq c\}$$

where c is a real number and f is a non-zero linear functional on R^k. The form of such a functional is

$$f(X) = a_1 x_1 + \cdots + a_k x_k$$

where $a_1, \ldots, a_k$ are real numbers, not all 0. The hyperplane $\{f = c\}$ determines two (closed) half-spaces, $\{f \leq c\}$ and $\{f \geq c\} = \{-f \leq -c\}$. So, a half-space is determined by a linear inequation (inequality)

$$a_1 x_1 + \cdots + a_k x_k \leq c$$

just as a hyperplane is determined by a linear equation

$$a_1 x_1 + \cdots + a_k x_k = c.$$

In Chapter 1 we discussed the "linear" or flat subsets of R^k (lines, planes, etc.). These can be described by the following equivalent conditions:

(a) If X, Y are in K, the line through X and Y is in K.

(b) K is an intersection of hyperplanes, *i.e.*, K can be described by a system of linear equations.

(c) K is a translate of a linear subspace of R^k.

We want to see that the following conditions on a *closed* set K are equivalent:

(a) K is convex.

(b) K is an intersection of half-spaces; i.e., K can be described by a (possibly infinite) system of linear inequalities.

Theorem 4. *Let* K *be a closed convex set in* R^k. *If* $X \in R^k$, *there is a unique point in* K *which is nearest to* X, *i.e., there is one and only one point* Y *in* K *such that*

$$|X - Y| = d(X, K).$$

Proof. If $|X - Y| = |X - Z|$ and if $Y \neq Z$, then $\frac{1}{2}(Y + Z)$ is closer to X than either Y or Z. If one wishes to verify that analytically, just do so in the case $X = 0$ and use the parallelogram law:

$$|Y - Z|^2 = 2(|Y|^2 + |Z|^2) - |Y + Z|^2$$
$$= 2(|Y|^2 + |Z|^2) - 4|\tfrac{1}{2}(Y + Z)|^2.$$

(You may feel that use of the triangle inequality is simpler.)

In case K is a linear subspace, the nearest point to X is the orthogonal projection of X onto K. (See Chapter 1.) In fact, any closed convex set K defines a "projection." Define a function

$$R^k \xrightarrow{\ P_K\ } K$$

by

$$Y = P_K(X) \quad \text{if} \quad Y \in K \quad \text{and} \quad |X - Y| = d(X, K).$$

Since Y is the unique best approximation to X by elements of K, the function P_K is well-defined. Several points X, together with their corresponding best approximations in K, are shown in Figure 11: $Y_j = P_K(X_j)$. We also have

(a) P_K is continuous;
(b) $P_K(P_K(X)) = P_K(X)$;
(c) $P_K(X) = X$ if and only if $X \in K$.

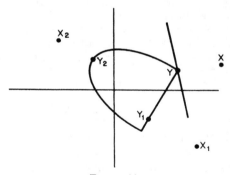

FIGURE 11

Take any X in R^k which is not in K. Let $Y = P_K(X)$. Then K is entirely on one side of the hyperplane through Y orthogonal to $X - Y$. (We have left the proof as an exercise.) That hyperplane is

$$Y + (X - Y)^\perp = \{Z; \langle Z, X - Y \rangle = \langle Y, X - Y \rangle\}.$$

In other words, the hyperplane is

$$\{Z; f(Z) = f(Y)\}$$

where f is the linear functional "inner product with $X - Y$":

$$f(Z) = \langle Z, X - Y \rangle.$$

Either K is in the half-space where $f \leq f(Y)$ or the half-space where $f \geq f(Y)$, according as $f(X) > f(Y)$ or $f(X) < f(Y)$.

Theorem 5. *Every closed convex set is an intersection of half-spaces.*

The only property of the projection P_K which we asserted but which is not obvious is the continuity. We end this section by indicating how this can be verified. If $Y = P_K(X)$ we want to choose $\delta > 0$ so that $|Y - P_K(Z)|$ is small for all vectors Z such that $|X - Z| < \delta$. Suppose $\delta > 0$ is small and Z satisfies $|X - Z| < \delta$. Let $W = P_K(Z)$. We obtain the following inequalities for the reasons indicated at the right.

$$|X - W| \geq |X - Y|, \qquad Y = P_K(X)$$
$$|X - W| \leq |Z - W| + \delta, \qquad |X - Z| < \delta$$
$$\leq |Z - Y| + \delta, \qquad W = P_K(Z)$$
$$\leq |X - Y| + 2\delta, \qquad |X - Z| < \delta.$$

Therefore,

$$|X - Y| \leq |X - W| \leq |X - Y| + 2\delta.$$

If δ is small, does this imply that W is close to Y? No, but it does if we add the one other piece of information we have, which is that (since K is convex) every point on the line segment from Y to W is at least as far from X as Y is. Since there are but three vectors involved, this need only be verified in the plane where it says: If X and Y are points in the plane, if W is a point such that the line segment from Y to W lies between the circles of radius $|X - Y|$ and $|X - Y| + 2\delta$, and if δ is small, then W is close to Y.

Exercises

1. True or false? If f is a continuous real-valued function on R_k and if K is a compact subset of R, then $f^{-1}(K)$ is compact.

2. True or false? If F is continuous from R^k to R^m and S is a subset of R^m, then $F^{-1}(\overline{S})$ is the closure of $F^{-1}(S)$.

3. If f and g are continuous complex-valued functions on R^k, then $\{X; f(X) = g(X)^2\}$ is closed.

4. Prove that the function
$$R^k \xrightarrow{\;\;F\;\;} R^m$$
is continuous if and only if, for every set $S \subset R^m$, the boundary of $F^{-1}(S)$ is contained in $F^{-1}(\text{boundary } S)$.

5. Let D be a connected subset of R^k and suppose that there is a continuous real-valued function on D which is nowhere 0. Show that the function is either everywhere positive or everywhere negative.

6. Let D be a disconnected subset of R^k. Prove that there exists on R^k a continuous real-valued function which is nowhere zero on D, yet takes on both positive and negative values at points of D. *Hint*: Use distance functions and Theorem 14 of Chapter 2.

7. Let f be a real function on R^k such that $f^{-1}(V)$ is closed, for every open set V in R. What can you say about f? Interchange open and closed.

8. Prove the converse of Theorem 4: If K is a subset of R^k such that every point of R^k has a unique best approximation by points of K, then K is closed and convex.

9. Let D be an open subset of R^k and let F

$$D \xrightarrow{\ F\ } R^m$$

be a continuous function on D. Let K be a closed subset of D. Prove that there exists a continuous function

$$R^k \xrightarrow{\ G\ } R^m$$

such that $G(X) = F(X)$ for all $X \in K$.

10. If E is a closed convex set and K is a compact convex set, then E and K can be strictly separated by a hyperplane, i.e., there is a non-zero linear f and a number c such that $f > c$ on E and $f < c$ on K. (*Hint*: What kind of a set is the algebraic sum $E + (-K)$?)

11. An **extreme point** of a convex set K is a point $Y \in K$ which does not lie on the line segment joining any two different points in K, i.e., if X_1, X_2 are in K and

$$Y = \tfrac{1}{2}(X_1 + X_2)$$

then $X_1 = X_2 = Y$.

  (a) A closed convex set need have no extreme points.

  (b) A non-empty compact convex set has an extreme point.

  *(c) A compact convex set K consists of all convex combinations

$$t_1 Y_1 + \cdots + t_n Y_n, \qquad t_j \geq 0, \qquad \sum_j t_j = 1$$

where $Y_1, \ldots, Y_n$ are extreme points of K.

*12. If S is any set in R^k, the smallest convex set which contains S consists of all convex combinations

$$\sum_{j=1}^{n} t_j Y_j, \qquad Y_j \in S, \qquad t_j \geq 0, \qquad \sum_j t_j = 1$$

with $n \leq k + 1$ (all convex combinations of $k + 1$ or less points of S).

3.3. The Limit of a Function

Continuity can be formulated in terms of the concept of the limit of a function at a point. This concept is similar to the idea of the limit of a sequence, as we shall discuss later.

Definition. *Let* F *be a function*

$$D \xrightarrow{F} R^m, \qquad D \subset R^k.$$

Let X_0 *be a cluster point of the set* D. *We say that* F **has the limit** L **at** X_0 *provided*

(i) $L \in R^m$;

(ii) *if* V *is any neighborhood of* L, *there exists a neighborhood* U *of the point* X_0 *such that*

$$f(X) \in V \quad \text{for every} \quad X \in D \cap U, X \neq X_0.$$

If (i), (ii) *hold, we write*

$$L = \lim_{X \to X_0} F(X).$$

We say that *F* **has a limit** at X_0 or that $\lim_{X \to X_0} F(X)$ **exists** provided there exists $L \in R^m$ such that

$$L = \lim_{X \to X_0} F(X).$$

(If the limit exists, it is unique. This follows from the fact that X_0 is a cluster point of *D*.)

Lemma. *Let* X_0 *be a cluster point of the domain* D, *and let* L *be a point in* R^m. *The following are equivalent.*

(i) $\lim_{X \to X_0} F(X) = L$;

(ii) *For each* $\epsilon > 0$, *there exists a number* $\delta > 0$ *such that* $|F(X) - L| < \epsilon$ *for every* X *in* D *such that* $0 < |X - X_0| < \delta$.

(iii) *If* $\{X_n\}$ *is a sequence of points in* D *such that* $X_n \neq X_0$ *and* $\{X_n\}$ *converges to* X_0, *then* $\{F(X_n)\}$ *converges to* L.

(iv) *If* $D_0 = D \cup \{X_0\}$ *and if* F_0 *is the function on* D_0 *defined by*

$$F_0(X) = \begin{cases} F(X), & \text{if } X \in D \text{ and } X \neq X_0 \\ L, & \text{if } X = X_0 \end{cases}$$

then F_0 *is continuous at the point* X_0.

Proof. The point X_0 is in the closure of *D*. It may or may not be in *D*. The definition of "the limit of *F* at X_0" makes no mention of whether or not $X_0 \in D$, and if $X_0 \in D$, the definition is completely independent of $F(X_0)$. This must be understood very clearly. Then, it should be understood clearly that (iv) is a reformulation of the definition. It is then apparent that (ii) and (iii) are reformulations of (i).

Theorem 6. *If* X_0 *is a cluster point of* D, *then* F *is continuous at* X_0 *if and only if*

$$\lim_{X \to X_0} F(X) = F(X_0).$$

Lemma. *Let* X_0 *be a cluster point of* D. *These are equivalent.*

(i) F *has a limit at* X_0.

(ii) *If* $\epsilon > 0$, *there exists* $\delta > 0$ *such that* $|F(X_1) - F(X_2)| < \epsilon$ *for all points* X_1, X_2 *in* D *such that*

$$0 < |X_1 - X_0| < \delta$$
$$0 < |X_2 - X_0| < \delta.$$

(iii) *If* $X = \lim X_n$, *where* $X_n \in D$ *and* $X_n \neq X_0$, *then* $\{F(X_n)\}$ *is a Cauchy sequence.*

Proof. We content ourselves with verifying that (i) follows from (iii). In other words, if (iii) holds, we must show that the limit of the sequence $\{F(X_n)\}$ is independent of the sequence $\{X_n\}$. So suppose we have two such sequences

$$X_0 = \lim_n X_n, \qquad X_n \in D, \quad X_n \neq X_0$$
$$X_0 = \lim_n Y_n, \qquad Y_n \in D, \quad Y_n \neq X_0.$$

Let

$$L = \lim_n F(X_n)$$
$$M = \lim_n F(Y_n).$$

The sequence X_1, Y_1, X_2, Y_2, X_3, $Y_3, \ldots$ converges to X_0. By (iii), the sequence $F(X_1)$, $F(Y_1)$, $F(X_2)$, $F(Y_2), \ldots$ is a Cauchy sequence. That is not likely to happen if $L \neq M$.

Evidently, we now have several elementary facts which we can state. The limit of a sum is the sum of the limits, etc. Such results follow from the corresponding results for sequences, or the corresponding results for continuous functions. But, a better way to think of it is that the formal reasoning is the same in the sequence case as it is in the case of the limit of a function at a point. Some people find it helpful to regard as analogous the facts that (1) in defining $\lim_{X \to X_0} F(X)$ we do not discuss $F(X_0)$; (2) a sequence $\{X_n\}$ has no last term.

One word of warning is in order. The domain of definition of a function is part of the function. The discussion of limits is one of the situations in which it is extremely important to bear this in mind. The same sort of rule may define functions on various sets, and a change in the domain of definition may affect the question of the existence of the limit at some point. Of course, we follow the usual convention that if a function is to be defined by a rule and if the domain is not explicitly mentioned, then the domain is the largest subset of R^k on which the rule makes sense.

EXAMPLE 9. The function f defined by

$$f(x) = \frac{x}{|x|}, \qquad x \neq 0$$

does not have a limit at 0. That is, we cannot extend f to a continuous function on R. The function g defined by

$$g(x) = \frac{x}{|x|}, \qquad x > 0$$

does have a limit at 0.

EXAMPLE 10. Define

$$f(x) = \sin\left(\frac{1}{x}\right), \qquad x \neq 0.$$

Then f does not have a limit at 0, because f oscillates near 0 between the values 1 and -1. (See Figure 12.) For instance, the sequence

$$x_n = \frac{2}{n\pi}$$

converges to 0, but the values $f(x_n)$ are $1, 0, -1, 0, \ldots$.

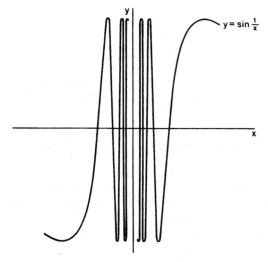

FIGURE 12

EXAMPLE 11. Let

$$g(x) = x \sin\left(\frac{1}{x}\right), \qquad x \neq 0.$$

Then

$$\lim_{x \to 0} g(x) = 0$$

because $|g(x)| \leq |x|$. (See Figure 13.) Therefore we can extend g to a continuous function on R, by assigning the value 0 to the point 0.

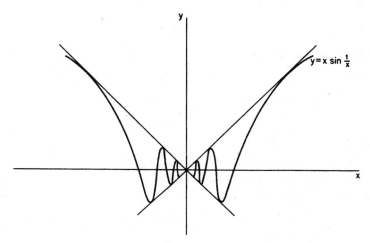

FIGURE 13

EXAMPLE 12. Let

$$f(x, y) = \frac{xy}{x^2 + y^2}, \qquad x^2 + y^2 \neq 0.$$

Then f does not have a limit at the origin. If $x = r \cos \theta$ and $y = r \sin \theta$, then

$$f(x, y) = \tfrac{1}{2} \sin 2\theta.$$

Thus, the restriction of f to a line through the origin has a limit at $(0, 0)$. The limits along those various lines tend to be different.

EXAMPLE 13. Let F be the inversion function on the set of invertible $k \times k$ matrices:

$$F(A) = A^{-1}, \qquad A \text{ invertible.}$$

Let B be a $k \times k$ matrix which is not invertible. Then B is a cluster point of the domain of F, because the set of invertible matrices is dense in the space of $k \times k$ matrices. Does F have a limit at B? No. In fact, there does not exist any sequence of invertible matrices A_n with $B = \lim_n A_n$ for which the sequence $\{A_n^{-1}\}$ converges. For, if we had

$$\lim_n A_n = B$$

$$\lim_n A_n^{-1} = C$$

then it would follow that

$$\lim_n I = BC$$

and hence that B is invertible.

Exercises

1. In the space of $k \times k$ matrices,

$$\lim_{A \to 0} e^A = I.$$

2. Let A be a fixed $k \times k$ matrix. Then

$$\lim_{t \to 0} \frac{e^{tA} - 1}{t} = A.$$

3. Let

$$f(x) = \exp\left(\frac{1}{x}\right), \qquad x \neq 0.$$

Then f does not have a limit at 0.

4. Let

$$f(x) = \exp\left(-\frac{1}{x^2}\right), \qquad x \neq 0.$$

If k is any positive integer

$$\lim_{x \to 0} x^{-k} f(x) = 0.$$

5. Let $D = \{z \in C; |z| < 1\}$ be the open unit disk in the plane. Let

$$f(z) = (1 - z) \exp\left(\frac{z + 1}{z - 1}\right), \qquad z \in D.$$

Let's prove that

$$\lim_{z \to 1} f(z) = 0.$$

   (i) If w is a complex number, then

$$|e^w| = e^{\mathrm{Re}(w)}.$$

   (ii)
$$\mathrm{Re}\left(\frac{z + 1}{z - 1}\right) = \frac{|z|^2 - 1}{|1 - z|^2}.$$

   (iii)
$$\left| \exp\left(\frac{z + 1}{z - 1}\right) \right| < 1, \qquad |z| < 1.$$

So

$$\lim_{z \to 1} f(z) = 0.$$

6. If

$$f(z) = (1 - z) \exp\left(\frac{z + 1}{z - 1}\right), \qquad z \in C, \quad z \neq 1$$

then f does not have a limit at $z = 1$.

7. Define a function f on the set of rational numbers, as follows. If x is a nonzero rational number, write $x = p/q$, where $p \in Z$, $q \in Z_+$, and p, q are relatively prime (no non-trivial common divisor); then let $f(x) = 1/q$. If x is irrational or zero, define $f(x) = 0$. Prove that

$$\lim_{t \to x} f(t) = 0$$

for each real number x. At which points is f continuous?

8. For functions on the real line, one can discuss left- and right-hand limits. Suppose

$$D \longrightarrow R^m, \qquad D \subset R$$

and let x be a point which is a cluster point of D from both sides. Define this type of cluster point, and then define (if they exist)

$$\lim_{t \to x_-} F(t) \quad \text{and} \quad \lim_{t \to x_+} F(t).$$

9. Let f be a real-valued function on the real line which is increasing:

$$f(x) \leq f(t), \qquad x \leq t.$$

At every point x the left- and right-hand limits of f exist. (See Exercise 8.) Such an f is continuous, except at a countable number of points. (Use Exercise 12 of Section 2.5.)

10. Let f be a real-valued function on D. Let X_0 be a cluster point of D. How would you define

$$\lim_{x \to X_0} \inf f(x) \quad \text{and} \quad \lim_{x \to X_0} \sup f(x)?$$

Every bounded function should have a lim sup and a lim inf at X_0; and, the function should have a limit at X_0 exactly when the lim sup equals the lim inf.

3.4. Infinite Limits and Limits at Infinity

In order to work effectively in analysis, it is important to know how to handle such statements as

(3.8) $$\lim_{x \to \infty} xe^{-x} = 0$$

(3.9) $$\lim_{X \to 0} \log |X| = -\infty.$$

Such things do not present much more difficulty than the limits we have discussed thus far. As with most things that smack of "the infinite," the secret of handling them rests not so much in knowing what to say as in knowing what not to say.

The most concise way of dealing with limits such as (3.8) and (3.9) is this. We introduce the **extended real number system,** which consists of the set of real numbers together with two additional objects, ∞ and $-\infty$. We do not say what those objects are, except that we expressly assume that $\infty \notin R$ and $-\infty \notin R$. The ordering of the real numbers is extended to the enlarged system by

$$-\infty < \infty$$
$$-\infty < x < \infty, \qquad x \in R.$$

We can extend addition and multiplication only partially to the enlarged system

$$x + \infty = \infty \quad \text{and} \quad x + (-\infty) = -\infty, \qquad x \in R$$
$$x\infty = \infty, \qquad\qquad\qquad\qquad\qquad\qquad x > 0$$
$$x\infty = -\infty, \qquad\qquad\qquad\qquad\qquad\qquad x < 0$$
$$x(-\infty) = -(x^\infty), \qquad\qquad\qquad\qquad\quad x \neq 0$$
$$\infty^2 = \infty$$
$$\infty(-\infty) = -\infty$$
$$(-\infty)^2 = \infty.$$

In short, $a + b$ is defined unless a, b are ∞ and $-\infty$ in one or the other order; and ab is defined unless $a = 0$ or $b = 0$. The usual rules of algebra hold, in the sense that "any algebraic calculation we might make will be correct, as long as we avoid $\infty + (-\infty)$ and $0 \cdot (\pm\infty)$". But, that sort of algebraic formalism is of relatively little use to us.

In the extended real number system, a **neighborhood of** ∞ is a subset which contains ∞ and also contains $\{x \in R; x > t\}$ for some $t \in R$. Replace ∞ by $-\infty$ and $x > t$ by $x < t$, and we have the definition of a **neighborhood of** $-\infty$.

Now we define limits as before, using neighborhoods; and statements such as (3.8) and (3.9) have a well-defined meaning. Suppose, for instance, that f is a real-valued function on a subset of R^k:

$$D \xrightarrow{\ f\ } R, \qquad D \subset R^k.$$

What does it mean to say that

(3.10) $$\lim_{X \to X_0} f(X) = \infty\,?$$

It means this: X_0 is a cluster point of D and, for every neighborhood V of ∞ there exists a neighborhood U of X_0 such that $f(U - \{X_0\}) \subset V$. In other words, it means that for each real number t, there exists $\delta > 0$ such that

$$f(X) > t, \qquad X \in D, \quad 0 < |X - X_0| < \delta.$$

For another illustration, suppose that F is a vector-valued function defined on a subset of R:

$$D \xrightarrow{\ F\ } R^m, \qquad D \subset R.$$

What does it mean to say that

(3.11) $$\lim_{x \to \infty} F(x) = L\,?$$

It means that "∞ is a cluster point of D" and that, for every $\epsilon > 0$, there exists $t \in R$ such that

$$|F(x) - L| < \epsilon, \qquad x \in D, \quad x > t.$$

The terminology "∞ is a cluster point of D" is not to be taken too seriously. It means that D is not bounded above. Notice that, in case D is the set of positive integers, F is a sequence and the limit reverts to the limit of a sequence, as previously defined.

From time to time, we shall use "infinite limits" as a piece of shorthand. *Keep this in mind:* If f is a real-valued function and if we say that f has a limit at X_0, that means that the limit exists in R, as in Section 3.3. If we wish to say that

$$\lim_{X \to X_0} f(X) = \infty$$

we shall say precisely that.

While we are on the subject of the extended real number system, we may as well formally extend the definitions of sup S and inf S to arbitrary sets S on the real line. In the extended real number system, every set S has a least upper bound and a greatest lower bound and they are denoted by sup S and inf S, respectively. Thus, for example,

$$\text{sup } S = \infty$$

means that S is non-empty but not bounded above by any real number. Some people prefer to apply sup and inf only to non-empty sets, because sup $\varnothing = -\infty$ and inf $\varnothing = \infty$.

Exercises

1. Let f, g, and h be real-valued functions on a set D in R^k. Let X_0 be a cluster point of D. Suppose that $f \le g \le h$ and that f, h have limits at X_0 which are equal. Show that g has the same limit at X_0.

2. Define what is meant by

$$\lim_{x \to \infty} f(x) = -\infty$$

without mentioning ∞ or $-\infty$.

3. If p is a polynomial with complex coefficients,

$$\lim_{x \to +\infty} p(x)e^{-x} = 0.$$

4. If f is an increasing function on the real line:

$$f(x) \le f(t), \qquad x \le t$$

and if f is bounded, then

$$\lim_{x \to \infty} f(x)$$

exists.

5. Find:

(a) $\lim_{x \to +\infty} \exp{(1/x)}$;

(b) $\lim_{x \to -\infty} \exp{(1/x)}$.

6. Let f be a real-valued continuous function on the real line. Suppose that, for every $\epsilon > 0$, the set

$$\{x; |f(x)| \geq \epsilon\}$$

is compact. Then

$$\lim_{|x| \to \infty} f(x) = 0.$$

7. We defined neighborhoods of ∞ and $-\infty$ in the extended real number system. With those neighborhoods, define open sets in the extended system, just as we defined them in R^k. Then prove that the extended real number system is compact.

3.5. *Continuous Mappings*

This section contains two theorems which are (now) quite easy to prove, but which have widespread application. The theorems state that a continuous mapping preserves compactness and connectedness. A continuous function F,

$$D \overset{F}{\longrightarrow} R^m, \qquad D \subset R^k$$

has the property that $F^{-1}(V)$ is open relative to D, for every open set V in R^m. As we have seen, this is the same as saying that $F^{-1}(S)$ is closed relative to D, for every closed set S in R^m. In general, the continuous function F does not preserve open sets or closed sets, as does F^{-1}. But, it is true that $F(K)$ is compact if K is compact and is connected if K is connected.

Theorem 7. *Let* F *be a continuous function. If* K *is a compact subset of the domain of* F, *then* F(K) *is compact.*

Proof. We may assume that $K = D$, because the restriction of F to K is a continuous function from K into R^m. Let $\{V_\alpha; \alpha \in A\}$ be an open cover of $F(D)$. Let

$$U_\alpha = F^{-1}(V_\alpha).$$

Then $\{U_\alpha; \alpha \in A\}$ is a cover of D. Since F is continuous, each U_α is open relative to D. Since D is compact, there exist indices $\alpha_1, \ldots, \alpha_n$ in A such that

$$D = U_{\alpha_1} \cup \cdots \cup U_{\alpha_n}.$$

Then

$$F(D) \subset V_{\alpha_1} \cup \cdots \cup V_{\alpha_n}.$$

We have extracted from each open cover of $F(D)$ a finite subcover: $F(D)$ is compact.

Corollary. *If* F

$$D \overset{F}{\longrightarrow} R^m$$

is continuous and D *is compact, then* F *is bounded and there exists a point* X_0 *in* D *such that*

$$|F(X_0)| = \sup |\{F(X)\}; X \in D|.$$

Proof. Since F is continuous, $|F|$ is continuous. Thus, the range of $|F|$ is compact and contains its supremum.

Corollary. *If* f *is a continuous real-valued function on a compact set* D, *there exist points* X_1, X_2 *in* D *such that*

$$f(X_1) = \sup \{f(X); X \in D\}$$
$$f(X_2) = \inf \{f(X); X \in D\}.$$

Notice that, if the domain of F is compact, then F does carry closed sets onto closed sets: Let K be a closed subset of D; then K is compact, because D is compact; so $F(K)$ is compact and therefore closed. It does not follow that F carries open subsets of D onto sets which are open relative to $F(D)$. Remember that F does not usually preserve intersections, and so we would not expect preservation of closed sets to imply preservation of open sets.

Theorem 7 can be proved another way. Suppose D is closed and bounded and F is continuous. We can show that $F(D)$ is closed and bounded in this way. Let $\{Y_n\}$ be any sequence of points in $F(D)$. For each n, choose some point X_n in D such that $Y_n = F(X_n)$. Since D is bounded, the Bolzano-Weierstrass theorem tells us that there is a subsequence $\{X_{n_k}\}$ which converges to some point X. Since D is closed, X is in D. Since F is continuous, the sequence $\{Y_{n_k}\}$ converges to $F(X)$. We have proved that every sequence in $F(D)$ has a subsequence which converges to a point *in* $F(D)$. Thus, $F(D)$ is bounded and closed.

Theorem 8. *Let* F *be a continuous function. If* K *is a connected subset of the domain* D, *then* F(K) *is connected.*

Proof. We may assume that $K = D$. Let S be a subset of $F(D)$ which is both open and closed relative to $F(D)$. Since F is continuous, $F^{-1}(S)$ is both open and closed relative to D. Since D is connected, either $F^{-1}(S)$ is empty or $F^{-1}(S) = D$. Therefore, either S is empty or $S = F(D)$.

Corollary. *Let* D *be an interval on the real line, and let* f *be a continuous real-valued function on* D. *Either* f *is constant or* f(D) *is an interval.*

Proof. The connected subsets of R are the empty set, sets with one member, and intervals.

Corollary. *Let* $D = [a, b]$ *be a closed (bounded) interval on the real line, and let* f *be a continuous non-constant real-valued function on* D. *Then* f(D) *is a closed (bounded) interval.*

Proof. The set $f(D)$ is compact, connected, and contains more than one point. Of course, $f(D) = [m, M]$, where

$$m = \inf\{f(x), a \leq x \leq b\}$$
$$M = \sup\{f(x), a \leq x \leq b\}.$$

The last corollary combines two theorems usually found in a calculus course. (1) A continuous real-valued function on a closed interval attains its maximum and minimum. (2) A continuous real-valued function on an interval has the intermediate value property: If $f(x_1) = y_1$ and $f(x_2) = y_2$, then, between x_1 and x_2, f takes on every value between y_1 and y_2.

The intermediate value property tells us this about a continuous real-valued function f on an interval. The function f cannot be $1:1$ unless f is **strictly increasing**

$$f(x) < f(y), \qquad x < y$$

or **strictly decreasing**

$$f(x) > f(y), \qquad x < y.$$

If we have, for instance, a strictly increasing continuous f on the interval $[a, b]$, we can define the inverse function f^{-1}:

$$[a, b] \xleftarrow{\ f^{-1}\ } [f(a), f(b)].$$

It is automatic that f^{-1} is continuous in this particular case.

Theorem 9. *If* f *is a continuous and strictly increasing function on an interval* $I \subset R$, *then* f(I) *is an interval and*

$$I \xleftarrow{\ f^{-1}\ } f(I)$$

is continuous.

Proof. Since f maps each closed interval $[a, b]$ in I onto the closed interval $[f(a), f(b)]$, the image of the open interval (a, b) must be $[f(a), f(b)]$ with $f(a), f(b)$ deleted:

$$f((a, b)) = (f(a), f(b)).$$

Thus f maps open sets onto open sets, i.e., f^{-1} is continuous.

For the case of a closed interval, Theorem 9 is a special case of the following very basic theorem.

Theorem 10. *Let* F *be a continuous function on a compact set* D. *If* F *is* $1:1$, *then the inverse function*

$$D \xleftarrow{\ F^{-1}\ } F(D)$$

is continuous.

Proof. Since D is compact, every closed subset of D is compact. Therefore, F carries closed sets onto closed sets. Since F is $1:1$ the function F^{-1} is defined, and if K is a closed subset of D,

$$(F^{-1})^{-1}(K) = F(K)$$

is closed. Thus F^{-1} is continuous.

EXAMPLE 14. Here is an illustration of Theorem 9. We could have proved the existence of nth roots this way. Let

$$f(x) = x^n, \qquad x \geq 0.$$

Since the domain $D = [0, \infty)$ is an interval, and since f is continuous, $f(D)$ is an interval. Observe that $f(x) \geq 0, f(0) = 0$, and f is not bounded. *Conclusion:* $f(D) = [0, \infty)$. So, every $t \geq 0$ is the nth power of some $x \geq 0$. On D, f is $1 : 1$, so the inverse function

$$f^{-1}(t) = t^{1/n}$$

is continuous.

For the next few examples, we need to know that the exponential function is continuous. We shall prove this now, as a lemma. We suggest that (perhaps) the reader should look at the examples and then return to the proof of the lemma.

Lemma. *The exponential function on* k $\times$ k *matrices is continuous.*

Proof. By definition

$$e^A = \sum_{n=0}^{\infty} \frac{1}{n!} A^n$$

$$= \lim_n p_n(A)$$

where

$$p_n(A) = \sum_{k=0}^{n} \frac{1}{k!} A^k.$$

Since p_n is a polynomial function, it is continuous. The p_n's approximate the exponential function, and so it seems plausible that exponentiation is continuous; however, such reasoning is not always valid. We need to be careful about how well $p_n(A)$ approximates e^A. We know that

$$|e^A - p_n(A)| \leq \sum_{n+1}^{\infty} \frac{1}{k!} |A|^k.$$

Look at

$$e^A - e^B = e^A - p_n(A) + p_n(A) - p_n(B) + p_n(B) - e^B.$$

We have

$$|e^A - e^B| \leq |e^A - p_n(A)| + |p_n(A) - p_n(B)| + |p_n(B) - e^B|$$

$$\leq \sum_{n+1}^{\infty} \frac{1}{k!} |A|^k + |p_n(A) - p_n(B)| + \sum_{n+1}^{\infty} \frac{1}{k!} |B|^k$$

$$= |p_n(A) - p_n(B)| + \sum_{n+1}^{\infty} \frac{1}{k!} [|A|^k + |B|^k].$$

This is valid for every n.

Fix A. Suppose we want to guarantee that $|e^A - e^B| < \epsilon$ if B is suffi-ciently close to A. If we at least keep $|A - B| < 1$, then $|B| < 1 + |A|$ and

$$|A|^k + |B|^k < 2(1 + |A|)^k.$$

Thus,

$$|e^A - e^B| \leq |p_n(A) - p_n(B)| + R_n$$

where

$$R_n < 2 \sum_{n+1}^{\infty} \frac{1}{k!}(1 + |A|)^k.$$

Now $1 + |A|$ is some fixed positive number and so

$$\sum_{k=1}^{\infty} \frac{1}{k!}(1 + |A|)^k < \infty.$$

If $\epsilon > 0$, we can choose N so that

$$(3.12) \qquad \sum_{N+1}^{\infty} \frac{1}{k!}(1 + |A|)^k < \frac{\epsilon}{4};$$

and, for that particular N

$$|e^A - e^B| \leq |p_N(A) - p_N(B)| + \frac{\epsilon}{2}, \qquad |A - B| < 1.$$

Retain the fixed A and the fixed N (which depends upon A). Since p_N is continuous, there exists $\delta > 0$ such that

$$|p_N(A) - p_N(B)| < \frac{\epsilon}{2}, \qquad |A - B| < \delta.$$

We may assume $\delta < 1$, and then we have

$$(3.13) \qquad |e^A - e^B| < \epsilon, \qquad |A - B| < \delta.$$

EXAMPLE 15. Let's look at the exponential function

$$f(x) = e^x, \qquad x \in R$$

on the real line. We know that f is continuous, and Example 9 in Section 2.3 showed us that

$$f(x + t) = f(x)f(t).$$

From the series definition, it is obvious that $f(x) > 1$ if $x > 0$. Since $e^x e^{-x} = 1$, clearly $f(x) < 1$ if $x < 0$. Therefore f is strictly increasing:

$$e^t = e^{t-x}e^x < e^x, \qquad t < x.$$

So the image of f is an interval contained in the interval $(0, \infty)$. Plainly f is not bounded, because

$$f(n) = f(1)^n$$
$$> 2^n.$$

Hence, the image of f is the interval $(0, \infty)$. *Conclusion:* If $y > 0$, there exists one and only one real number x such that

$$y = e^x.$$

The inverse function f^{-1} is called the (natural) **logarithm function**

$$x = \log y,$$

i.e.,

$$e^{\log y} = y,$$

and that function is continuous.

EXAMPLE 16. Suppose we did not know that every complex number of absolute value 1 had the form e^{it}, for some real t. Let's see how we could deduce that from things we have developed in this book. Let F be the map from R into C defined by

$$F(t) = e^{it}, \qquad t \in R.$$

Since $e^{it}e^{-it} = e^0 = 1$ and $e^{-it} = (e^{it})^*$, we have

$$|e^{it}| = 1.$$

Thus F maps R into the unit circle. Why *onto* the circle? Certainly F is not constant, because $F(0) = 1$ and

$$F(2) = 1 + \sum_{n=1}^{\infty} \frac{2^n}{n!} i^n$$

so that

$$Re\, F(2) = \sum_{k=0}^{\infty} (-1)^k \frac{4^k}{(2k)!}$$
$$= 1 - 2 + \tfrac{2}{3} - \cdots$$
$$< 0.$$

What do we know about the image of F? It is a connected subset of the unit circle. It contains 1, and it contains the number $w = F(2)$ which satisfies Re $w < 0$. Therefore, either $-w$ or $-w^*$ is in the image of F. (Otherwise, the line through $-w$ and $-w^*$ would disconnect the image. See Figure 14.) But, the image of F is symmetric about the real axis because $F(-t) = F(t)^*$. Thus $-w$ is in the image of F:

$$-w = F(a), \qquad a \in R.$$

Since $F(2)F(a - 2) = F(a) = -F(2)$ we have $F(a - 2) = -1$. The fact that both 1 and -1 are in the (connected) image makes it apparent that the image of F is the full unit circle.

The function $F(t) = e^{it}$ is certainly not 1 : 1. In fact (as you know) it is periodic. If $F(b) = -1$, then $F(2b) = 1$ so that the set

$$S = \{t \in R; F(t) = 1\}$$

contains a non-zero number. Let

$$c = \inf \{t \in S; t > 0\}$$

and it is not too difficult to show that $c > 0$ and that S consists of all integer multiples of c. The number c has come to be called 2π.

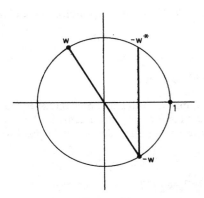

FIGURE 14

EXAMPLE 17. We cannot delete from Theorem 10 the hypothesis that D is compact. Look at the map

(3.14) $$F(t) = e^{it}, \qquad 0 \le t < 2\pi.$$

Then F is continuous and $1:1$, mapping the interval $[0, 2\pi)$ onto the unit circle in the plane. But F^{-1} is not continuous because it "jumps" at $z = 1$.

EXAMPLE 18. We have the exponential map

$$F(A) = e^A$$

from $k \times k$ matrices into the invertible $k \times k$ matrices. Is it onto? i.e., does every invertible matrix have a logarithm? Not in the real case. The image of F is connected, because the space of all $k \times k$ matrices is connected. (It is R^{k^2}.) But the set of invertible $k \times k$ real matrices is disconnected by the determinant function. It is the union of the matrices with positive determinant and those with negative determinant. Clearly then $\det e^A > 0$, for all real $k \times k$ matrices A. In the complex case, F does map onto the set of all invertible matrices, but we'll worry about that later.

Exercises

1. True or false? If S is a subset of R and

$$S^2 = \{x^2; x \in S\}$$

is compact, then S is compact.

2. If w is a complex number and $w \ne 0$, then $w = e^z$ for some $z \in C$.

3. The matrix A is called skew-symmetric if $A^t = -A$. Show that, if A is skew-symmetric, then e^A is an orthogonal matrix. Is every orthogonal matrix of that form?

4. The set of points $(x, y) \in R^2$ such that $x^2 + y^3$ is irrational is disconnected.

5. True or false? The set of points $(x, y) \in R^2$ such that either x or y is rational is a disconnected set.

6. Prove Theorem 10 using sequences.

7. Prove that every continuous function

$$[0, 1] \xrightarrow{f} [0, 1]$$

has a fixed point (a point x such that $f(x) = x$).

8. Prove that every continuous map of $[0, 1)$ *onto* $[0, 1)$ has a fixed point.

9. Prove that there does not exist any $1{:}1$ continuous map of $(0, 1)$ onto $[0, 1]$.

10. Let I be the closed interval $[0, 1]$ and let S be the closed square $I \times I$ in R^2. It is a (somewhat painful) fact that there exists a continuous map of I *onto* S. You are not expected to prove that, but prove that there does *not* exist a $1 : 1$ continuous map of I onto S. *Hint*: What happens to I when you throw away $\frac{1}{2}$?

11. True or false? If

$$R^k \xrightarrow{F} R^m$$

and if $F^{-1}(K)$ is compact for all compact K, then F is continuous.

12. A function is called a **proper map** if it is continuous and the inverse image of every compact set is compact.

(a) True or false? If F is a proper map and K is closed relative to the domain of F, then $F(K)$ is closed.

(b) True or false? If F is a proper map and U is open, then $F(U)$ is open.

13. If K_1 and K_2 are compact sets in R^m, then the algebraic sum

$$K_1 + K_2 = \{X_1 + X_2; X_j \in K_j\}$$

is compact.

14. True or false? If T is a linear transformation from R^k into R^m and if U is an open set in R^k, then $T(U)$ is open relative to the image of T.

15. Let D be a compact set in R^k; let F be a function from R^k into R^m, and suppose F is *bounded*. Then F is continuous if and only if the graph of F is a closed subset of R^{k+m}.

16. Let f be a non-zero continuous function from the real line into the complex numbers which satisfies

$$f(x + t) = f(x)f(t)$$

for all x, t. Prove that there exists a complex number z such that $f(t) = e^{tz}$ for all t.

3.6. *Uniform Continuity*

The motivation for our definition of continuity was the idea of a function

$$(3.15) \qquad D \xrightarrow{\ F\ } R^m, \qquad D \subset R^k$$

such that $F(X)$ is near $F(X_0)$ provided X is sufficiently near X_0. But, as we defined continuity, the "sufficiently near" could vary from one locality to the other in the domain D. Given $\epsilon > 0$ and *given* $X_0 \in D$, we can find $\delta > 0$ such that

$$|F(X) - F(X_0)| < \epsilon, \qquad |X - X_0| < \delta.$$

The "sufficiently near" is measured by δ, which presumably depends on X_0 as well as on ϵ. For some functions, the δ can always be chosen so as not to depend on X_0.

Definition. *The function* F (3.15) *is* **uniformly continuous** *if, for each* $\epsilon > 0$, *there exists a number* $\delta > 0$ *such that*

$$|F(X_1) - F(X_2)| < \epsilon, \qquad X_1, X_2 \in D, \quad |X_1 - X_2| < \delta.$$

EXAMPLE 19. Let us look at (what is perhaps) the simplest example of a continuous function which is not uniformly continuous. Let

$$f(x) = x^2, \qquad x \in R.$$

Now

$$f(x) - f(t) = x^2 - t^2$$
$$= (x + t)(x - t).$$

Suppose $\epsilon > 0$ and we wish to find δ so that

$$(3.16) \qquad |f(x) - f(t)| < \epsilon, \qquad |x - t| < \delta.$$

If we fix t, we can do that because $x + t$ does not become too large near any fixed t. But, if we try to do it for all points x, t simultaneously, that factor $x + t$ is not bounded and we get into trouble. For instance, if (3.16) holds, then it holds for every $t > 0$ with $x = t + \frac{1}{2}\delta$; hence

$$\left(2t + \frac{1}{2}\delta\right)\frac{\delta}{2} < \epsilon, \qquad t > 0.$$

That implies $\delta = 0$, which won't do.

At times it is convenient to reword the condition of uniform continuity in the following way. Suppose we have a function F with non-empty domain D. For each positive number η define

$$\omega(F, \eta) = \sup \{|F(X) - F(T)|; X, T \in D, |X - T| \le \eta\}.$$

This function ω is called the **modulus of continuity** of F (whether or not F is continuous). Evidently we have

$$0 \leq \omega(F, \eta) \leq \infty$$

for each η. Roughly speaking, the value $\omega(F, \eta)$ measures how far apart F can send any two points in its domain which are themselves not more than η apart. *The function* F *is uniformly continuous if and only if*

(3.17) $$\lim_{\eta \to 0} \omega(F, \eta) = 0.$$

For instance, if F is uniformly continuous and $\epsilon > 0$, there exists $\delta > 0$ such that

$$|F(X) - F(T)| < \epsilon, \qquad X, T \in D, \quad |X - T| < \delta.$$

Thus we have $\omega(F, \eta) \leq \epsilon$ for all $\eta < \delta$. It is equally easy to see that, if (3.17) holds, then F is uniformly continuous.

The next theorem provides us with many examples of continuous functions which are not uniformly continuous. The subsequent theorem is more positive.

Theorem 11. *If* F *is uniformly continuous and if the domain* D *has compact closure (i.e., if* D *is bounded), then* F *is a bounded function.*

Proof. Apply the definition of uniform continuity with $\epsilon = 5$. There exists δ such that

$$|F(X_1) - F(X_2)| < 5, \qquad X_j \in D, \quad |X_1 - X_2| < \delta.$$

The family of open balls $B(X; \delta)$, as X varies over D, is an open cover of $\bar{D}$. Thus there exist points $X_1, \ldots, X_n$ in D such that the balls $B(X_j; \delta)$ cover $\bar{D}$. In particular, if $X \in D$ then X is within distance δ of one of the points $X_1, \ldots, X_n$. Because of the way δ was chosen we then have

$$|F(X) - F(X_j)| < 5$$

for some j. Clearly $|F| \leq M$, where

$$M = \max_j (|F(X_j)| + 5).$$

Here is (the sketch of) another proof. Suppose F is not bounded. Choose points X_n in D such that

(3.18) $$\lim |F(X_n)| = \infty.$$

Some subsequence of $\{X_n\}$ converges in R^k. The limit point need not be in D, but that is irrelevant. After a while, all the points in the subsequence are less than δ (3.17) apart. Thus, after a while, all the corresponding values of F are less than 5 apart. That contradicts (3.18).

Theorem 12. *Let* F *be a continuous function. If the domain of* F *is compact, then* F *is uniformly continuous.*

Proof. Let $\epsilon > 0$. Let $X \in D$. Since F is continuous at X, there exists a number $\delta_X > 0$ such that

$$|F(Y) - F(X)| < \tfrac{1}{2}\epsilon, \qquad Y \in D, \quad |Y - X| < \delta_X.$$

Let $U_X = B(X; \tfrac{1}{2}\delta_X)$. Then $\{U_X; X \in D\}$ is an open cover of D. Since D is compact, there exist $X_1, \ldots, X_n$ in D such that the sets U_{X_j} cover D. The fact that these sets cover D means that if $X \in D$ there exists j, $1 \leq j \leq n$, such that

(3.19)
$$|X - X_j| < \tfrac{1}{2}\delta_{X_j}.$$

Let

$$\delta = \min_j \tfrac{1}{2}\delta_{X_j}.$$

Now, let X and Y be any two points in D such that $|X - Y| < \delta$. Choose an index j such that (3.19) holds. Then

$$|Y - X_j| \leq |Y - X| + |X - X_j|$$
$$< \delta + \tfrac{1}{2}\delta_{X_j}$$
$$\leq \delta_{X_j}.$$

By (3.19)

$$|F(Y) - F(X_j)| < \tfrac{1}{2}\epsilon$$
$$|F(X) - F(X_j)| < \tfrac{1}{2}\epsilon.$$

Thus

$$|F(X) - F(Y)| < \epsilon.$$

This holds for any two points in D which are less than δ apart. We have proved that F is uniformly continuous.

The proof of Theorem 12 is not trivial; but, it is a very important theorem and the proof must be understood thoroughly. Here is another proof: Let $\epsilon > 0$. Suppose we cannot find a suitable δ. Then, for each n, we can find points X_n, Y_n in D with

$$|X_n - Y_n| < \frac{1}{n} \quad \text{and} \quad |F(X_n) - F(Y_n)| \geq \epsilon.$$

Since D is compact, there exists in D a cluster point X of the sequence $\{X_n\}$. Now F is continuous at the point X. So there exists $\alpha > 0$ such that

$$|F(Y) - F(X)| < \tfrac{1}{2}\epsilon, \qquad |Y - X| < \alpha.$$

Pick some $n > 2/\alpha$ such that

$$|X - X_n| < \frac{\alpha}{2}.$$

We then have

$$|F(X_n) - F(X)| < \tfrac{1}{2}\epsilon$$
$$|F(Y_n) - F(X)| < \tfrac{1}{2}\epsilon$$
$$|F(Y_n) - F(X_n)| \geq \epsilon ?$$

Before we look at some additional specific examples, let us tie together Theorems 11 and 12. The following result makes Theorem 11 rather obvious.

Theorem 13. *Let* F *be a uniformly continuous function on* D. *Then* F *can be extended to a continuous function on the closure of* D, *i.e., there exists a function* $\bar{F}$

$$\bar{D} \xrightarrow{\bar{F}} R^m$$

such that

(a) $\bar{F}$ *is continuous;*
(b) $\bar{F}(X) = F(X)$ *for each* X *in* D.

The extension $\bar{F}$ *is unique, and* $\bar{F}$ *is uniformly continuous.*

Proof. Obviously there could not be more than one such extension: Suppose $\bar{F}$ satisfies (*a*) and (*b*). Let X be any point in $\bar{D}$. There exists in D a sequence $\{X_n\}$ which converges to X. Since $\bar{F}$ is continuous,

$$\bar{F}(X) = \lim_n \bar{F}(X_n)$$
$$= \lim_n F(X_n).$$

That shows that $\bar{F}(X)$ is uniquely determined by F.

But, why does any such extension exist? We shall define $\bar{F}$. Motivated by the previous paragraph, we proceed as follows. Let $X \in \bar{D}$. Choose a sequence $\{X_n\}$ in D which converges to X. We'll prove that the sequence $\{F(X_n)\}$ converges; and then, we'll define

(3.20) $$\bar{F}(X) = \lim_n F(X_n).$$

Since F is continuous, we shall have $\bar{F}(X) = F(X)$ for $X \in D$. All that we must do then is to prove that $\bar{F}$ is uniformly continuous.

Everything is taken care of by this observation: If $\{X_n\}$ is a Cauchy sequence in D, then $\{F(X_n)\}$ is a Cauchy sequence. That is where we use the *uniform* continuity of F. We have left the proof to the exercises. Suppose we have verified that fact about Cauchy sequences. The existence and uniform continuity of $\bar{F}$ then go like this.

(i) Pick any $X \in \bar{D}$. Let $\{X_n\}$ be a sequence in D which converges to X. Then $\{X_n\}$ is a Cauchy sequence; so $\{F(X_n)\}$ is a Cauchy sequence in R^m and therefore converges to a point in R^m. Define $\bar{F}(X)$ to be the limit of $F(X_n)$. In case X is in D, we will have $\bar{F}(X) = F(X)$ because F is continuous at X.

(ii) Let $\epsilon > 0$. Choose $\delta > 0$ such that

(3.21) $$|F(X) - F(Y)| < \epsilon, \qquad X, Y \in D, \quad |X - Y| < \delta.$$

Let X, Y be in $\bar{D}$. We have the corresponding sequences $\{X_n\}$ and $\{Y_n\}$ in

D which were used to define $\bar{F}(X)$. Suppose $|X - Y| < \delta$ (3.21). Then

$$|X_n - Y_n| < \delta, \qquad n > N.$$

Consequently (3.21),

$$|F(X_n) - F(Y_n)| < \epsilon, \qquad n > N.$$

By (3.20)

$$|\bar{F}(X) - \bar{F}(Y)| \leq \epsilon.$$

Thus $\bar{F}$ is uniformly continuous.

We should remark that the uniqueness of the extension tells us that the vector $\bar{F}(X)$ defined in (i) did not depend upon the particular sequence $\{X_n\}$ which we used. The fact that $\bar{F}(X)$ is independent of $\{X_n\}$ is also clear from the special case of (ii) in which $X = Y$.

EXAMPLE 20. Let's look at three functions on the interval $(0, 1)$:

$$f(x) = \frac{1}{x}, \qquad 0 < x < 1$$

$$g(x) = \sin\frac{1}{x}, \qquad 0 < x < 1$$

$$h(x) = x^2 \sin\frac{1}{x}, \qquad 0 < x < 1.$$

Each of these functions is continuous. The function f is not uniformly continuous. This can be seen several ways. For instance, $(0, 1)$ is bounded but f is not bounded. The function g is bounded, but g is not uniformly continuous. For one thing we obviously cannot extend g to a continuous function on the closed interval $[0, 1]$, because g does not have a limit at 0. Now h is uniformly continuous, because

$$\lim_{x\to 0} h(x) = 0 \quad \text{and} \quad \lim_{x\to 1} h(x) = \sin 1$$

so, h has a continuous extension to $[0, 1]$.

EXAMPLE 21. Let f_1, g_1, h_1 be the functions on $(0, \infty)$ defined by the formulas for f, g, h in Example 20. Then f_1 and g_1 cannot be uniformly continuous, because f and g are not. What about h_1? We can extend h_1 continuously to the closure of the domain, i.e., to $[0, \infty)$; however, that doesn't necessarily guarantee that h_1 is uniformly continuous, because $[0, \infty)$ is not compact. Now

$$\lim_{x\to\infty} h_1(x) = \infty$$

because

$$\lim_{x\to\infty} x \sin\frac{1}{x} = \lim_{t\to 0}\frac{\sin t}{t}$$
$$= 1.$$

Thus $h_1(x)$ behaves like x for large x. No function (other than a constant) could be more uniformly continuous than "x":

$$|x - t| < \epsilon \quad \text{if} \quad |x - t| < \epsilon.$$

Thus, it seems likely that h_1 is an (unbounded) uniformly continuous function. Such is the case. We omit the details.

EXAMPLE 22. Suppose T

$$R^k \xrightarrow{\ T\ } R^m$$

is a linear transformation: $T(cX_1 + X_2) = cT(X_1) + T(X_2)$. In Example 7 we saw that every linear transformation is uniformly continuous. This is easy to see. Let $\epsilon > 0$. Since T is continuous at the origin, there exists $\delta > 0$ such that

$$(3.22) \qquad\qquad |T(X)| < \epsilon. \qquad |X| < \delta.$$

(Remember that $T(0) = 0$.) That same δ will work at every other point X_0. Why? Well, since T is linear,

$$(3.23) \qquad\qquad T(X) - T(X_0) = T(X - X_0)$$

so that (3.22) tells us

$$|T(X) - T(X_0)| < \epsilon, \qquad |X - X_0| < \delta.$$

Essentially the same argument establishes the uniform continuity of any **affine** transformation:

$$A(X) = Y_0 + T(X)$$

where T is linear and Y_0 is a fixed vector in R^m.

For linear transformations, there is a concise way to describe the uniform continuity. Suppose we determine the δ for some ϵ, as in (3.22). Let X be any non-zero vector in R^k. Then

$$\left| \frac{\delta}{|X|} X \right| = \delta$$

so

$$\left| T\left(\frac{\delta}{|X|} X \right) \right| \le \epsilon$$

$$\frac{\delta}{|X|} |T(X)| \le \epsilon$$

$$|T(X)| \le \frac{\epsilon}{\delta} |X|.$$

That is, T does not change length by a factor of more than ϵ/δ. The **norm** of T is defined to be

$$(3.24) \qquad\qquad \|T\| = \sup_{X \ne 0} \frac{|T(X)|}{|X|}$$

$$= \sup \{|T(X)|; |X| = 1\}.$$

Then we have

$$(3.25) \qquad\qquad |T(X)| \le \|T\| |X|, \qquad X \in R^k.$$

And $\|T\|$ is the smallest positive constant which will suffice on the right-hand side of (3.25). The delta-epsilonics is now trivially described:

$$|T(X)| < \epsilon, \qquad |X| < \frac{\epsilon}{\|T\|}.$$

EXAMPLE 23. Suppose $t > 0$. How do we define t^x, where $x \in R$? That's easy:

$$t^x = \exp(x \log t).$$

For purposes of illustration, suppose we did it this way. After we have nth roots of positive numbers, it is easy to define t^x when x is rational:

$$t^{m/n} = (t^{1/n})^m.$$

Then, for rational b, a

$$t^{a+b} = t^a t^b$$

$$t^{ab} = (t^a)^b.$$

It is also easy to verify that

$$\lim_{x \to 0} t^x = 1.$$

Since x is only rational, we should say

(3.26) $$\lim_{x \to 0} f(x) = 1$$

where

$$f(x) = t^x, \qquad x \in Q.$$

Is f uniformly continuous on Q? Well

(3.27)
$$f(b) - f(a) = t^b - t^a$$
$$= t^b(1 - t^{a-b}).$$

That factor t^b suggests that f is not uniformly continuous on all of Q; however, on any bounded interval the uniform continuity is apparent from (3.26) and (3.27). So we can extend f continuously to that full interval. That defines t^x, for all real x.

EXAMPLE 24. Let K be the Cantor set (Example 15 of Chapter 2). We obtain K by deleting the open interval $(\frac{1}{3}, \frac{2}{3})$ from $[0, 1]$, and then deleting the middle thirds of the two remaining intervals, etc. Let D be the union of those open intervals. We define a function f on D by

$$f = \tfrac{1}{2} \quad \text{on} \quad (\tfrac{1}{3}, \tfrac{2}{3})$$
$$f = \tfrac{1}{4} \quad \text{on} \quad (\tfrac{1}{9}, \tfrac{2}{9})$$
$$f = \tfrac{3}{4} \quad \text{on} \quad (\tfrac{7}{9}, \tfrac{8}{9})$$
$$f = \tfrac{1}{8} \quad \text{on} \quad (\tfrac{1}{27}, \tfrac{2}{27})$$
$$f = \tfrac{3}{8} \quad \text{on} \quad (\tfrac{7}{27}, \tfrac{8}{27})$$
$$\vdots \qquad \qquad \vdots$$

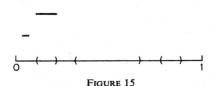

FIGURE 15

(See Figure 15.) It should be clear that

$$|f(x) - f(y)| < 2^{-n}, \quad \text{if } |x - y| < 3^{-n}.$$

Thus, f is a uniformly continuous function on D, and so has a continuous extension to $\bar{D} = [0, 1]$. The extension $\bar{f}$ is called the **Cantor function.** Notice that its graph is flat practically everyplace, but $\bar{f}(x)$ still manages to climb from 0 up to 1, continuously.

EXAMPLE 25. One nice class of uniformly continuous functions consists of the contractions:

$$|F(X) - F(Y)| \leq |X - Y|.$$

For such functions "we can always take $\delta = \epsilon$". A **rigid motion** (or **congruence**) of R^k is a function

$$R^k \xrightarrow{\;T\;} R^k$$

such that

(3.28) $$|T(X_1) - T(X_2)| = |X_1 - X_2|.$$

Euclidean geometry is essentially the study of those properties of sets which are invariant under (unaffected by) rigid motions. One example of a rigid motion is **translation**

$$T(X) = X_0 + X.$$

Another example is an **orthogonal (linear) transformation:**

    (a) T is linear;
    (b) $\langle T(X_1), T(X_2) \rangle = \langle X_1, X_2 \rangle$.

Such a linear transformation on R^k is one which can be represented by an orthogonal $k \times k$ matrix A, i.e., one for which the rule is $T(X) = AX$ where $AA^t = I$.

Theorem 14. *Let* Y_0 *be a fixed vector in* R^k *and let* T *be an orthogonal linear transformation on* R^k. *Then the (affine) transformation*

$$(3.29) \qquad S(X) = Y_0 + T(X)$$

is a rigid motion of R^k. *Furthermore every rigid motion of* R^k *has this form.*

Proof. We have already remarked that a transformation of the form (3.29) is a rigid motion, provided T is an orthogonal linear transformation. It is the converse which must be verified, that every rigid motion is an orthogonal transformation followed by a translation. So, let S be any rigid motion of R^k. Let $Y_0 = S(0)$ and define T by

$$T(X) = S(X) - Y_0.$$

Since S preserves distance, T does also. Therefore T is a rigid motion *and* $T(0) = 0$. If we apply the "rigidity" property

$$(3.30) \qquad |T(X_1) - T(X_2)| = |X_1 - X_2|$$

to the case $X_2 = 0$ and use the fact that $T(0) = 0$, we obtain

$$(3.31) \qquad |T(X)| = |X|.$$

Now we can verify easily that T preserves inner products. We have

$$|X_1 - X_2|^2 = |X_1|^2 + |X_2|^2 - 2\langle X_1, X_2 \rangle$$
$$|T(X_1) - T(X_2)|^2 = |T(X_1)|^2 + |T(X_2)|^2 - 2\langle T(X_1), T(X_2) \rangle$$

which, if we use (3.30) and (3.31) yields immediately

$$\langle T(X_1), T(X_2) \rangle = \langle X_1, X_2 \rangle.$$

A transformation which preserves inner products is necessarily linear. We have left the proof as an exercise.

Corollary. *Every rigid motion of* R^k *maps* R^k *onto* R^k, *and its inverse map is a rigid motion of* R^k.

Proof. A linear transformation from R^k into R^k which is $1:1$ necessarily maps R^k onto R^k.

Exercises

1. Let F be a uniformly continuous function. If $\{X_n\}$ is a Cauchy sequence in the domain of F, then $\{F(X_n)\}$ is a Cauchy sequence.

2. If F is a function from R^k to R^m, we say that F **vanishes at infinity** if, for every $\epsilon > 0$, the set $\{X; |F(X)| \geq \epsilon\}$ is compact. If F is continuous and vanishes at infinity, then F is uniformly continuous.

3. True or false? The map $F = (f_1, \ldots, f_m)$ is uniformly continuous if and only if each f_j is uniformly continuous.

4. True or false? The continuous function F is uniformly continuous if and only if F has a limit at each cluster point of its domain.

5. True or false? If f is a uniformly continuous function on the real line then the difference quotient

$$\frac{f(x) - f(t)}{x - t}, \qquad x \neq t$$

is bounded.

6. Let F be a uniformly continuous function

$$D \xrightarrow{\ F\ } R^m$$

and let $\bar{F}$ be the extension of F to $\bar{D}$. True or false? The graph of $\bar{F}$ is the closure of the graph of F.

7. True or false? If f and g are bounded uniformly continuous functions on D, then fg is uniformly continuous.

8. True or false? If f is a uniformly continuous function on R, then $g(x, y) = f(x) - f(y)$ defines a uniformly continuous function on R^2.

9. True or false? If

$$D \xrightarrow{\ F\ } R^m$$

is uniformly continuous and $S \cap D$ is bounded, then $F(S)$ is bounded.

10. Let T be a linear transformation from R^k to R^m. Let $E_1, \ldots, E_k$ be the standard basis for R^k. Let

$$T(E_j) = (a_{1j}, a_{2j}, \ldots, a_{mj}), \qquad 1 \leq j \leq k.$$

The $m \times k$ matrix $A = [a_{ij}]$ is the **standard representing matrix for** T. It completely determines T, as follows. If $Y = T(X)$, then

$$Y^t = AX^t$$

where X^t is the transpose of X:

$$X^t = \begin{bmatrix} x_1 \\ \cdot \\ \cdot \\ \cdot \\ x_k \end{bmatrix}, \qquad \text{etc.}$$

Verify this and then verify that T is orthogonal from R^k to R^k if and only if A is an orthogonal matrix. (It might be more natural to let A^t be the representing matrix, so that $Y = T(X)$ means $Y = XA$; but, analysts write functions on the left, and so we do it the other way.)

11. Recall the definition of the norm of a linear transformation, in Example 22. Let's look at linear transformations from R^k into R^k.

   (a) $\| T_1 \circ T_2 \| \leq \| T_1 \| \, \| T_2 \|$.
   (b) $\| T \| \leq |A|$, where A is the standard representing matrix (Exercise 10).
   (c) Give an example where $\| T \| < |A|$.
   (d) If $\| T \| < 1$, then $I - T$ is invertible.

(Think about the relation to Example 9 of Chapter 2.)

12. Let T be a linear transformation from R^k into R^k.

(a) T is an orthogonal transformation if and only if the standard representing matrix A is an orthogonal matrix.

(b) T is an orthogonal transformation if and only if T is invertible and
$$\|T\| = \|T^{-1}\| = 1.$$

13. Show that any transformation from R^k to R^k which preserves inner products is linear.

14. Let T be a function from R^k into R^m which is additive
$$T(X_1 + X_2) = T(X_1) + T(X_2).$$

Suppose that there exists some neighborhood of the origin in R^k such that the restriction of T to that neighborhood is a bounded function. Prove that T is a linear transformation.

4. Calculus
Revisited

4.1. Differentiation on Intervals

This chapter is a review of what one might call the intellectual skeleton of calculus. It is not intended to be a comprehensive review of the subject because it does not deal extensively with the techniques or applications of calculus. The focus is on the basic concepts and theorems, some of which we will need to extend beyond the level of generality customarily found in a calculus course.

Definition. *Let* I *be an interval on the real line, and let* F *be a function from* I *into* $\mathbf{R}^m$. *If* x *is a point of* I, *we say that* F *is* **differentiable** *at* x *if the limit*

(4.1)
$$\lim_{t \to x} \frac{F(t) - F(x)}{t - x}$$

exists. If F *is differentiable at* x, *the limit* (4.1) *is called the* **derivative** *of* F *at* x *and is denoted by* F'(x) *or* (DF)(x). *The function* F *is called* **differentiable** *if* F *is differentiable at each point of* I.

If F is any function from I to R^m and if we fix a point $x \in I$, the difference quotient

$$\frac{F(t) - F(x)}{t - x}$$

is a function defined at all points $t \in I$ such that $t \neq x$. Thus, it makes sense to ask whether the limit (4.1) exists. Suppose it does, and let G be another function from I to R^m which is differentiable at x. From the corresponding properties of limits, we see that $F + G$ is differentiable at x

119

and

(4.2) $(F + G)'(x) = F'(x) + G'(x).$

If F and G are (complex) matrix-valued functions, then FG is differentiable at x and

(4.3) $(FG)'(x) = F(x)G'(x) + F'(x)G(x).$

The product rule (4.3) is derived by adding and then subtracting a term to the difference quotient for FG:

$$\frac{(FG)(t) - (FG)(x)}{t - x} = \frac{F(t)G(t) - F(x)G(x)}{t - x}$$

$$= F(t)\frac{G(t) - G(x)}{t - x} + \frac{F(t) - F(x)}{t - x}G(x).$$

As t approaches x, $F(t)$ approaches $F(x)$ and the difference quotients approach $G'(x)$ and $F'(x)$, respectively. The result follows from the rules for limits of sums and products.

In the last paragraph we used the fact that, if F is differentiable at x, then

$$\lim_{t \to x} [F(t) - F(x)] = 0.$$

This is apparent, since in (4.1) the denominator of the difference quotient is tending to 0 and the quotient could not approach a limit unless the numerator were tending to 0 as well. What this says is that *differentiability at x implies continuity at x*. The reader should be aware that the converse is false, as the example $f(x) = |x|$ at the point $x = 0$ shows.

If F is differentiable at each point of I, then $DF = F'$ is a function on I, called the **derivative** of F, and it makes sense to ask whether DF is differentiable. If it is, we denote its derivative by D^2F or F''. The successive derivatives (if they exist) are denoted by $F^{(n)} = D^nF$. We say that the function F is **of class C^k** provided the kth derivative D^kF exists and is continuous. If F has derivatives of all orders, then F is **of class C^∞**.

If I is a semi-closed interval, then, at the closed end, the derivative is only a one-sided derivative; e.g., from within $I = [a, b)$ approach to a is possible only from the right. At times, it is useful to have left- and right-hand derivatives at interior points of the interval. Let us discuss this briefly.

Suppose F is defined on a neighborhood of x. We consider the difference quotient

$$\frac{F(x) - F(t)}{x - t}, \qquad x < t < x + \delta.$$

If that quotient has a limit at x, it is the **right-hand derivative** of F at x:

(4.4) $(D^+F)(x) = \lim_{t \to x^+} \dfrac{F(x) - F(t)}{x - t}.$

The **left-hand derivative** of F at x (if it exists) is defined similarly

$$(4.5) \qquad (D^-F)(x) = \lim_{t \to x^-} \frac{F(x) - F(t)}{x - t}.$$

Either of the one-sided derivatives may fail to exist. They may both exist and be different, as the example $f(x) = |x|$ at $x = 0$ shows. Evidently, $(DF)(x)$ *exists if and only if* $(D^+F)(x)$ *exists,* $(D^-F)(x)$ *exists, and* $(D^+F)(x) = (D^-F)(x)$.

Any map

$$I \xrightarrow{F} R^m$$

is described by an m-tuple of real-valued functions, $F = (f_1, \ldots, f_m)$. From the corresponding property of limits, it is immediate that F is differentiable at x precisely when every f_j is differentiable at x. In this case $F'(x) = (f'_1(x), \ldots, f'_m(x))$. Therefore, the entire discussion of differentiation could be carried out using only real-valued functions; however, this is not an advisable way to proceed. It is much better to learn to handle differentiation of vector-valued functions directly, without using coordinates.

This seems an appropriate time at which to discuss the special features of differentiation of real-valued functions. If f is a differentiable function, then $f'(x)$ is the slope of the tangent line to the graph of f at the point $(x, f(x))$. It measures the rate at which f is increasing at x. Thus, the tangent line to the graph is horizontal at points where f has a maximum or minimum. This simple observation has a number of important consequences.

Theorem 1. *Let* f *be a real-valued function on an interval* (a, b). *Let* x $\in$ (a, b) *and suppose that*

(i) f *has a local maximum (or local minimum) at* x;
(ii) f *is differentiable at* x.

Then f$'$(x) $= 0$.

Proof. To say that f has a local maximum at x means that there exists $\delta > 0$ such that

$$(4.6) \qquad f(t) \le f(x), \qquad |x - t| < \delta.$$

From (4.6), it is clear that $(D^+f)(x) \le 0$, if $(D^+f)(x)$ exists. Similarly $(D^-f)(x) \ge 0$ if $(D^-f)(x)$ exists. Hence, if they both exist and are equal, their common value must be 0.

Corollary (*Mean Value Theorem*). *If* f *is continuous on* [a, b] *and differentiable on* (a, b), *there exists an* x $\in$ (a, b) *such that*

$$(4.7) \qquad f'(x) = \frac{f(b) - f(a)}{b - a}.$$

Proof. Let

$$g(x) = f(b) - f(x) + \frac{f(b) - f(a)}{b - a}(x - b).$$

Then g satisfies the same conditions as f and $g(a) = g(b) = 0$. The continuous function g attains its maximum and minimum on $[a, b]$. One of those must occur at a point x in the open interval (a, b). Then $g'(x) = 0$, which gives us (4.7).

Corollary. *Suppose* f *is differentiable on* (a, b). *Then* f *is an increasing function if and only if* f$' \geq 0$*;* f *is constant if and only if* f$' = 0$.

Proof. The mean value theorem for derivatives.

Corollary (Chain Rule). *Let* F, g *be differentiable functions such that the image of* g *is contained in the domain of* F:

$$(a, b) \xrightarrow{\ g\ } (c, d) \xrightarrow{\ F\ } R^m.$$

Then the composition F $\circ$ g *is differentiable and* $(F \circ g)' = g'(F' \circ g)$*, i.e.,* $(F \circ g)'(x) = g'(x)F'(g(x))$.

Proof. Fix $x \in (a, b)$. From the definition of derivative

$$F(y) - F(g(x)) = [y - g(x)]F'(g(x)) + (y - g(x))R(y, g(x))$$

where

$$\lim_{y \to g(x)} R(y, g(x)) = 0.$$

Thus,

$$\left| \frac{F(g(t)) - F(g(x))}{t - x} - \frac{g(t) - g(x)}{t - x}F'(g(x)) \right| \leq \left| \frac{g(t) - g(x)}{t - x} \right| |R(g(t), g(x))|.$$

Corollary. *If* g *is a differentiable real-valued function on the interval* (a, b) *such that* g$'(x) \neq 0$ *for each* $x \in (a, b)$*, then* g *is* 1:1*, the inverse function* g^{-1} *is differentiable, and*

$$(Dg^{-1})(g(x)) = \frac{1}{g'(x)}$$

Proof. If we had $g(x) = g(t)$ with $x \neq t$, the mean value theorem would provide a point c between x and t with $g'(c) = 0$. Hence g is $1:1$, that is, g is either strictly increasing or strictly decreasing. As we observed in Chapter 3 (Theorem 9) this means that the image of (a, b) is an open interval and g^{-1} is continuous. Thus,

$$(Dg^{-1})(g(x)) = \lim_{u \to g(x)} \frac{g^{-1}(u) - g^{-1}(g(x))}{u - g(x)} = \lim_{t \to x} \frac{t - x}{g(t) - g(x)} = \frac{1}{g'(x)}.$$

EXAMPLE 1. Let A be a fixed $k \times k$ matrix with complex entries. Let

$$F(t) = e^{tA}, \qquad t \in R.$$

Then F is a differentiable function. We have

$$\frac{F(t) - F(x)}{t - x} = \frac{e^{tA} - e^{xA}}{t - x}$$

$$= \frac{e^{(t-x)A} - I}{t - x} \cdot e^{xA}$$

What we must investigate is

Now

$$\lim_{t \to 0} \frac{e^{tA} - I}{t}.$$

$$e^{tA} - I = \sum_{n=1}^{\infty} \frac{1}{n!} t^n A^n$$

so that

$$t^{-1}(e^{tA} - I) - A = \sum_{n=2}^{\infty} \frac{1}{n!} t^{n-1} A^n$$

$$|t^{-1}(e^{tA} - I) - A| \leq \sum_{n=2}^{\infty} \frac{1}{n!} |t|^{n-1} |A|^n$$

$$\leq \frac{|A|}{2} \cdot \frac{|t||A|}{1 - |t||A|}.$$

Now it is clear that F is differentiable and

(4.8) $$F'(x) = Ae^{xA} = AF(x).$$

Since $F' = AF$, it follows that F is a function of class C^∞.

We remark that the differential equation in (4.8) determines the exponential function. If G is a differentiable function from R into $C^{k \times k}$ and if

$$G'(x) = AG(x)$$

where A is some fixed matrix, then $e^{-xA}G(x)$ has derivative 0. Thus,

$$G(x) = e^{xA}B$$

where $B = G(0)$. Notice that, if $G(x) = Be^{xA}$, then in order that $G' = AG$ be satisfied, B must commute with A, i.e., $AB = BA$.

The most important special case is, of course, the real-valued function

$$f(t) = e^t, \qquad t \in R$$

which satisfies $f' = f$. Since $f'(t) \neq 0$, the last corollary tells us that the inverse function f^{-1}, which we defined to be the log function, satisfies

$$(D \log)(e^t) = e^{-t} \quad \text{or} \quad (D \log)(x) = \frac{1}{x}.$$

EXAMPLE 2. Let F be the complex-valued function on the real line: $F(x) = e^{ix}$. According to Example 1, this is a function of class C^∞ and $F^{(n)}(x) = i^n F(x)$. Notice that, if we did not know about sines and cosines,

We could define the sine and cosine functions by:

$$\sin x = \text{Im}\,(F(x))$$

$$= \text{Im} \sum_{0}^{\infty} \frac{1}{n!}(ix)^n$$

$$= x - \frac{x^3}{3!} + \frac{x^5}{5!} - \cdots$$

$$= \sum_{n=0}^{\infty} \frac{(-1)^n}{(2n+1)!} x^{2n+1};$$

$$\cos x = \text{Re}\,(F(x))$$

$$= \text{Re} \sum_{n=0}^{\infty} \frac{1}{n!}(ix)^n$$

$$= 1 - \frac{x^2}{2!} + \frac{x^4}{4!} - \cdots$$

$$= \sum_{n=0}^{\infty} \frac{(-1)^n}{(2n)!} x^{2n}.$$

These functions are of class C^∞ and

$$D(\cos) = -\sin$$
$$D(\sin) = \cos.$$

Each of these functions satisfies the differential equation

(4.9) $f'' + f = 0.$

Furthermore, any f which satisfies (4.9) on an interval is a linear combination of sin and cos. The proof of this should be very familiar to every serious student of mathematics. For convenience, suppose f is real-valued and (4.9) holds on a neighborhood of 0. Let

$$g(x) = f(x) - f(0) \cos x - f'(0) \sin x.$$

Then $g'' + g = 0$ and $g(0) = g'(0) = 0$. Thus,

$$g'g'' + gg' = 0$$
$$D((g')^2 + g^2) = 0$$
$$(g')^2 + g^2 = \text{constant}.$$

Since $g(0) = g'(0) = 0$, the constant is 0; so,

$$(g')^2 + g^2 = 0$$
$$g' = g = 0 \qquad \text{(since } g \text{ is real-valued)}$$
$$f(x) = f(0) \cos x + f'(0) \sin x.$$

EXAMPLE 3. Let f be a real-valued function on an interval. We call f a **convex function** if

(4.10) $f[cx + (1-c)u] \leq cf(x) + (1-c)f(u)$

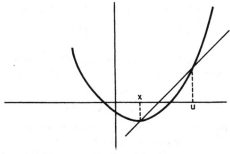

FIGURE 16

for all x, u in the domain and all c such that $0 \le c \le 1$. This says that, between x and u, the graph of f lies below the line joining the point $(x, f(x))$ to the point $(u, f(u))$. (See Figure 16.) Thus $f(x) = x^2$ and $f(x) = e^x$ define convex functions on the real line, while $f(x) = \sin x$ is a convex function on the interval $[-\pi, 0]$.

The defining property (4.1) can be rewritten in another useful way by observing that if $x < t < u$, then $t = cx + (1 - c)u$ where

$$c = \frac{u - t}{u - x}.$$

The condition (4.10) then says that

$$f(t) \le \frac{u - t}{u - x} f(x) + \frac{t - x}{u - x} f(u)$$

which is more conveniently expressed as

$$\frac{f(t) - f(x)}{t - x} \le \frac{f(u) - f(x)}{u - x}.$$

In other words, if $x < t < u$, the chord joining $(x, f(x))$ to $(u, f(u))$ has slope at least as great as the chord from $(x, f(x))$ to $(t, f(t))$. This suggests that for smooth functions convexity amounts to the fact that the derivative f' is an increasing (non-decreasing) function. This is the case, and we have left the proof to the Exercises.

As a matter of fact, the convexity property alone implies that f has a derivative at most points and that (where defined) f' is an increasing function. We sketch a proof. We know that if $x < t < u$, then

$$\frac{f(u) - f(x)}{u - x} \ge \frac{f(t) - f(x)}{t - x}.$$

Thus, unless x is the right end point of the interval of definition, the right-hand derivative $(D^+f)(x)$ exists and it is given by

$$(D^+f)(x) = \inf_{>x} \frac{f(t) - f(x)}{t - x}.$$

Similarly, unless x is the left end point, the left-hand derivative exists and

$$(D^-f)(x) = \sup_{t<x} \frac{f(t) - f(x)}{t - x}.$$

Since both $(D^+f)(x)$ and $(D^-f)(x)$ exist where they make sense, f is continuous at each such point x. More than that can be said. A little thought shows that

$$(D^+f)(t) \geq (D^-f)(t) \geq (D^+f)(x) \geq (D^-f)(x), \qquad t > x.$$

So, the left derivative D^-f is (as is D^+f) an increasing function. At any point where f fails to be differentiable, $(D^+f)(x) > (D^-f)(x)$ and so D^-f has a jump. The sum of all those jumps cannot exceed $(D^-f)(b) - (D^+f)(a)$, on the interval $[a, b]$. Therefore, there can be only a countable number of jumps. *Conclusion:* A convex function f on an open interval is continuous; it is differentiable except at a countable number of points; and the derivative Df (where defined) is an increasing function.

Exercises

1. Suppose that F is differentiable at a point x. Near x we approximate $F(t)$ by the affine function

$$A(t) = F(x) + (t - x)F'(x).$$

Show that the approximation satisfies

$$\lim_{t \to x} \frac{F(t) - A(t)}{t - x} = 0.$$

2. Let F be defined on an open interval about x. Prove that, if $(D^+F)(x)$ and $(D^-F)(x)$ both exist, then F is continuous at x.

3. Let

$$f_n(x) = \begin{cases} x^n \sin \dfrac{1}{x}, & x \neq 0 \\ 0, & x = 0. \end{cases}$$

Prove that f_1 is continuous but not differentiable at 0. Prove that f_2 is differentiable but not of class C^1. What is the corresponding fact about f_n?

4. Let $f_n(x) = x^n |x|$, $n = 0, 1, 2, \ldots$. Compute f_n' and find the largest integer k such that f_n is of class C^k.

5. Let F be a differentiable function from an interval I into R^m and suppose that $F' = 0$. Prove that F is constant.

6. Let f be the real-valued function on the real line

$$f(x) = \exp(-x^{-2}), \qquad x \neq 0$$

$$f(0) = 0.$$

Prove that f is a function of class C^∞ and that $f^{(n)}(0) = 0$ for every n.

7. Let f be a complex-valued function on an interval such that $f + f'' = 0$. Find the form of f by using the fact that $g = f - if'$ and $h = f + if'$ satisfy $g' = ig$, $h' = -ih$.

8. Let f, g be differentiable functions on R such that

$$f^2 + g^2 = 1$$
$$Df = g$$
$$Dg = -f$$
$$f(0) = 0$$
$$g(0) = 1.$$

Prove that f, g are what you think they are.

9. Let f be the Cantor function on $[0, 1]$, which was defined in Example 24 of Chapter 3. Prove that the derivative of f exists and is 0 at each point not in the Cantor set. Prove that f is not differentiable at any point of the Cantor set.

10. Show that the exponential function on the real line is a convex function and hence that

$$e^x \geq e^t(1 + x - t)$$

for all real numbers x and t.

11. Let f be a continuous real-valued function on an interval I which is differentiable at each interior point of I. Show that f is convex if and only if the derivative f' is an increasing (non-decreasing) function. (It is the "if" half of the result which we left unproved in the text.) What does this result tell you about the form of convex functions of class C^2?

12. If f and g are convex functions, is max (f, g) a convex function? What about min (f, g)?

13. Let f be a real-valued function on an interval. The following are equivalent.

   (a) f is convex.

   (b) For each x, there is a line through the point $(x, f(x))$ such that the graph of f lies above that line.

   (c) The set of all points in R^2 which lie above the graph of f is a convex set.

   (d) f is continuous and

$$f\left(\frac{x+t}{2}\right) \leq \frac{1}{2}[f(x) + f(t)].$$

14. Let f be a continuous complex-valued function on the real line, which satisfies the functional equation

$$f(x + t) = f(x)f(t).$$

If $f \neq 0$, show that there is a complex number α such that $f(t) = e^{\alpha t}$. (You might ask first what kind of a function $\log |f|$ is.)

15. What about Exercise 14 for matrix-valued functions?

16. Let F be a differentiable function on the real line such that $|F'| \leq |F|$ and $F(0) = 0$. Prove that $F = 0$.

17. We know how to solve the differential equation $f'(x) = ax$. Suppose we have a system of differential equations

$$f'_1 = a_{11}f_1 + \cdots a_{1k}f_k$$

$$\cdot \qquad \cdot \qquad \cdot$$
$$\cdot \qquad \cdot \qquad \cdot$$
$$\cdot \qquad \cdot \qquad \cdot$$

$$f'_k = a_{k1}f_1 + \cdots + a_{kk}f_k$$

involving k unknown functions f_i and k^2 known constants a_{ij}. Rewrite the system in the form

$$F' = AF$$

where A is the matrix of constants and

$$F = \begin{bmatrix} f_1 \\ \cdot \\ \cdot \\ \cdot \\ f_k \end{bmatrix}.$$

Prove that every solution for F has the form $F(t) = e^{tA}X_0$, where X_0 is some constant column matrix.

4.2. *Integration on Intervals*

Let F be a continuous function on a closed interval $I = [a, b]$. We want to define and discuss the integral of F over I:

$$\int_I F = \int_a^b F(x)\, dx.$$

In case F is a non-negative (and continuous) real-valued function, this integral is a *number* which measures the area under the graph of F and it is defined by a *process* known as integration, which gives a systematic method for approximating the area.

Since these ideas were first exploited systematically by Newton and Leibnitz 300 years ago, the range of applications of integration has been broadened enormously. This has been made possible in part by the modern definitions of the integral, in which the integration process is described in precise analytic terms, without appeal to any of its particular applications. In this chapter, we shall use a modified version of the formulation of integration given by Riemann. In Chapter 7 we will discuss the concept of integral given by Lebesgue, which is more general than Riemann's because it allows one to integrate many more functions. But we have no need of such generality at this point because our primary goal is to understand the basic theorems of calculus. Furthermore, it is important for every student of analysis to know that Riemann's process of integration converges for continuous functions.

In order to present the Riemann integral it is convenient to introduce two terms. A **partition** of the interval $[a, b]$ is any $(n + 1)$-tuple of the form

(4.11)
$$P = (x_0, \ldots, x_n)$$
$$a = x_0 < x_1 < \cdots < x_n = b.$$

Such a partition divides the interval into n subintervals, and the greatest of the lengths of these subintervals is called the **mesh** of the partition

(4.12) $$\|P\| = \max_k (x_k - x_{k-1}) \qquad (1 \leq k \leq n).$$

Suppose F is a function from the interval $I = [a, b]$ into R^m. The integral of F over I, if it exists, will be a limit of sums

(4.13) $$S(F, P, T) = \sum_{k=1}^{n} (x_k - x_{k-1})F(t_k)$$

where
$$P = (x_0, \ldots, x_n)$$

is a partition of $[a, b]$ and

$$T = (t_1, \ldots, t_n), \qquad t_k \in [x_{k-1}, x_k]$$

is a choice of points t_k, one from each subinterval defined by the partition. Such sums are frequently referred to as "Riemann sums", although we shall not use this terminology. The "limit" of these sums is to be taken as the mesh of the partition goes to 0; loosely speaking

$$\int_I F = \lim_{\|P\| \to 0} S(F, P, T).$$

There are so many (uncountably many) different partitions and different choices of points at which to evaluate the function that we will need to be careful how we describe the "limit" involved.

For any function which is integrable by the process we shall describe, it will always be possible (theoretically) to calculate the integral using any particular sequence of partitions $\{P_n\}$ for which the meshes tend to zero and any corresponding sequence of choices T_n. Any two such sequences will provide a sequence of vectors (approximating sums)

$$S_n = S(F, P_n, T_n)$$

which converges to the integral of F. For example, if

(4.14) $$S_n = \sum_{k=1}^{n} \frac{b - a}{n} F\left(\frac{k(b - a)}{n}\right)$$

then for each integrable function F we shall have

$$\lim_n S_n = \int_a^b F(x)\, dx.$$

One might ask, therefore, why we do not use the sequence in (4.14) to define the integral, since it seems so much simpler than involving all sorts of partitions P and choices T, as in (4.13). There are two reasons, one practical and one theoretical.

(i) Even for very smooth functions F the particular sequence (4.14) may not provide the most convenient way of calculating the integral. We want to be able to use any sequence of partitions with meshes tending to 0 and know that the sums converge to the same limit. In other words, we want to be able to choose the sequence most convenient for each particular function F.

(ii) We need to deal with integrals of some functions which are not smooth and, therefore, we must be precise about the class of functions to which the integration process applies. Simple examples show that, were we to have the definition of this class of functions depend upon the values of the function at a particular sequence of points such as $(k/n)(b - a)$, we would generate chaos.

EXAMPLE 4. Let f be the real-valued function defined by

$$f(x) = \begin{cases} 0, & \text{if } x \text{ is rational} \\ 1, & \text{if } x \text{ is irrational.} \end{cases}$$

Suppose $a < b$ and let $P = (x_0, x_1, \ldots, x_n)$ be any partition of $[a, b]$. Each interval $[x_{k-1}, x_k]$ has positive length and, accordingly, it contains both rational and irrational points. For each k, choose a rational point $t_k \in [x_{k-1}, x_k]$ and an irrational point $t'_k \in [x_{k-1}, x_k]$. From the definition of f we have $f(t_k) = 0$ and $f(t'_k) = 1$ for each k. Accordingly,

$$S(F, P, T) = 0$$
$$S(F, P, T') = b - a.$$

Thus we can choose sequences (P_n, T_n) with $\|P_n\| \to 0$ for which $S(F, P_n, T_n) = 0$ and we can choose other sequences of this type for which $S(F, P_n, T_n) = b - a$. This leaves us a bit perplexed as to what the integral of f ought to be. In order to avoid this confusion, we shall define the class of functions which are "integrable" in such a way as to exclude this function f.

Definition. Let F be a function from the interval $[a, b]$ into $\mathbf{R}^m$. We say that F is **Riemann-integrable** if there exists a vector S in $\mathbf{R}^m$ with this property: For each $\epsilon > 0$ there exists $\delta > 0$ such that if P is any partition of $[a, b]$ with $\|P\| \leq \delta$ and if T is any choice of points in the subintervals defined by P, then

$$|S - S(F, P, T)| < \epsilon.$$

Although this definition sounds somewhat complicated, it expresses a rather simple idea: The function F will be called Riemann-integrable if there exists a vector S (which will be the integral of F) to which the approximating sums $S(F, P, T)$ converge as $\|P\| \to 0$. Converge in which sense? In the sense that $S(F, P, T)$ will be close to S provided only that $\|P\|$ is small, and with no condition on T, the choice of points in the subintervals of P.

It should be clear that this basic definition can be reformulated as follows. The function F is Riemann-integrable if (and only if) there exists a vector S in R^m such that

$$\lim_n S(F, P_n, T_n) = S$$

for every sequence of partitions P_n with $\| P_n \| \to 0$ and any corresponding choice(s) T_n of points in the subintervals defined by P_n. Thus the point of Example 4 can be described by saying that the function f which occurs there is not Riemann-integrable.

If F is Riemann-integrable, the vector S which occurs in the definition is obviously unique. It is called the **integral of F over the interval** $I = [a, b]$ and is denoted by $\int_I F$ or $\int_a^b F(x)\, dx$.

A word about this notation. Part of the power of calculus as a tool derives from the very useful notation which Leibnitz invented for derivatives and integrals. For real-valued functions on an interval $[a, b]$ the integral was denoted by

$$\int_a^b f(x)\, dx$$

as a reminder of the process of integration

$$\int_a^b f(x)\, dx = \lim_{\|P\| \to 0} \sum_{k=1}^n f(x_k)(x_k - x_{k-1})$$
$$= \lim_{\max \Delta x_k \to 0} \sum_k f(x_k)\, \Delta x_k$$
$$= \lim_{\Delta x \to 0} \sum f(x)\, \Delta x.$$

We have not used the $\Delta x_k = x_k - x_{k-1}$ notation, in part because for vector-valued functions it would then be natural to write

$$\lim_{\|P\| \to 0} \sum_{k=1}^n \Delta x_k F(x_k) = \int_a^b dx\, F(x)$$

which is potentially confusing.

We want to direct our efforts immediately to proving that various well-behaved functions (e.g., continuous functions) are Riemann-integrable. In other words, we want to show that the process of integration "converges" for various well-behaved functions. We need a criterion for convergence analogous to the Cauchy criterion for sequences, i.e., a criterion which does not explicitly involve the limit vector.

If F is any function from $[a, b]$ into R^m, then, for each $\delta > 0$, we define the set $\Sigma_\delta = \Sigma_\delta(F)$ as follows. We consider all partitions $P = (x_0, \ldots, x_n)$ of the interval $[a, b]$ such that $\|P\| \leq \delta$ and all possible corresponding choices T of points $t_k \in [x_{k-1}, x_k]$. Each such pair (P, T) determines a sum

$$S(F, P, T) = \sum_{k=1}^n (x_k - x_{k-1})F(t_k)$$

and $\Sigma_\delta(F)$ is defined to be the set of all such sums:

(4.15) $\sum_\delta (F) = \{S(F, P, T); \|P\| \le \delta, t_k \in [x_{k-1}, x_k]\}.$

If F is Riemann-integrable then, for small δ, all the sums $S(F, P, T)$ with $\|P\| \le \delta$, i.e., all the vectors in $\Sigma_\delta(F)$, will be close to the integral of F. Thus, when δ is small, Σ_δ must be a set of small diameter clustered near the integral of F. And the integral is approximable by vectors in Σ_δ, that is, the integral lies in the closure $\bar{\Sigma}_\delta$. What happens to the set Σ_δ as we decrease δ? If $0 < \eta < \delta$, then it is immediate from the definition (4.15) that $\Sigma_\eta \subset \Sigma_\delta$. This is true for any F. For any integrable F it must be that, as δ tends to 0, the closed sets $\bar{\Sigma}_\delta$ close down on the single vector $\int_I F$. In other words, the diameters of the sets Σ_δ converge to 0 and $\int_I F$ is the single vector contained in the intersection

$$\bigcap_{\delta > 0} \bar{\Sigma}_\delta.$$

Theorem 2. *Let* F *be a function from the interval* [a, b] *into* R^m, *and define the set* $\Sigma_\delta(F)$ *by* (4.15). *Then* F *is Riemann-integrable if and only if*

(4.16) $\lim_{\delta \to 0} \text{diam } \Sigma_\delta = 0.$

Proof. It is the "if" statement which remains to be proved. Assume (4.16) and consider the intersection of the closures of the sets Σ_δ:

$$\bigcap_{\delta > 0} \bar{\Sigma}_\delta.$$

By the condition on the diameters, there cannot be more than one vector in this intersection. Furthermore, the intersection is nonempty for this reason. We know that $\Sigma_\eta \subset \Sigma_\delta$ if $\eta < \delta$. Hence the closed sets $\bar{\Sigma}_\delta$ are nested:

$$\bar{\Sigma}_\eta \subset \bar{\Sigma}_\delta, \qquad \eta < \delta.$$

Furthermore, since the diameters of these sets tend to 0, they are bounded (sets) for all sufficiently small δ. Thus their intersection is non-empty. (See Theorem 9 of Chapter 2.) Let S be the single vector in the intersection. If $\epsilon > 0$, choose $\delta > 0$ such that diam $\Sigma_\delta(F) < \epsilon$. If P is any partition of [a, b] with $\|P\| \le \delta$ and if T is any choice of points in the subintervals defined by P, then the sum $S(F, P, T)$ is in the set $\Sigma_\delta(F)$ and the vector S is in $\bar{\Sigma}_\delta(F)$. Since diam $\Sigma_\delta(F) < \epsilon$,

$$|S - S(F, P, T)| < \epsilon.$$

Thus F is Riemann-integrable (and $S = \int_I F$).

In order to use the diameter criterion (4.16) to show that a given function F is Riemann-integrable, one needs to estimate, for a given δ, what the maximum distance is between any two approximating sums, formed by partitions of mesh at most δ. To carry out such estimates, we need to see what happens when we pass from a given partition to a more refined one.

Let $P = (x_0, \ldots, x_n)$ and $Q = (y_0, \ldots, y_p)$ be partitions of the interval $[a, b]$. We say that Q is a **refinement** of P if $\{x_0, \ldots, x_n\}$ is a subset of $\{y_0, \ldots, y_p\}$. In other words Q is a refinement of P if Q can be obtained by starting with the points $x_0 < x_1 < \cdots < x_n$ and interposing some $p - n$ points between them to form a finer subdivision. The fact that Q is a refinement of P guarantees among other things that $\|Q\| \leq \|P\|$, but note that it is a much stronger condition than the (mere) fact that Q has a smaller mesh than does P.

Lemma. *Let* F *be a function from* [a, b] *into* $\mathbf{R}^m$. *Let* $P = (x_0, \ldots, x_n)$ *and* $Q = (y_0, \ldots, y_p)$ *be partitions of* [a, b] *such that* Q *is a refinement of* P. *If* T *is any choice of points* $t_k \in [x_{k-1}, x_k]$, $k = 1, \ldots, n$, *there exists points* $u_1, \ldots, u_p$ *in* [a, b] *such that*

(i) $y_{j-1} - \|P\| \leq u_j \leq y_j + \|P\|$, $j = 1, \ldots, p$;

(ii) $\sum_{k=1}^{n} (x_k - x_{k-1})F(t_k) = \sum_{j=1}^{p} (y_j - y_{j-1})F(u_j)$.

Proof. Let j be an index, $1 \leq j \leq p$. We shall define the corresponding point u_j. Since Q is a refinement of P, none of the points x_i lies in the open interval (y_{j-1}, y_j). Therefore the closed interval $[y_{j-1}, y_j]$ must lie wholly within one of the intervals $[x_{i-1}, x_i]$. In fact, since the interiors of the different intervals $[x_{i-1}, x_i]$ do not overlap, there is precisely one index i, $1 \leq i \leq n$, such that $[y_{j-1}, y_j] \subset [x_{i-1}, x_i]$. Call this index $i(j)$ and define $u_j = t_{i(j)}$.

Now observe that

$$\sum_{j=1}^{p} (y_j - y_{j-1})F(u_j) = \sum_{k=1}^{n} \sum_{i(j)=k} (y_j - y_{j-1})F(u_j)$$

$$= \sum_{k=1}^{n} \sum_{i(j)=k} (y_j - y_{j-1})F(t_k).$$

For a particular k, the indices j such that $i(j) = k$ are those for which $[y_{j-1}, y_j] \subset [x_{k-1}, x_k]$. Since Q is a refinement of P, these intervals form a partition of $[x_{k-1}, x_k]$, that is,

$$\sum_{i(j)=k} (y_j - y_{j-1}) = x_k - x_{k-1}.$$

This proves the lemma.

Theorem 3. *If* F *is a continuous function from the closed interval* [a, b] *into* $\mathbf{R}^m$, *then* F *is Riemann-integrable.*

Proof. The idea of the proof is to use the uniform continuity of F to estimate the diameters of the sets $\Sigma_\delta(F)$. Let $\delta > 0$. Let P, P' be partitions of $[a, b]$ of mesh at most δ and let $T = (t_1, \ldots, t_n)$, $T' = (t'_1, \ldots, t'_{n'})$

be choices of points in the subintervals determined by P and P', respectively. We want to estimate the distance from $S(F, P, T)$ to $S(F, P', T')$. To do this, we need to rewrite them as sums over the same partition. Accordingly, let $Q = (y_0, \ldots, y_p)$ be the partition which is made up of the *distinct* points in the sequence $x_0, \ldots, x_n, x'_0, \ldots, x'_n$, arranged in increasing order. Then Q is a refinement of P and also a refinement of P'. By the preceding lemma, there exist points $u_1, \ldots, u_p$ and $u'_1, \ldots, u'_p$ in $[a, b]$ such that

$$S(F, P, T) = S(F, Q, U)$$
$$S(F, P', T') = S(F, Q, U').$$

We do not know that u_j and u'_j lie in the interval $[y_{j-1}, y_j]$. But, from the lemma, we do know that both u_j and u'_j lie in the interval $[y_{j-1} - \delta, y_j + \delta]$. Thus,

$$S(F, P, T) - S(F, P', T') = \sum_{j=1}^{p} (y_j - y_{j-1})[F(u_j) - F(u'_j)]$$

and $|u_j - u'_j| \leq 2\delta$, $j = 1, \ldots, p$. Let

$$M = \max_j |F(u_j) - F(u'_j)|$$

and we have

(4.17)
$$|S(F, P, T) - S(F, P', T')| \leq M \sum_{j=1}^{p} (y_j - y_{j-1})$$
$$= M(b - a).$$

The idea now is to show that M is small whenever $\|P\|$ and $\|P'\|$ are small. The critical observation is that F is continuous on the closed interval $[a, b]$, and is therefore uniformly continuous, (Theorem 12 of Chapter 3). This guarantees that, when δ is small, each of the numbers $|F(u_j) - F(u'_j)|$ will be small, because u_j, u'_j are not more than 2δ apart. To be precise, if ω is the modulus of continuity of F (see (3.17)):

$$\omega(F, \eta) = \sup_{|x-t| \leq \eta} |F(x) - F(t)|$$

then (4.17) tells us that

$$|S(F, P, T) - S(F, P', T')| \leq (b - a)\omega(F, 2\delta).$$

Thus,

$$\text{diam } \Sigma_\delta(F) \leq (b - a)\omega(F, 2\delta).$$

Now

$$\lim_{\eta \to 0} \omega(F, \eta) = 0$$

because this is what it means to say that F is uniformly continuous. We conclude that F is Riemann-integrable.

Corollary. *If* F *is a continuous function on the interval* [a, b], *if* P = $(x_0, \ldots, x_n)$ *is a partition of* [a, b] *and* T = $(t_1, \ldots, t_n)$ *any choice of points* $t_k \in [x_{k-1}, x_k]$ *then*

$$\left| \int_a^b F(x)\, dx - \sum_{k=1}^n (x_k - x_{k-1}) F(t_k) \right| \le (b-a)\omega(F, 2\|P\|)$$

where ω is the modulus of continuity of F.

Proof. Let $\delta = \|P\|$. In the proof of the theorem we verified that

$$\text{diam } \Sigma_\delta(F) \le (b, a)\omega(F, 2\delta).$$

Since the integral of F is in the closure of $\Sigma_\delta(F)$, its distance from the sum $S(F, P, T)$ cannot exceed $(b - a)\omega(F, 2\delta)$.

Exercises

1. Let F be a Riemann-integrable function on the closed interval I. Show that, if c is any real number, then cF is Riemann-integrable and

$$\int_I cF = c \int_I F.$$

(Work directly with $c \int F - S(cF, P, T)$ rather than $\Sigma_\delta(cF)$.)

2. In Exercise 1, if F is a complex-valued function and c is a complex number, is the same assertion valid? (Of course, we regard a complex-valued function as a map into R^2.)

3. Let F and G be functions from the closed interval I into R^m. Show that, if F and G are Riemann-integrable, then $F + G$ is and

$$\int_I (F + G) = \int_I F + \int_I G.$$

4. If f and g are real-valued Riemann-integrable functions on the closed interval I such that $f \ge g$, then

$$\int_I f \ge \int_I g.$$

5. If f is a non-negative continuous function on $[a, b]$ and

$$\int_a^b f(x)\, dx = 0$$

then $f = 0$.

6. Use the result of Exercise 5 to prove the following. If f is a continuous real-valued function on $[a, b]$ and

$$\int_a^b f(x)\, dx = 0$$

then there exists a point c, $a \le c \le b$, such that $f(c) = 0$.

7. Use the result of Exercise 6 to prove the *first mean value theorem of integral calculus*: If f is a continuous real-valued function on the interval $[a, b]$, there exists a point c, $a \le c \le b$, such that

$$\int_a^b f(x)\, dx = (b - a) f(c).$$

8. Give an example which shows that the mean value theorem of Exercise 6 is false for (some) complex-valued continuous functions.

9. Let f be a real-valued continuous function on $[a, b]$ such that

(i) $|f| \leq 1$;

(ii) $\int_a^b f(x)\, dx = b - a$.

Prove that $f = 1$.

10. Prove Exercise 9 for complex-valued continuous functions f.

11. Let f be a complex-valued continuous function on a closed interval I such that

$$\int_I f = \int_I |f|.$$

Prove that f is real-valued and $f \geq 0$.

12. Prove the Cauchy-Schwartz inequality for integrals of continuous functions: If f, g are continuous real-valued functions on $[a, b]$, then

$$\left(\int_I fg\right)^2 \leq \int_I f^2 \int_I g^2.$$

4.3. *Fundamental Properties of the Integral*

We shall summarize the properties of integration which are used repeatedly.

Theorem 4. *Let* F *be a function from the closed interval* $I = [a, b]$ *into* R^m *and suppose* F *is Riemann-integrable.*

(i) (**Positivity**). *If* F *is non-negative real-valued, i.e., if* $m = 1$ *and* $F \geq 0$, *then*

$$\int_I F \geq 0.$$

(ii) (**Linearity**). *If* G *is another Riemann-integrable function from* I *into* R^m *and if* c *is any real number, then* $(cF + G)$ *is Riemann-integrable and*

$$\int_I (cF + G) = c \int_I F + \int_I G.$$

(iii) (**Boundedness**). *The function* F *is bounded and*

$$\left| \int_a^b F(x)\, dx \right| \leq (b - a) \sup_{a \leq x \leq b} |F(x)|.$$

(iv) *The function* $|F|$ *is Riemann-integrable and*

$$\left| \int_I F \right| \leq \int_I |F|.$$

Proof. (i) and (ii) were to be verified as part of the exercises for Section 4.2. They are repeated here for emphasis.

(iii) Let $P = (x_0, \ldots, x_n)$ be any partition of $[a, b]$ and $T = (t_1, \ldots, t_n)$ any choice of points $t_k \in [x_{k-1}, x_k]$. The sum

$$S(F, P, T) = \sum_{k=1}^{n} (x_k - x_{k-1})F(t_k)$$

satisfies

(4.18) $$|S(F, P, T)| \leq \sum_{k=1}^{n} (x_k - x_{k-1})|F(t_k)|.$$

If

$$M = \sup_{a \leq x \leq b} |F(x)|$$

then it is apparent from (4.18) that

$$|S(F, P, T)| \leq M \sum_{k=1}^{n} (x_k - x_{k-1})$$

$$= M(b - a).$$

Since this is true for every sum $S(F, P, T)$, we also have

$$\left| \int_a^b F(x)\, dx \right| \leq M\,(b - a).$$

But how do we know that $M < \infty$, i.e., that F is bounded? Suppose that F is *not* bounded. Let $P = (x_0, \ldots, x_n)$ be any partition of $[a, b]$. Then there must be (at least) one index k such that F is not bounded on the interval $[x_{k-1}, x_k]$. Fix such a value of k, and select a sequence of points $\{c_j\}$ in the interval $[x_{k-1}, x_k]$ such that

(4.19) $$\lim_j |F(c_j)| = \infty.$$

Let T_j be the choice of points in the subintervals defined by P for which $t_i = x_i, i \neq k$, and $t_k = c_j$. That is, let $T_j = (x_1, \ldots, x_{k-1}, c_j, x_{k+1}, \ldots, x_n)$. Then

$$S(F, P, T_j) = \sum_{i \neq k} (x_i - x_{i-1})F(x_i) + (x_k - x_{k-1})F(c_j).$$

From (4.19) we see that

$$\lim_j |S(F, P, T_j)| = \infty.$$

Since there is such a sequence $\{T_j\}$ for each partition P, the set $\Sigma_\delta(F)$ is unbounded, for every $\delta > 0$. Thus, F is not Riemann-integrable.

(iv) The inequality

$$\left| \sum_{k=1}^{n} (x_k - x_{k-1})F(t_k) \right| \leq \sum_{k=1}^{n} (x_k - x_{k-1})|F(t_k)|$$

states that

$$|S(F, P, T)| \leq S(|F|, P, T)$$

for all partitions P and choices T. Thus, it is apparent that

(4.20) $$\left| \int_I F \right| \le \int_I |F|$$

if $|F|$ is known to be Riemann-integrable. For instance, if F is continuous we know that $|F|$ is continuous and therefore Riemann-integrable; hence the norm inequality (4.20) holds. For most of this book the functions which we integrate will be piecewise-continuous (defined later) and it is a trivial fact that if F is such a function, then $|F|$ is as well. Thus on first reading one should omit the proof that the integrability of $|F|$ follows from the integrability of F. It is given in the next section, where we discuss some technical facts about Riemann-integrability of real-valued functions.

Theorem 5. *Let* F *be a function from the interval* [a, b] *into* R^m *and let* c *be a point such that* a < c < b. *If* F *is Riemann-integrable on the interval* [a, c] *and on the interval* [c, b] *as well, then* F *is Riemann-integrable on* [a, b] *and*

$$\int_a^b F(x)\,dx = \int_a^c F(x)\,dx + \int_c^b F(x)\,dx.$$

Proof. The idea here is relatively simple, although it does take a few lines to write down the details. Let $P = (x_0, \ldots, x_n)$ be any partition of $[a, b]$. Then there is a unique index k such that $x_{k-1} < c \le x_k$. Accordingly, $Q = (x_0, \ldots, x_{k-1}, c)$ and $R = (c, x_k, \ldots, x_n)$ are partitions of $[a, c]$ and $[c, b]$, respectively. Let $T = (t_1, \ldots, t_n)$ be any choice of points $t_i \in [x_{i-1}, x_i]$. Then $U = (t_1, \ldots, t_{k-1}, c)$ and $V = (c, t_k, \ldots, t_n)$ are choices of points within the subintervals of Q and R. Now

$$S(F, Q, U) = \sum_{i=1}^{k-1} (x_i - x_{i-1})F(t_i) + (c - x_{i-1})F(c)$$

$$S(F, R, V) = \sum_{i=k+1}^{n} (x_i - x_{i-1})F(t_i) + (x_k - c)F(c)$$

$$S(F, P, T) = \sum_{i \ne k} (x_i - x_{i-1})F(t_i) + (x_k - x_{k-1})F(t_k).$$

Therefore,

(4.21) $$\begin{aligned} S(F, P, T) = &\ S(F, Q, U) + S(F, R, V) \\ &+ (x_k - x_{k-1})[F(t_k) - F(c)]. \end{aligned}$$

The idea is that, when $\|P\|$ is small, the first term on the right will be near $\int_a^c F$, the second will be near $\int_c^b F$, and the third will be small. To be precise, let $\epsilon > 0$. Since F is integrable on $[a, c]$, there exists $\delta > 0$ such that

(4.22) $$\left| \int_a^c F(x)\,dx - S(F, Q, U) \right| < \frac{\epsilon}{3}$$

for all partitions Q of $[a, c]$ such that $\|Q\| \le \delta$ (and for all choices U).

Similarly, there exists $\eta > 0$ such that

(4.23) $$\left| \int_c^b F(x)\,dx - S(F, R, V) \right| < \frac{\epsilon}{3}$$

for all partitions R of $[c, b]$ such that $\|R\| \leq \eta$ (and for all V). Now

(4.24) $$|(x_k - x_{k-1})[F(t_k) - F(c)]| \leq 2\|Q\|M$$

where

$$M = \sup_{a \leq x \leq b} |F(x)|.$$

Let P be any partition of $[a, b]$ such that

$$\|P\| \leq \min\left(\delta, \eta, \frac{\epsilon}{6\|Q\|M} \right)$$

and let T be a corresponding choice of points t_i. Use the point c to construct the corresponding Q, R, U, and V. Then (4.22), (4.23) are applicable and the term on the right side of (4.24) is less than $\epsilon/3$. From (4.21) we have

$$\left| \int_a^c F(x)\,dx + \int_c^b F(x)\,dx - S(F, P, T) \right| < \epsilon.$$

It follows that F is Riemann-integrable on $[a, b]$ and that its integral is the sum of the integrals on the two subintervals.

It is conventional and extremely useful to define

$$\int_a^b F(x)\,dx = - \int_b^a F(x)\,dx$$

whenever $b < a$ and F is integrable on $[b, a]$. The addition formula

$$\int_a^b = \int_a^c + \int_c^b$$

is then valid with a, b, c in any order, as long as the function is integrable on the largest of the three intervals.

Definition. *Let* F *be a function from the closed interval* [a, b] *into* R^m. *We say that* F *is* **piecewise-continuous** *if there is a partition* P = (x₀, ..., xₙ) *of* [a, b] *such that* F *is uniformly continuous on each of the open intervals* (x_{i-1}, x_i) i = 1, ..., n.

Theorem 6. *Every piecewise-continuous function on the interval* [a, b] *is Riemann-integrable.*

Proof. Let F be piecewise continuous and let $P = (x_0, \ldots, x_n)$ be a partition of $[a, b]$ such that F is uniformly continuous on each subinterval (x_{k-1}, x_k). By Theorem 5, it suffices to show that F is Riemann-integrable on each subinterval $[x_{k-1}, x_k]$. Fix an index k. Since F is uniformly continuous on (x_{k-1}, x_k) we can extend F to a continuous function on the closed interval $[x_{k-1}, x_k]$, that is, there is a continuous function G on

$[x_{k-1}, x_k]$ such that $G(x) = F(x)$, $x_{k-1} < x < x_k$. We know that G is Riemann-integrable and, since $F(x)$ differs from $G(x)$ for at most two points $x \in [x_{k-1}, x_k]$, F is Riemann-integrable. (See Exercise 2.)

Theorem 7 (*Fundamental Theorem of Calculus*). *Let* F *be a function from* [a, b] *into* $\mathbb{R}^m$ *which is Riemann-integrable. Let* G *be the function defined by*

$$G(x) = \int_a^x F(u)\, du, \qquad a \le x \le b.$$

Then G *is a continuous function on* [a, b]. *If* x *is any point of* [a, b] *at which* F *is continuous, then* G *is differentiable at* x *and* $G'(x) = F(x)$.

Proof. We have

$$G(t) - G(x) = \int_a^t F(u)\, du + \int_a^x F(u)\, du$$

$$= \int_x^t F(u)\, du.$$

Thus,

$$|G(t) - G(x)| \le |t - x| \sup_{a \le u \le b} |F(u)|.$$

Since F is bounded, this shows (directly) that G is uniformly continuous. Now, let x be a point at which F is continuous. We have

$$\frac{G(t) - G(x)}{t - x} = \frac{1}{t - x} \int_x^t F(u)\, du.$$

Therefore,

$$\frac{G(t) - G(x)}{t - x} - F(x) = \frac{1}{t - x} \int_x^t F(u)\, du - F(x)$$

$$= \frac{1}{t - x} \int_x^t [F(u) - F(x)]\, du$$

and

$$(4.25) \qquad \left| \frac{G(t) - G(x)}{t - x} - F(x) \right| \le |t - x|^{-1} \left| \int_x^t [F(u) - F(x)]\, du \right|.$$

How large can the last integral be? Not larger than $|t - x|\,\omega(x, |x - t|)$, where

$$\omega(x, \delta) = \sup \{|F(u) - F(x)|; |u - x| \le \delta\}.$$

From (4.25)

$$(4.26) \qquad \left| \frac{G(t) - G(x)}{t - x} - F(x) \right| \le \omega(x, |x - t|).$$

Since F is continuous at the point x,

$$\lim_{\delta \to 0} \omega(x, \delta) = 0.$$

From (4.26) we see, therefore, that

$$\lim_{t \to x} \frac{G(t) - G(x)}{t - x} = F(x)$$

i.e., we see that G is differentiable at x and $G'(x) = F(x)$.

Corollary. *Let* I *be any interval on the real line which has a non-empty interior and let* F *be any continuous function from* I *into* $\mathbf{R}^m$. *There exists a function* G *from* I *into* $\mathbf{R}^m$ *such that* (G *is differentiable and*)

$$G' = F.$$

Proof. The only purpose in assuming I has a non-empty interior is to avoid wondering what derivatives might be on degenerate intervals, i.e., points. Other than for this condition, I may be any type of interval: open, closed, semi-closed, bounded or unbounded. For the proof, let a be a point in the interior of I and define

$$G(x) = \int_a^x F(u)\, du, \qquad x \in I.$$

Corollary. *Let* F *be a piecewise-continuous function from the closed interval* [a, b] *into* $\mathbf{R}^m$. *If* G *is any continuous function from* [a, b] *into* $\mathbf{R}^m$ *such that* G'(x) = F(x) *except at a finite number of points* x, *then*

$$\int_a^b F(x)\, dx = G(b) - G(a).$$

Proof. Define

$$H(x) = G(a) - G(x) + \int_a^x F(u)\, du.$$

Then H is continuous on $[a, b]$ and, except at a finite number of points, its derivative exists and is 0. Therefore $H = 0$. The corollary says that $H(b) = 0$.

This last corollary is also referred to at times as the fundamental theorem of calculus. It is the result of combining Theorem 7 with the fact that a differentiable function on an open interval which has derivative 0 is constant. It is a powerful tool which says, for example, that if F is continuous and we either know or can dig up a function G such that $G' = F$, then we can evaluate the integral of F over any closed interval within its domain by subtracting the values of G at the end points. All of this is quite familiar to the reader, of course. We repeat it here to emphasize the beauty and power of this result relating differentiation and integration. Once we know the derivatives of x^{n+1}, $\sin x$, $\cos x$, e^x, $\tan^{-1} x$ we can reduce the problem of calculating

$$\int_a^b x^n\, dx, \quad \int_a^b \cos x\, dx, \quad \int_a^b \sin x\, dx, \quad \int_a^b e^x\, dx, \quad \int_a^b \frac{1}{1 + x^2}\, dx$$

to calculating the values of the "antiderivative" function at b and a. It is worthwhile for the reader to pause a few moments to be impressed once again with such formulae as

$$\sin x = \int_0^x \cos t \, dt, \qquad e^b - e^a = \int_a^b e^t \, dt, \qquad \frac{\pi}{4} = \int_0^1 \frac{1}{1 + x^2} \, dx$$

and their interpretations in terms of areas.

The first corollary is unquestionably one of the most fundamental theorems in mathematics: If F is continuous on an interval, the differential equation $G'(x) = F(x)$ has a solution G. If we specify the value of G at some point x_0, then the G is unique. The process of integration describes how to go about calculating the value of this G at any given point.

Theorem 8 (*Change of Variable Theorem*). *Let* g *be a real-valued function of class* C^1 *from the interval* [c, d] *into the interval* [a, b] *such that* a $=$ g(c) *and* b $=$ g(d). *If* F *is any continuous function from* [a, b] *into* R^m, *then*

$$\int_a^b F(x) \, dx = \int_c^d g'(t) F(g(t)) \, dt.$$

Proof. Let $G(x) = \int_a^x F(u) \, du$. Then G is differentiable and $G' = F$. Thus the composition $H(t) = G(g(t))$ is a function of class C^1 on the interval [c, d]. Furthermore, the chain rule tells us that

$$H'(t) = g'(t) G'(g(t))$$
$$= g'(t) G(g(t)).$$

Thus

$$\int_c^d g'(t) F(g(t)) \, dt = \int_c^d H'(t) \, dt$$
$$= H(d) - H(c)$$
$$= G(b) - G(a)$$
$$= \int_a^b F(x) \, dx.$$

Exercises

1. If F is a piecewise-continuous function on [a, b] and

$$\int_a^b |F(x)| \, dx = 0,$$

what can you say about F?

2. If G is a Riemann-integrable function on [a, b] and F is a function such that $F(x) = G(x)$ except at a finite number of points x, then F is Riemann-integrable and

$$\int_a^b F(x) \, dx = \int_a^b G(x) \, dx.$$

3. Show that

$$\int_{-\pi}^{\pi} e^{inx}\, dx = 0, \qquad n = \pm 1, \pm 2, \ldots .$$

4. Show that for any fixed a, b

$$\lim_{x \to \infty} \int_a^b e^{itx}\, dt = 0$$

5. Let f be the function on the real line defined by

$$f(x) = \begin{cases} 0, & \text{if } x \text{ is irrational or zero} \\ \dfrac{1}{q}, & \text{if } x = \dfrac{p}{q} \text{ (lowest terms).} \end{cases}$$

(The second description means: If x is rational and non-zero, write $x = p/q$ where p is an integer, q is a positive integer, p and q have no common factors other than ± 1; then define $f(x) = 1/q$.) Prove that f is Riemann-integrable on every interval $[a, b]$ and has integral zero.

6. Let f be the characteristic function of the Cantor set K

$$f(x) = \begin{cases} 1, & x \in K \\ 0, & x \notin K. \end{cases}$$

(The Cantor set is Example 24 of Chapter 3.) Prove that f is Riemann-integrable on the interval $[0, 1]$ and

$$\int_0^1 f(x)\, dx = 0.$$

7. Let F be a function which is bounded and continuous on the open interval (a, b). Prove that F is Riemann-integrable on $[a, b]$. (According to Exercise 2, it doesn't matter how $F(a)$ and $F(b)$ are defined.)

8. Let $\{x_n\}$ be a convergent sequence of (distinct) points in the interval $[a, b]$. Let F be a function from $[a, b]$ into R^m such that

   (i) F is bounded;
   (ii) $F(x) = 0$ except (possibly) at the points $x_1, x_2, \ldots$.

Prove that F is Riemann-integrable and

$$\int_a^b F(x)\, dx = 0.$$

9. Let A be a fixed $k \times k$ matrix and define

$$F(t) = e^{tA}, \qquad t \in R.$$

Apply the fundamental theorem of calculus to obtain the function G with $G' = F$. Wouldn't you guess that G is given by

$$G(t) = A^{-1} e^{tA}?$$

What if A is not an invertible matrix?

10. If f is a real-valued Riemann integrable function on $[a, b]$ and F is a Riemann-integrable function from $[a, b]$ into R^m then fF is Riemann-integrable.

4.4. *Integrability and Real-Valued Functions*

In this short section, we reformulate the criterion for Riemann-integrability of real-valued functions, along the lines that Riemann himself followed. The essential point is that the ordering of the real numbers allows one to test the integrability of a real function by comparing sums $S(f, P, T)$ for different choices T but the same partition P. In fact, as the mesh of the partition P decreases to 0, one obtains for each bounded function an upper and a lower integral to compare. These will be equal if and only if the function is Riemann-integrable.

Definition. *Let* f *be a bounded real-valued function on the interval* [a, b], *and let* $P = (x_0, \ldots, x_n)$ *be a partition of* [a, b]. *The* **upper** *and* **lower Riemann sums** *associated with* f *and* P *are*

$$\bar{S}(f, P) = \sum_{k=1}^{n} M_k(x_k - x_{k-1})$$

$$\underline{S}(f, P) = \sum_{k=1}^{n} m_k(x_k - x_{k-1})$$

where
$$M_k = \sup_{x \in I_k} f(x) \quad and \quad m_k = \inf_{x \in I_k} f(x), \qquad I_k = [x_{k-1}, x_k].$$

Note that the suprema and infima involved in these definitions are finite because we have assumed that f is bounded. One purpose in introducing the upper and lower Riemann sums should be apparent: If $T = (t_1, \ldots, t_k)$ is any choice of points $t_k \in [x_{k-1}, x_k]$, then

$$\underline{S}(f, P) \leq S(f, P, T) \leq \bar{S}(f, P)$$

and the gap between $\bar{S}(f, P)$ and $\underline{S}(f, P)$ indicates the range of values for the approximating sums derived from different choices T. Thus the sizes of $\bar{S}(f, P) - \underline{S}(f, P)$ for the various partitions P of small mesh provide some measure of the integrability of f. The theorem below will show that the Riemann-integrability of f can indeed be measured in this manner. We urge the reader to draw a picture for the case in which $f \geq 0$, to see that $\bar{S}(f, P)$ and $\underline{S}(f, P)$ are the upper and lower estimates which P provides for the area under the graph of f and to keep this picture in mind throughout this section.

Lemma. *Let* P *and* Q *be partitions of* [a, b] *such that* Q *is a refinement of* P. *Then*
$$\bar{S}(f, Q) \leq \bar{S}(f, P)$$
$$\underline{S}(f, Q) \geq \underline{S}(f, P).$$

Proof. We leave the proof as an exercise, modulo this remark. Each subinterval I defined by Q is contained in precisely one subinterval J

defined by P and

$$\sup_I f \le \sup_J f$$

$$\inf_I f \ge \inf_J f.$$

Lemma. *If* P *and* P′ *are any partitions of* [a, b], *then*

$$\underline{S}(f, P) \le \bar{S}(f, P').$$

Proof. Let Q be the common refinement of P and P' obtained by using the *distinct* points of subdivision provided by P and P' jointly. By the lemma above applied to (P, Q) and (P', Q), we have

$$\underline{S}(f, P) \le \underline{S}(f, Q) \le \bar{S}(f, Q) \le \bar{S}(f, P').$$

Definition. *If* f *is a bounded real-valued function on the interval* [a, b], *the* **upper** *and* **lower** **(Riemann)** **integrals** *of* f *are, respectively,*

$$U(f) = \inf_P \bar{S}(f, P)$$

$$L(f) = \sup_P \underline{S}(f, P)$$

where in each case P *ranges over all partitions of the interval* [a, b].

By the last lemma, we have

$$L(f) \le U(f)$$

for every bounded function f. If f is Riemann-integrable, then clearly

$$\int_a^b f(x)\, dx = L(f) = U(f).$$

Theorem 9. *Let* f *be a bounded real-valued function on the closed interval* [a, b]. *The following are equivalent.*

(i) f *is Riemann-integrable.*
(ii) $U(f) = L(f)$.
(iii) *For each* $\epsilon > 0$ *there exists a partition* P *of* [a, b] *such that* $\bar{S}(P) - \underline{S}(P) < \epsilon$.
(iv) $\lim_{\|P\| \to 0} [\bar{S}(P) - \underline{S}(P)] = 0$.

Proof. (i) $\Rightarrow$ (ii). Suppose f is Riemann-integrable. Let $\epsilon > 0$. There exists $\delta > 0$ such that the set $\Sigma_\delta(f) = \{S(f, P, T); \|P\| \le \delta\}$ has diameter less than ϵ. This certainly shows that $U(f) - L(f) < \epsilon$.

(ii) $\Rightarrow$ (iii). Suppose $U(f) = L(f)$. Let $\epsilon > 0$. There exist partitions P, P' such that

$$\bar{S}(P) < U(f) + \frac{\epsilon}{2}$$

$$\underline{S}(P') > L(f) - \frac{\epsilon}{2}.$$

Then, since $U(f) = L(f)$, we have $\bar{S}(P) - \underline{S}(P') < \epsilon$. Now, let Q be the common refinement of P and P' which we have formed several times previously. Since $\bar{S}(Q) \leq \bar{S}(P)$ and $\underline{S}(Q) \geq \underline{S}(P')$, we have $\bar{S}(Q) - \underline{S}(Q) < \epsilon$.

(iii) $\Rightarrow$ (iv). Assume (iii), and let $\epsilon > 0$. Choose a partition $Q = (y_0, \ldots, y_m)$ of the interval $[a, b]$ such that $\bar{S}(Q) - \underline{S}(Q) < \epsilon/2$. We will use this fixed partition Q to find $\delta > 0$ such that $\bar{S}(P) - \underline{S}(P) < \epsilon$ for all partitions P of mesh at most δ. The idea is this. Given any partition $P = (x_0, \ldots, x_n)$, define P^* to be the common refinement of P and Q. Let's compare $\bar{S}(f, P)$ and $\bar{S}(f, P^*)$. In the sum

$$\bar{S}(f, P) = \sum_{k=1}^{n} M_k(x_k - x_{k-1})$$

every term will appear as a term in $\bar{S}(f, P^*)$ except for those values of k such that some y_j lands in the interior of $[x_{k-1}, x_k]$. There are at most $(m - 1)$ such values of k; hence the sum of the terms $M_k(x_k - x_{k-1})$ corresponding to such values of k cannot exceed

$$(m - 1)M \| P \|$$

where $M = \sup_{[a,b]} f$. Consequently,

$$\bar{S}(f, P) \leq \bar{S}(f, P^*) + (m - 1)M \| P \|.$$

By closely analogous reasoning,

$$\underline{S}(f, P) \geq \bar{S}(f, P^*) - (m - 1)M \| P \|.$$

Now, let $\delta = \epsilon/4(m - 1)M$. Then the inequalities above tell us that

$$\bar{S}(f, P) - \underline{S}(f, P) < \bar{S}(f, P^*) - \underline{S}(f, P^*) + \frac{\epsilon}{2}.$$

Furthermore, P^* is a refinement of Q, so that

$$\bar{S}(f, P^*) - \underline{S}(f, P^*) \leq \bar{S}(f, Q) - \underline{S}(f, Q)$$

$$< \frac{\epsilon}{2}.$$

We conclude that $\bar{S}(f, P) - \underline{S}(f, P) < \epsilon$, for all partitions P with $\| P \| \leq \delta$.

(iv) $\Rightarrow$ (i). We need to show that (iv) implies that the diameter of the set $\Sigma_\delta(f) = \{S(f, P, T); \| P \| \leq \delta\}$ tends to 0 as δ tends to 0. What we shall show is that

$$\tfrac{1}{2} \operatorname{diam} \Sigma_\delta(f) \leq \sup_{\| P \| \leq \delta} [\bar{S}(f, P) - \underline{S}(f, P)] \leq \operatorname{diam} \Sigma_\delta(f).$$

The right-hand inequality is trivial because

$$\operatorname{diam} \Sigma_\delta(f) = \sup | S(f, P, T) - S(f, Q, U) |$$

where the supremum is extended over all partitions P and Q of mesh at most δ and over all corresponding choices T and U of points in the subintervals defined by P and Q.

To verify the left-hand inequality, let P and Q be any partitions of $[a, b]$. By the last lemma, $\bar{S}(f, Q) - \underline{S}(f, P) \geq 0$ and hence

$$\bar{S}(f, P) - \underline{S}(f, Q) \leq \bar{S}(f, P) - \underline{S}(f, Q) + [\bar{S}(f, Q) - \underline{S}(f, P)]$$

$$= \bar{S}(f, P) - \underline{S}(f, P) + \bar{S}(f, Q) - \underline{S}(f, Q)$$

$$= [\bar{S}(f, Q) - \underline{S}(f, P)] + [\bar{S}(f, Q) - \underline{S}(f, Q)]$$

$$\leq 2 \sup_{\|P\| \leq \delta} [\bar{S}(f, P) - \underline{S}(f, P)].$$

Corollary. *If* f *is a monotone increasing (or decreasing) function on the interval* [a, b], *then* f *is Riemann-integrable.*

Proof. Let $P = (x_0, x_1, \ldots, x_n)$ be any partition of $[a, b]$. Then, since f is an increasing function, for each subinterval $L_k = [x_{k-1}, x_k]$ we have

$$\sup_{I_k} f = f(x_k)$$

$$\inf_{I_k} f = f(x_{k-1}).$$

Accordingly,

$$\bar{S}(f, P) = \sum_{k=1}^{n} f(x_k)(x_k - x_{k-1})$$

$$\underline{S}(f, P) = \sum_{k=1}^{n} f(x_{k-1})(x_k - x_{k-1})$$

and so

$$\bar{S}(f, P) - \underline{S}(f, P) = \sum_{k=1}^{n} [f(x_k) - f(x_{k-1})](x_k - x_{k-1})$$

$$\leq \|P\| \sum_{k=1}^{n} [f(x_k) - f(x_{k-1})]$$

$$= \|P\|[f(b) - f(a)].$$

Hence it is apparent that

$$\lim_{\|P\| \to 0} [\bar{S}(f, P) - \underline{S}(f, P)] = 0.$$

Corollary. *If* f *is a real-valued Riemann-integrable function on the interval* I = [a, b], *then* |f| *is Riemann-integrable and*

$$\left| \int_I f \right| \leq \int_I |f|.$$

Proof. We remarked earlier that the inequality on integrals follows as soon as $|f|$ is proved to be Riemann-integrable. This proof consists of showing that

$$\bar{S}(|f|, P) - \underline{S}(|f|, P) \leq \bar{S}(f, P) - \underline{S}(f, P)$$

for every partition P. This is easy to see because, on each subinterval

$I_k = [x_{k-1}, x_k]$ defined by P,

$$\sup_{I_k} |f| - \inf_{I_k} |f| = \sup_{s,t \in I_k} [|f(t)| - |f(s)|]$$
$$\leq \sup_{s,t \in I_k} |f(t) - f(s)|$$
$$= \sup_{s,t \in I_k} [f(t) - f(s)]$$
$$= \sup_{I_k} f - \inf_{I_k} f.$$

The critical step occurred between the second and third lines above, where we used the fact that f is real-valued and the consequent fact that $|f(t) - f(s)|$ is either $f(t) - f(s)$ or $f(s) - f(t)$.

Corollary. *If* F *is a Riemann-integrable function from the interval* I = [a, b] *into* $\mathbf{R}^m$, *then* |F| *is Riemann-integrable and*

$$\left| \int_I F \right| \leq \int_I |F|.$$

Proof. Once again, all we need to show is that $|F|$ is Riemann-integrable. Let $P = (x_0, \ldots, x_n)$ be any partition of $[a, b]$. Then

$$\bar{S}(|F|, P) - \underline{S}(|F|, P) = \sum_{k=1}^{n} \eta_k (x_k - x_{k-1})$$

where

$$\eta_k = \sup_{s,t \in I_k} [|F(t)| - |F(s)|], \qquad I_k = [x_{k-1}, x_k].$$

If $Y = (y_1, \ldots, y_m)$ and $Z = (z_1, \ldots, z_m)$ are any two vectors in $\mathbf{R}^m$, then

$$|Z| - |Y| \leq \sum_{j=1}^{m} |z_j - y_j|.$$

Thus, if $F = (f_1, \ldots, f_m)$, we have

$$|F(t)| - |F(s)| \leq \sum_{j=1}^{m} |f_j(t) - f_j(s)|.$$

Therefore,

$$\eta_k \leq \sup_{s,t \in I_k} \sum_{j=1}^{m} |f_j(t) - f_j(s)|$$
$$\leq \sum_{j=1}^{m} \sup_{s,t \in I_k} |f_j(t) - f_j(s)|$$
$$= \sum_{j=1}^{m} \sup_{s,t \in I_k} [f_j(t) - f_j(s)]$$

and consequently,

$$\bar{S}(|F|, P) - \underline{S}(|F|, P) \leq \sum_{j=1}^{m} [\bar{S}(f_j, P) - \underline{S}(f_j, P)].$$

Since F is Riemann-integrable, each f_j is; hence, it is apparent from the last inequality (and Theorem 9) that $|F|$ is Riemann-integrable.

Exercises

1. Complete the proof of the first lemma of this section.

2. Prove that if the real-valued function f is Riemann-integrable on $[a, b]$ then f^2 is Riemann-integrable.

3. Use the result of Exercise 2 to prove that the product of two real-valued Riemann-integrable functions is Riemann-integrable.

4. If F and G are Riemann-integrable functions from $[a, b]$ into R^m, then the function $\langle F, G \rangle$ is Riemann-integrable.

5. Establish the result of Exercise 3 for complex-valued functions.

6. A real-valued function is called **piecewise-monotone** on $[a, b]$ if there is a partition P of $[a, b]$, on each subinterval of which f is either non-decreasing or non-increasing. Show that every piecewise monotone function is Riemann-integrable.

7. If f is a non-decreasing real-valued function on $[a, b]$, then

$$g(x) = \int_a^x f(t)\, dt$$

is a convex function on $[a, b]$.

8. If f is a non-negative Riemann-integrable function, then $\sqrt{f}$ is Riemann-integrable.

4.5. Differentiation and Integration in $\mathbf{R}^n$

We shall not engage in a thorough discussion of either derivatives or integrals in n-space at this point. But there are a few basic facts which are used so often and which follow so naturally upon the 1-dimensional case that it would be artificial to delay their discussion.

If we start from differentiation in R^1, one natural thing to consider is the differentiation of functions along lines in R^n. Such derivatives measure rates of change in various directions. If we start from, say, the origin, the different (straight-line) directions in which we can strike out are represented by the rays which emanate from the origin. A convenient way to enumerate those rays is by the points of the unit sphere. Each ray emanating from the origin contains exactly one vector V such that $|V| = 1$.

Suppose that F is a function from (a subset of) R^n into R^m. Let X be a point in the interior of the domain of F. If $|V| = 1$, we can inquire whether the limit

$$\lim_{t \to 0} \frac{F(X + tV) - F(X)}{t}$$

exists. If it exists, we call it the **derivative of** F **at** X **in the direction of the** (unit) **vector** V and we denote it by $(D_V F)(X)$. The point of insisting that X be in the interior of the domain of F is (of course) that $X + tV$ will then be in the domain for all sufficiently small $t \in R$.

We can (and shall) discuss

(4.27) $$(D_V F)(X) = \lim_{t \to 0} t^{-1}[F(X + tV) - F(X)]$$

for *all* vectors V in R^n. It is easy to see that if $(D_V F)(X)$ exists, then $(D_{cV} F)(X)$ exists for every $c \in R$ and

$$(D_{cV} F)(X) = c(D_V F)(X).$$

If $V \neq 0$, the unique unit vector which has the same direction as does V is $|V|^{-1} V$ and thus (if it exists) the derivative of F at X in the direction of V is

$$\frac{1}{|V|}(D_V F)(X).$$

If they exist, the derivatives of F in the directions of the standard basis vectors are called the **partial derivatives** of F. We write $D_j F$ rather than the cumbersome $D_{E_j} F$. Thus

$$(D_j F)(X) = \lim_{t \to 0} t^{-1}[F(x_1, \ldots, x_j + t, \ldots, x_n)$$

$$- F(x_1, \ldots, x_j, \ldots, x_n)].$$

At times, we shall also use the notation

$$\frac{\partial F}{\partial x_j} = D_j F.$$

A **function** (or map) **of class** C^1 is a function F such that

(i) the domain of F is an open subset of R^n;

(ii) the partial derivatives $D_1 F, \ldots, D_n F$ exist and are continuous functions.

We call F a **function of class** C^k ($k \geq 2$) if the partial derivatives $D_1 F, \ldots, D_n F$ are functions of class C^{k-1}. If F is of class C^k for every $k = 1, 2, 3, \ldots$, then we say that F is **of class** C^∞.

Theorem 10. *Let* F *be a map of class* C^1 *on the open set* U *in* R^n. *If* X $\in$ U *and* V $\in$ R^n, *then the derivative* $(D_V F)(X)$ *exists and*

(4.28) $$(D_V F)(X) = v_1(D_1 F)(X) + \cdots + v_n(D_n F)(X).$$

Proof. The map F is given by an m-tuple of real-valued functions on U, $F = (f_1, \ldots, f_m)$. It should be clear that

$$D_j F = (D_j f_1, \ldots, D_j f_m)$$

and that (therefore) we need only prove this theorem for real-valued functions.

So, let

$$U \xrightarrow{\ f\ } R$$

be a function of class C^1 on U. Let $X \in U$ and $V \in R^n$. There is a $\delta > 0$ such that $(X + tV) \in U$, $|t| < \delta$. For any t with $|t| < \delta$ consider the difference quotient

$$\frac{f(X + tV) - F(X)}{t} = \frac{f(x_1 + tv_1, \ldots, x_n + tv_n) - f(x_1, \ldots, x_n)}{t}.$$

We rewrite that difference quotient by adding and then subtracting a few terms. Let us write it out fully in the case $n = 3$:

$$\frac{f(X+tV)-f(X)}{t} = \frac{f(x_1+tv_1,\, x_2+tv_2,\, x_3+tv_3) - f(x_1,\, x_2+tv_2,\, x_3+tv_3)}{t}$$

$$(4.29) \qquad\qquad + \frac{f(x_1,\, x_2+tv_2,\, x_3+tv_3) - f(x_1,\, x_2,\, x_3+tv_3)}{t}$$

$$+ \frac{f(x_1,\, x_2,\, x_3+tv_3) - f(x_1,\, x_2,\, x_3)}{t}.$$

Fix t. Then $P_t = (x_1, x_2 + tv_2, x_3 + tv_3)$ is a point in the open ball $B(X; \delta |V|)$ as is $Q_t = X + tV$. We apply the mean value theorem to the function which f defines along the line segment from P_t to Q_t, i.e., we apply it to the function

$$g(x) = f(P_t + xE_1).$$

Since $D_1 f$ exists on U, g is a differentiable function on the interval between 0 and tv_1. The mean value theorem tells us that

$$\frac{g(tv_1) - g(0)}{tv_1} = g'(\theta)$$

where θ is between 0 and tv_1. Thus

$$\frac{f(Q_t) - f(P_t)}{t} = v_1(D_1 f)(X_t)$$

where X_t is some point between P_t and Q_t. Now apply the same sort of reasoning to the second and third terms in (4.29) and it should be clear that

$$(4.30) \qquad \frac{f(X + tV) - f(X)}{t} = v_1(D_1 f)(X_t) + v_2(D_2 f)(Y_t) + v_3(D_3 f)(Z_t)$$

where X_t, Y_t, and Z_t are points in the ball $B(X; |t| |V|)$.

As t goes to 0, the points X_t, Y_t, Z_t converge to X. Since the derivatives $D_j f$ are continuous, we obtain from (4.30):

$$(D_V f)(X) = v_1(D_1 f)(X) + v_2(D_2 f)(X) + v_3(D_3 f)(X).$$

If f is a real-valued function of class C^1, then the **gradient** of f is the (continuous) vector-valued function

$$\operatorname{grad} f = (D_1 f, \ldots, D_n f).$$

In this case, (4.28) becomes

$$D_V f = \langle V, \operatorname{grad} f \rangle$$

$$= v_1 \frac{\partial f}{\partial x_1} + \cdots + v_n \frac{\partial f}{\partial x_n}.$$

In other words the derivative of f at X in the direction of the vector V is the inner product of V with the gradient of f at X.

This notation for the gradient is a bit awkward. At times it is more convenient to employ the notation

$$f' = \operatorname{grad} f.$$

The reformulation of (4.28) then becomes

$$(D_V f)(X) = \langle V, f'(X) \rangle$$

In employing this notation, remember that f' is a vector-valued function.

We have not yet said anything about higher order derivatives in R^n. If f is of class C^2, then $D_1 f, \ldots, D_n f$ are of class C^1 and so we have the second order derivatives

$$D_i D_j f = \frac{\partial^2 f}{\partial x_i \, \partial x_j}, \qquad 1 \le i \le n, \quad 1 \le j \le n.$$

In case $i = j$ it is customary to write

$$D_i D_j f = \frac{\partial^2 f}{\partial x_i^2}.$$

One basic fact here is that, if f is of class C^2, then $D_i D_j f = D_j D_i f$, that is

(4.31) $$\frac{\partial^2 f}{\partial x_i \, \partial x_j} = \frac{\partial^2 f}{\partial x_j \, \partial x_i}.$$

We shall defer the proof for a couple of pages, because it comes easily out of the basic observations on integration which we are about to make. From (4.31) we see that, if f is of class C^k, each kth partial derivative of f has the form

$$D_1^{k_1} \cdots D_n^{k_n} f = \frac{\partial^k f}{\partial x_1^{k_1} \cdots \partial x_n^{k_n}}$$

where $k_1 + \cdots + k_n = k$.

Let f be a real-valued function defined on an open set U in R^n. We say that f has a **local maximum** at the point X in U if there is a $\delta > 0$ such that $f(X) \ge f(T)$ for all T such that $|X - T| \le \delta$. In short, f has a local maximum at X if there is a neighborhood of X on which the value of f is never greater than $f(X)$. The term "local minimum" is defined similarly.

Theorem 11. *Let* f *be a real-valued function of class* C^1 *on an open set* U *in* R^n. *If* f *has a local maximum at the point* X *in* U, *then* $f'(X) = 0$, *that is*, $(D_j f)(X) = 0, 1 \leq j \leq n$.

Proof. Exercise.

Let us review very quickly the definition of the integral of a (continuous) function over a closed box in R^n. We shall be brief because the reasoning involved is essentially the same as for integrals over closed intervals.

A **closed box** in R^n is a Cartesian product of n closed intervals

$$B = I_1 \times \cdots \times I_n$$
$$= \{X; x_k \in I_k, 1 \leq k \leq n\}.$$

If $I_k = [a_k, b_k]$, then

$$B = \{X; a_k \leq x_k \leq b_k, 1 \leq k \leq n\}.$$

A **partition** of such a closed box B is a set $P = \{B_1, \ldots, B_N\}$ of (closed) boxes B_k such that

(i) $P = B_1 \cup \cdots \cup B_N$;

(ii) if $i \neq j$ the intersection of the interior of B_i with the interior of B_j is empty.

The **mesh** of such a partition P is

$$\|P\| = \max_k \text{ diam } (B_k).$$

The partition $P' = \{B'_1, \ldots, B'_{N'}\}$ is a **refinement** of the partition $P = \{B_1, \ldots, B_N\}$ if each box B'_j is contained in one of the boxes B_k.

We should remark that when $n = 1$ the last three definitions reduce to the definitions we have already given in relation to partitions of an interval. Let us also note the following. One way to construct a partition of the box

(4.32) $B = I_1 \times \cdots \times I_n,$ $I_k = [a_k, b_k]$

is by choosing partitions of each of the intervals $I_1, \ldots, I_n$ and using the associated Cartesian products of all the subintervals. For instance suppose that $n = 2$, $I = [a, b]$, $J = [c, d]$ and

$$B = I \times J = \{(x, y); a \leq x \leq b, c \leq y \leq d\}.$$

Let $P = (x_0, \ldots, x_n)$ be a partition of $[a, b]$ and $Q = [y_0, \ldots, y_p]$ be a partition of $[c, d]$. The product boxes

$$I_k \times J_\ell = \{(x, y); x_{k-1} \leq x \leq x_k, y_{\ell-1} \leq y \leq y_\ell\}$$

partition the box B. There are np such boxes since k ranges from 1 to n and ℓ from 1 to p. The reasoning by which one obtains a partition of the box B (4.32) from partitions of the intervals I_k is similar; it is notationally a bit more complicated when $n > 2$.

It is not true that every partition of B is obtained from partitions of $I_1, \ldots, I_n$ in the manner just described. But it is true that every partition has a refinement which is of this special type.

In order to define integrals in n-dimensions, we need to use the n-dimensional volume of a box, i.e., that number which extends the idea of length in R^1, area in R^2, volume in R^3. We shall use the term "measure", rather than "n-dimensional volume". Thus, the **measure** of the box $B = I_1 \times \cdots \times I_n$ is

$$m(B) = \prod_{k=1}^{n} \text{length } (I_k).$$

Now let B be a closed box in R^n and let F be a function

$$B \xrightarrow{F} R^m.$$

If $P = \{B_1, \ldots, B_N\}$ is any partition of B and if $T = (T_1, \ldots, T_N)$ is any choice of points $T_k \in B_k$, define

(4.33) $$S(F, P, T) = \sum_{k=1}^{N} m(B_k) F(T_k).$$

The function F is called **Riemann-integrable** on the box B if the limit

$$\lim_{\|P\| \to 0} S(F, P, T)$$

exists. For each $\delta > 0$ define $\Sigma_\delta (F) = \{S(F, P, T); \|P\| \leq \delta\}$. Then F is Riemann-integrable if and only if

$$\lim_{\delta \to 0} \text{diam } \Sigma_\delta (F) = 0.$$

When F is integrable, the **integral of F over B** is

$$\int_B F = \bigcap_{\delta > 0} \overline{\Sigma_\delta (F)}.$$

Other notations employed for this integral are

$$\int_B F(X) \, dX, \qquad \int_B F(x_1, \ldots, x_n) \, dx_1 \ldots dx_n.$$

The latter notation comes from the notation $\sum F(x_1, \ldots, x_n) \Delta x_1 \ldots \Delta x_n$ for the approximating sums.

Each continuous function on a closed box B is Riemann-integrable. The proof proceeds just as it did in the case $n = 1$. The essential step exploits the uniform continuity.

As in the Corollary to Theorem 3, the integral of a continuous function F satisfies

$$\left| \int_B F - S(F, P, T) \right| \leq m(B) \omega(F, 2\delta)$$

for every partition P of mesh at most δ (and for every choice T of points in the subboxes of the partition). From the result about continuous functions it follows that every piecewise continuous function on B is Riemann-integrable. The function F

$$B \xrightarrow{\; F \;} R^m$$

is called **piecewise-continuous** if there is a partition $P = \{B_1, \ldots, B_N\}$ of the box B such that F is uniformly continuous on the interior of each box B_k.

The basic properties of integration summarized in Theorem 4 follow by precisely the same argument as in the case $n = 1$. In the analogue of property (iii) the role of $(b - a)$ is played by $m(B)$:

$$\left| \int_B F \right| \le m(B) \sup_{X \in B} |F(X)|.$$

The only technical lemma which must be verified to employ the arguments given when $n = 1$ is this: If $P = \{B_1, \ldots, B_N\}$ is a partition of the box B, then

$$m(B) = m(B_1) + \cdots + m(B_N).$$

We also obtain the extension of this additivity property which is analogous to Theorem 5: If $P = \{B_1, \ldots, B_N\}$ is a partition of B and

$$B \xrightarrow{\; F \;} R^m,$$

then F is integrable if and only if F is integrable on each B_k, and when F is integrable

$$\int_B F = \int_{B_1} F + \cdots + \int_{B_N} F.$$

If F is an integrable function on the box $B = I_1 \times \cdots \times I_n$, then the integral of F over B can be calculated as an iterated integral, i.e., we can calculate

$$\int_B F(x_1, \ldots, x_n)\, dx_1 \ldots dx_n$$

by integrating with respect to the variables x_i, one at a time. We will verify this only when F is continuous, because the more general result gets into technicalities which will take us too far afield.

Theorem 12 (Iterated Integrals Theorem). *Let* F *be a continuous function on the closed box* B. *Suppose that*

$$B = B_1 \times B_2$$

where B_1 *and* B_2 *are closed boxes. Then*

$$\int_B F = \int_{B_2} \left\{ \int_{B_1} F(X, Y)\, dX \right\} dY$$

$$= \int_{B_1} \left\{ \int_{B_2} F(X, Y)\, dY \right\} dX.$$

Proof. The hypothesis is that B_j is a closed box in R^{n_j}, where $n_1 + n_2 = n$ and that

$$B = \{X \in R^n; (x_1, \ldots, x_{n_1}) \in B_1 \quad \text{and} \quad (x_{n_1+1}, \ldots, x_n) \in B_2\}.$$

The only thing which is at all non-trivial in the proof is to avoid becoming lost in a maze of notation and detail. The point of the proof is that $m(B_1 \times B_2) = m(B_1)m(B_2)$, for any boxes B_1, B_2. Let us write out the proof when $n = 2$:

$$B = [a, b] \times [c, d].$$

Choose the partitions

$$P = (x_0, \ldots, x_n), \qquad a = x_0 < x_1 < \cdots < x_n = b$$
$$Q = (y_0, \ldots, y_k), \qquad c = y_0 < \cdots < y_k = d.$$

Let $B_{ij} = [x_{i-1}, x_i] \times [y_{j-1}, y_j]$. Let

$$\begin{aligned}
S &= \sum_{i,j} m(B_{ij})F(x_i, y_j) \\
&= \sum_{i,j} (x_i - x_{i-1})(y_j - y_{j-1})F(x_i, y_j) \\
&= \sum_{i=1}^{n} (x_i - x_{i-1}) \sum_{j=1}^{k} (y_j - y_{j-1})F(x_i, y_j).
\end{aligned}$$

Let δ be the mesh of the partition $\{B_{ij}; 1 \leq i \leq n, 1 \leq j \leq k\}$. Then $\|P\| \leq \delta$ and $\|Q\| \leq \delta$. The idea is that if δ is small,

$$\sum_{j=1}^{k} (u_j - y_{j-1})F(x_i, y_j) \approx \int_c^d F(x_i, y)\, dy$$

and we can get an estimate of how close the approximation is which is uniform in x_i (does not depend upon x_i). For any fixed index i, the function G_i defined by $G_i(y) = F(x_i, y)$ is a continuous function on the interval $[c, d]$. According to the Corollary of Theorem 3,

$$(4.34) \qquad \left| \int_c^d F(x_i, y)\, dy - \sum_{j=1}^{k} (y_j - y_{j-1})F(x_i, y) \right| \leq (d - c)\omega(G_i, 2\delta).$$

The modulus of continuity $\omega(G_i, \delta)$ is defined by

$$\begin{aligned}
\omega(G_i, \delta) &= \sup \{|G_i(y) - G_i(z)|; y, z \in [c, d], |y - z| \leq \delta\} \\
&= \sup \{|F(x_i, y) - F(x_i, z)|; y, z \in [c, d], |y - z| \leq \delta\}
\end{aligned}$$

If $|y - z| \leq \delta$, then the points (x_i, y) and (x_i, z) are not more than δ apart. Therefore,

$$\omega(G_i, \delta) \leq \omega(F, \delta).$$

By the inequality (4.34) we have

$$(4.35)$$
$$\begin{aligned}
\left| S - \sum_{i=1}^{n} (x_i - x_{i-1}) \int_c^d F(x_i, y)\, dy \right| &\leq \sum_{i=1}^{n} (x_i - x_{i-1})(d - c)\omega(G_i, 2\delta) \\
&\leq (d - c)\omega(F, 2\delta) \sum_{i=1}^{n} (x_i - x_{i-1}) \\
&= (d - c)(b - a)\omega(F, 2\delta) \\
&= m(B)\omega(F, 2\delta).
\end{aligned}$$

Define

$$G(x) = \int_c^d F(x, y)\, dy, \qquad a \le x \le b.$$

Then G is a continuous function on $[a, b]$ because

$$|G(x) - G(t)| = \left| \int_c^d [F(x, y) - F(t, y)]\, dy \right|$$

$$\le (d - c) \sup_{y \in [c, d]} |F(x, y) - F(t, y)|$$

If $|x - t| \le \delta$, we have

$$|G(x) - G(t)| \le (d - c)\omega(F, \delta).$$

Hence,

$$\omega(G, \delta) \le (d - c)\omega(F, \delta).$$

Let

$$\tilde{S} = \sum_{i=1}^n (x_i - x_{i-1})G(x_i).$$

By the Corollary to Theorem 3,

$$\left| \int_a^b G(x)\, dx - \tilde{S} \right| \le (b - a)\omega(G, 2\delta)$$

$$\le (b - a)(d - c)\omega(F, 2\delta)$$

$$= m(B)\omega(F, 2\delta)$$

and (4.35) says that

$$|S - \tilde{S}| \le m(B)\omega(F, 2\delta).$$

Therefore,

$$(4.36) \qquad \left| \int_a^b G(x)\, dx - S \right| \le 2m(B)\omega(F, 2\delta).$$

Let $\delta \to 0$. We have

$$\lim_{\delta \to 0} \omega(F, 2\delta) = 0$$

$$\lim_{\delta \to 0} S = \int_B F.$$

From (4.36) it follows that

$$\int_B F = \int_a^b G(x)\, dx$$

$$= \int_a^b \left\{ \int_c^d F(x, y)\, dy \right\} dx.$$

Corollary. *If* F *is a continuous function on the box*

$$B = I_1 \times \cdots \times I_n, \qquad I_k = [a_k, b_k]$$

then

$$\int_B F = \int_{a_n}^{b_n} \cdots \int_{a_1}^{b_1} F(x_1, \ldots, x_n)\, dx_1 \ldots dx_n$$

where the symbol on the right denotes the result of integrating $F(x_1, \ldots, x_n)$

first with respect to x_1, *then integrating the resulting function with respect to* x_2, *etc. The same result holds if the* n *integrations are carried out in any order.*

Theorem 13. *If* F *is a function of class* C^2 *on an open subset of* R^n, *then* $D_iD_jF = D_jD_iF$, *that is*

$$\frac{\partial^2 F}{\partial x_i \, \partial x_j} = \frac{\partial^2 F}{\partial x_j \, \partial x_i}, \qquad 1 \le i \le n, \quad 1 \le j \le n.$$

Proof. We need only prove the theorem for $n = 2$. Let

$$B = [a, b] \times [c, d]$$

be a box inside the domain of F. The derivative $D_1 D_2 F$ is continuous. By Theorem 12 and the fundamental theorem of calculus

$$
\begin{aligned}
\int_B D_1 D_2 F &= \int_c^d \int_a^b \frac{\partial^2 F}{\partial x_1 \, \partial x_2} \, dx_1 \, dx_2 \\
&= \int_c^d [(D_2 F)(b, x_2) - (D_2 F)(a, x_2)] \, dx_2 \\
&= F(b, d) - F(b, c) + F(a, c) - F(a, d).
\end{aligned}
$$

If we replace b by any number x, $a \le x \le b$, and replace d by any number y, $c \le y \le d$, we have a similar conclusion which, for convenience, we rewrite as

$$F(x, y) = \int_c^y \int_a^x (D_1 D_2 F)(x_1, x_2) \, dx_1 \, dx_2 - F(x, c) - F(a, y) + F(a, c)$$

$$- \int_a^x \int_c^y (D_1 D_2 F)(x_1, x_2) \, dx_2 \, dx_1 - F(x, c) - F(a, y) + F(a, c).$$

From this equation, calculate $D_1 F$. We have

$$(D_1 F)(x, y) = \int_c^y (D_1 D_2 F)(x, x_2) \, dx_2 - (D_1 F)(x, c).$$

If we differentiate this equation with respect to y, we obtain

$$(D_2 D_1 F)(x, y) = (D_1 D_2 F)(x, y).$$

This step involves use of the fundamental theorem of calculus and the fact that $D_1 D_2 F$ is continuous.

Corollary. *Let* F *be a map from an open subset of* R^n *into* R^m *and let* i, j *be indices,* $1 \le i \le n$, $1 \le j \le n$. *If the partial derivative* D_iD_jF *exists and is continuous, then* D_jD_iF *also exists and*

$$D_j D_i F = D_i D_j F.$$

Proof. In the proof of Theorem 13 we used the existence and continuity of one of the mixed partial derivatives to show that the other exists and is the same.

Exercises

1. Let F be a continuous function on an open set U in R^n. If

$$\int_B F = 0$$

for every closed box $B \subset U$, then $F = 0$.

2. Define

$$f(X) = |x_1|^{1/2} |x_2|^{1/2}, \qquad X \in R^2.$$

(a) Show that the partial derivatives of f at the origin exist.

(b) Find a vector V such that $(D_V f)(0)$ does not exist.

3. True or false? If f is a function of class C^1 on $R^n - \{0\}$ and if the partial derivatives of f are uniformly continuous functions, then f can be extended to a function of class C^1 on R^n.

4. Let f be a (real-valued) function of class C^1 on an open set U. If $X \in U$, show that (at X) the direction in which f is increasing most rapidly is the direction of $f'(X)$, the gradient of f at X.

5. If B is a box, let $X + B$ be the X-translate of B:

$$X + B = \{X + Y; \ Y \in B\}.$$

Let f be a continuous function on R^n and let B be a fixed box in R^n. Let

$$g(X) = \int_{X+B} f.$$

Is it true that g is of class C^1?

6. If f is of class C^1 and X is a point in the domain of f, then $(D_V f)(X)$ is a linear function of V.

7. (Mean Value Theorem) Recall that a set is convex provided it contains the line segment joining each pair of its points. Let U be a convex open set in R^n and let f be a (real-valued) function of class C^1 on U. If X, Y are in U, show that

$$f(Y) - f(X) = \langle Y - X, f'(P) \rangle$$

where P is some point on the line segment between X and Y. *Hint*: Let

$$g(t) = f[X + t(Y - X)], \qquad 0 \le t \le 1.$$

8. If $k = (k_1, \ldots, k_n)$ is an n-tuple of positive integers, let

$$k! = k_1! \cdots k_n!$$
$$D^k = D_1^{k_1} \cdots D_n^{k_n}.$$

Let f be a polynomial function on R^n. Show that

$$f(x_1, \ldots, x_n) = \sum_k \frac{1}{k!} (D^k f)(0) x_1^{k_1} \cdots x_n^{k_n}.$$

9. Let F be a continuous function on the box B and let K be a compact subset of B. Suppose you wanted to define

$$\int_K F.$$

Here is one thing to try. Let g be a continuous function on B such that $0 \le g \le 1$, $g = 1$ on K and $g < 1$ off K. Prove that

$$\lim_k \int_B g^k F$$

exists. *Hint*: Prove it first when F is a non-negative real-valued function.

10. Let F, B, and K be as in Exercise 9. Here is another reasonable way to attempt to define the integral of F over K. If $P = \{B_1, \ldots, B_N\}$ is a partition of B, consider those boxes B_k which intersect K. For each such k, choose a point $T_k \in B_k \cap K$. Form the sum

$$S_K = \sum F(T_k) m(B_k)$$

where the sum is extended over those k's such that B_k meets K. Prove that

$$\lim_{\|P\| \to 0} S_K$$

exists. *Hint*: Prove that it exists by showing that, if $\|P\|$ is small, then

$$S_K \approx \lim_k \int_B g^k F \qquad \text{(the limit from Exercise 9)}.$$

4.6. *Riemann-Stieltjes Integration*

Strictly speaking, this section does not constitute part of a review of the usual calculus course. It deals with a type of integral which is slightly more general, but technically no more involved, than the integral in R^1 which we discussed in Section 4.2. One example of a more general integral will be

$$\int_I f \, dG$$

where f is a continuous real-valued function on the interval I and G is a (vector-valued) function of "bounded variation". Such an integral is a limit of sums of the type

$$\sum_{k=1}^n f(t_k)[G(x_k) - G(x_{k-1})].$$

Let us define carefully the class of functions G which we shall use.
Suppose that

$$[a, b] \xrightarrow{G} R^k$$

is a (not necessarily continuous) map from $[a, b]$ into R^k. If

$$P = (x_0, \ldots, x_n)$$
$$a = x_0 < x_1 < \cdots < x_n = b$$

is a partition of $[a, b]$, we form the sum

$$(4.37) \qquad V(P; G) = \sum_{k=1}^n |G(x_k) - G(x_{k-1})|.$$

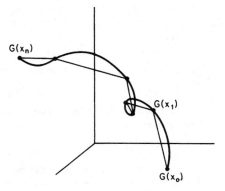

$G(x_n)$

$G(x_1)$

$G(x_0)$

FIGURE 17

This is the sum of the distances between the successive points $G(x_k)$. If G is $1:1$, it should provide a better approximation to the length of the image of G in R^k, as the mesh of the partition tends to 0. (See Figure 17.)

We say that the function G is of **bounded variation** if the sums $V(P; G)$ are bounded, i.e., if there is a constant M such that $V(P; G) \leq M$ for all partitions P. If G is of bounded variation, the **total variation** of G is

(4.38) $$V_a^b(G) = \sup_P V(P; G).$$

If G is a function of bounded variation on $[a, b]$ and if $a \leq x \leq b$, then obviously

(4.39) $$V_a^x(G) + V_x^b(G) = V_a^b(G).$$

The function V_G defined by

(4.40) $$V_G(x) = V_a^x(G), \qquad a \leq x \leq b$$

is called the **variation** of G. It is an increasing function on $[a, b]$.

Before we finish the discussion of integrals, let us look at some examples.

EXAMPLE 5. If G is continuously differentiable (of class C^1), then G is of bounded variation:

$$\sum_k |G(x_k) - G(x_{k-1})| = \sum_k \left| \int_{x_{k-1}}^{x_k} G'(t)\, dt \right|$$
$$\leq \sum_k (x_k - x_{k-1}) \sup |G'|$$
$$= (b - a) \sup |G'|.$$

The total variation of G is $\int_a^b |G'(x)|\, dx$, although that takes a little proving.

There is a natural generalization of the example just given. The function G is said to satisfy a **Lipschitz condition** if there exists a constant M such that

$$|G(x) - G(y)| \leq M|x - y|.$$

Evidently, such a function is uniformly continuous and of bounded variation.

EXAMPLE 6. Arc length is related to functions of bounded variation. A **path** in R^k is a continuous map

$$[a, b] \xrightarrow{G} R^k$$

from some closed interval into R^k. Paths can be pretty horrible, but not if G is of bounded variation. If that is the case, we call G a **rectifiable path** (or, path with finite length) and define

$$\text{length } (G) = V_a^b(G).$$

The motivation for this definition is to be found in Figure 17. Rectifiability makes the image of G look like a curve, and as t ranges over $[a, b]$ the point $G(t)$ traces out that curve. In general the curve may cross itself and $G(t)$ may double back and trace some segments of the curve several times. This is the reason for calling G the path rather than referring to the image of G as being the path. In the event that G is $1:1$, the length of the path (i.e., the variation of G) is what we define to be the length of the image curve $G([a, b])$.

EXAMPLE 7. Let g be an increasing (non-decreasing) real-valued function. Then g is of bounded variation and

$$V_a^b(g) = g(b) - g(a).$$

This is because any sum (4.37) telescopes and has the value $g(b) - g(a)$. Similarly, any decreasing function is a function of bounded variation. *Every real-valued function of bounded variation is the sum of an increasing function and a decreasing function:*

(4.41) $$g = V_g + (g - V_g).$$

We remarked that the variation of g (4.40) is an increasing function; and, the fact that $(g - V_g)$ is a decreasing function simply says that

$$g(t) - g(x) \leq V_x^t(g), \qquad x \leq t.$$

EXAMPLE 8. An example of a continuous function not of bounded variation is

$$g(x) = x \sin \frac{1}{x}, \qquad 0 < x \leq b$$

$$g(0) = 0.$$

The graph of this function oscillates too much to have a finite length. See Figure 13 on p. 95. At the points $2/\pi$, $2/3\pi$, $2/5\pi$, ... the function g has the values $2/\pi$, $-(2/3\pi)$, $2/5\pi$, $-(2/7\pi)$, ... Compute the distances between the successive points and we see that

$$V_0^b(g) \geq \frac{2}{\pi}\Big[\Big(1 + \frac{1}{3}\Big) + \Big(\frac{1}{5} + \frac{1}{7}\Big) + \cdots\Big].$$

Thus g is not of bounded variation.

Now, suppose G is a function on $[a, b]$ and f is a numerical function on $I = [a, b]$. If G maps I into R^k, take f real-valued. If G maps into C^k, take f complex-valued. For a partition $P = (x_0, \ldots, x_n)$, choose points $t_k \in [x_{k-1}, x_k]$ and look at the sum

(4.42) $$S(f, P, T, G) = \sum_{k=1}^n f(t_k)[G(x_k) - G(x_{k-1})].$$

If the limit

$$\lim_{\|P\| \to 0} S(f, P, T, G)$$

exists, we denote it by $\int_I f \, dG$ and say that **the Riemann-Stieltjes integral** $\int_I f \, dG$ **exists.** If $\Sigma_\delta(f, G)$ is the set of all sums (4.42) obtained from partitions of mesh at most δ, the integral exists if and only if

$$\lim_{\delta \to 0} \text{diam } \Sigma_\delta(f, G) = 0$$

and, when this condition is satisfied, we have

$$\int_I f \, dG = \bigcap_{\delta > 0} \bar{\Sigma}_\delta.$$

This integral, when it exists, is also written

$$\int_a^b f(x) \, dG(x).$$

Theorem 14. *If* f *is continuous and* G *is of bounded variation, the Riemann-Stieltjes integral*

$$\int_I f \, dG = \int_a^b f(x) \, dG(x)$$

exists. In fact, if P *is a partition of mesh at most* δ *any sum* (4.38) *satisfies*

$$\Big|\int_I f \, dG - S(f, P, T, G)\Big| \leq \omega(f, 2\delta) V_a^b(G).$$

Proof. The proof is virtually identical with the proof when $G(x) = x$, except that in the estimate $(x_k - x_{k-1})$ is replaced by $|G(x_k) - G(x_{k-1})|$.

The same sort of result is valid for the integral

$$\int_I F \, dg = \int_a^b F(x) \, dg(x)$$

where F is continuous from $[a, b]$ into R^m $[C^m]$ and g is a real [complex]-valued function of bounded variation.

Theorem 15 (*Integration by Parts, Weak Form*). *If* f *and* G *are continuous functions of bounded variation on* [a, b], *then*

(4.43) $$\int_a^b f(x) \, dG(x) = f(x)G(x) \Big|_a^b - \int_a^b G(x) \, df(x).$$

Proof. Let $P = (x_0, \ldots, x_n)$ be a partition of $[a, b]$. Then

$$f(x_k)[G(x_k) - G(x_{k-1})] + [f(x_k) - f(x_{k-1})]G(x_{k-1})$$
$$= f(x_k)G(x_k) - f(x_{k-1})G(x_{k-1}).$$

Sum from $k = 1, \ldots, n$. Let the mesh of the partition go to 0.

EXAMPLE 9. If G is a function of class C^1, Riemann-Stieltjes integration with respect to G does not introduce anything particularly new. In this case we have

$$\int_a^b f(x) \, dG(x) = \int_a^b f(x)G'(x) \, dx$$

for each continuous function f. This is apparent by looking at approximating sums. If $P = (x_0, \ldots, x_n)$ is a partition of $[a, b]$ and $T = (t_1, \ldots, t_n)$ a choice of points $t_k \in [x_{k-1}, x_k]$, then

$$\sum f(t_k)[G(x_k) - G(x_{k-1})] = \sum_{k=1}^n f(t_k) \int_{x_{k-1}}^{x_k} G'(t) \, dt$$
$$= \int_a^b f_{PT}(t)G'(t) \, dt$$

where f_{PT} is the step function which has the (constant) value $f(t_k)$ on the kth subinterval of the partition P:

$$f_{PT}(t) = f(t_k), \qquad x_{k-1} \le t \le x_k.$$

For a continuous f, f_{PT} converges uniformly to f as $\|P\| \to 0$:

$$\lim_{\|P\| \to 0} \|f - f_{PT}\|_\infty = 0$$

Hence the Riemann-Stieltjes sums converge to $\int f(t)G'(t) \, dt$ as the mesh of P goes to 0.

EXAMPLE 10. Let

$$f(x) = \frac{1}{2} \frac{x}{|x|}, \qquad x \ne 0.$$
$$f(0) = 0.$$

On any interval $[-b, b]$, f is of bounded variation. If G is continuous, then

(4.44) $$\int_{-b}^b G(x) \, df(x) = G(0).$$

This has nothing to do with $f(0)$. We could have defined $f(0)$ to be any-thing. Just use partitions which do not involve 0. The "derivative" of f is what Dirac called the "delta function". Of course, f is not differentiable at 0. If f had a derivative, it would satisfy

$$\int_{-b}^{b} G(x) f'(x)\, dx = G(0)$$

for all continuous G. Now $f'(x) = 0$ for $x \neq 0$. But $\int_{-b}^{b} f' = 1$. So $f'(0) = \infty$. So, there is no such function f'. The Dirac delta "function" is the process of integration against df.

EXAMPLE 11. Look at the Cantor function from Example 24 of Chap-ter 3. It is a continuous function f on the interval $[0, 1]$ such that

(i) f is increasing (non-decreasing);
(ii) $f(0) = 0$; $f(1) = 1$;
(iii) f is constant on each of the intervals which were deleted from $[0, 1]$ to form the Cantor set.

Thus any integral $\int G\, df$ depends only upon the behavior of G on the Cantor set. But, the dependence is much more involved than in Example 9.

Suppose that we fix a function G

$$[a, b] \xrightarrow{\;G\;} R^m$$

which is of bounded variation. For each continuous real-valued function f on the interval $I = [a, b]$ the Riemann-Stieltjes integral $\int f\, dG$ exists and has these properties

(i) $\displaystyle\int_I (cf + g)\, dG = c \int_I f\, dG + \int_I g\, dG$

(ii) $\displaystyle\left| \int_I f\, dG \right| \leq V_a^b(G) \sup_I |f|$.

There is a particular form of converse to this which is useful in several parts of mathematics. Suppose we have a function L which assigns to each continuous function f on I a vector $L(f)$ in R^m, and suppose that L has these two properties:

(i) L is linear, i.e., $L(cf + g) = cL(f) + L(g)$;
(ii) there is a constant $M > 0$ such that $|L(f)| \leq M \sup_I |f|$ for every f continuous on I.

Then there exists a function G from I into R^m which is of bounded varia-tion such that

$$L(f) = \int_I f\, dG$$

for every f. We shall not stop to prove this here.

The reader has probably noticed a pattern to the various Riemann-Stieltjes integrals which we have discussed, and has probably observed that we could introduce other integrals by the same method. For instance, if F and G are $k \times k$ matrix-valued functions, if F is continuous and G is of bounded variation, the matrix

$$\int F \, dG$$

is defined. (Be careful; it's not the same as $\int (dG)F$.) We don't want to make a fuss about all that; however, let us say this much. We can define the integral $\int F \, dG$ just as we did, using a variety of different multiplications, that is, using a variety of different interpretations of the products in the sums

$$\sum_{k=1}^{n} F(t_k)[G(x_k) - G(x_{k-1})].$$

Suppose

$$[a, b] \xrightarrow{F} R^m$$

$$[a, b] \xrightarrow{G} R^k.$$

and suppose we are given any "multiplication" of vectors in R^m by vectors in R^k:

$$R^m \times R^k \xrightarrow{M} R^p.$$

Assume that

   (i) M is bilinear; i.e., $M(X, Y)$ is a linear function of X for fixed Y and a linear function of Y for fixed X.

   (ii) $|M(X, Y)| \leq |X| |Y|$.

If F is continuous and G is of bounded variation, the integral

$$(4.45) \qquad \int_a^b M(F(x), dG(x)) = \lim_{\|P\| \to 0} \sum_k M(F(x_k), G(x_k) - G(x_{k-1}))$$

exists with the proof which we gave. The integral $\int M(F, dG)$ is a bilinear function of F and G and

$$(4.46) \qquad \left| \int M(F, dG) \right| \leq \int |F| \, dV_G.$$

Exercises

1. If G is of bounded variation on $[a, b]$, then G has left- and right-hand limits at every point of $[a, b]$. (Prove first for real-valued functions.)

2. If G is of bounded variation, then G is continuous except at a countable number of points.

3. If G is of bounded variation, then $\int f\,dG$ is independent of the values of G at the points of discontinuity interior to $[a, b]$.

4. If f is a function of class C^1 on $[a, b]$, then

$$\int G\,df = \int Gf'.$$

5. True or false? If f is a continuous function of bounded variation on $[a, b]$ and if $f(a) = 0$, then

$$f(x) = \int_a^x df(t).$$

6. Show that, if Q is a refinement of P, then

$$V(Q; G) \leq V(P; G)$$

for all functions G on $[a, b]$.

7. Show that, if G is continuous and of bounded variation, then

$$\lim_{\|P\|\to 0} V(P; G) = V_a^b(G).$$

8. True or false? Every convex function on $[a, b]$ is of bounded variation.

9. True or false? If f

$$[a, b] \xrightarrow{\ f\ } [c, d]$$

is a $1:1$ function of bounded variation, then the inverse function f^{-1} is of bounded variation.

10. Let f be a differentiable function on the interval $[a, b]$. Prove that, if f' is of bounded variation, then f' is continuous.

11. Let g be the function on $[0, 1]$ defined as follows: If x is irrational or zero, $g(x) = 0$. If $x \neq 0$ is rational and $x = p/q$ (lowest terms), then $g(x) = q^{-3}$. Then g is of bounded variation. What is $\int f\,dg$?

12. If f is of bounded variation and G is both continuous and of bounded variation, the Riemann-Stieltjes integral $\int f\,dG$ exists.

13. Let g be an increasing function on $[a, b]$ such that

$$\int_a^b dg = 1.$$

If F is continuous from $[a, b]$ into R^m, then $\int F\,dg$ lies in the closed convex hull of the range of F. (The closed convex hull of a set is the smallest closed convex set which contains that set.)

*14. (Mean-value theorem) Let F, g be as in Exercise 13. There exist points $t_1, \ldots, t_k$ in $[a, b]$ and numbers $c_1, \ldots, c_k$ with

$$\int F\,dg = c_1 F(t_1) + \cdots + c_k F(t_k), \qquad c_j \geq 0, \quad \sum_j c_j = 1.$$

15. (Jensen) Let g be an increasing function on $[a, b]$ such that $g(b) - g(a) = 1$, and let ϕ be a continuous convex function on the real line. If f is any continuous

real-valued function on $[a, b]$, then

$$\int (\phi \circ f) \, dg \geq \phi\left(\int f \, dg\right).$$

(*Hint*: Let $t = \int f \, dg$. There is a line through $(t, \phi(t))$ such that the graph of ϕ is above the line. Write that down and integrate.)

16. According to Exercise 15

$$\int e^f \, dg \geq \exp\left(\int f \, dg\right)$$

under certain conditions. Use this to prove that, for positive numbers $a_1, \ldots, a_n$, the arithmetic mean exceeds the geometric mean:

$$\frac{1}{n}(a_1 + \cdots + a_n) \geq (a_1 \cdots a_n)^{1/n}.$$

(*Hint*: Pick n points $t_1, \ldots, t_n$; let g jump by $1/n$ at each t_k.)

5. Sequences
of Functions

5.1. Convergence

Suppose that we have a sequence of functions F_n, with a common domain of definition, D. We have in mind that D may be any set whatsoever. We want to talk about what it means for the sequence $\{F_n\}$ to converge. Basically, this will mean that the values of the functions converge. Therefore, we had better have all the functions map into the same (Euclidean) space:

$$(5.1) \qquad D \xrightarrow{F_n} R^m.$$

If we fix any point X in D, then $\{F_n(X)\}$ is a sequence of points in R^m, and we can ask whether or not that sequence converges.

Definition. *The sequence of functions* $\{F_n\}$ *(5.1)* **converges pointwise** *if, for each* X *in* D, *the sequence* $\{F_n(X)\}$ *converges. If* $\{F_n\}$ *converges pointwise and if*

$$F(X) = \lim_n F_n(X), \qquad X \in D$$

then (F is a function on D *and) we say that* $\{F_n\}$ **converges pointwise to (the function)** F *and we write*

$$F = \lim_n F_n.$$

The reader is familiar with the convergence of various sequences of functions, e.g.,

169

$$f_n(x) = x^{1/n}, \qquad x \in R_+$$
$$\lim_n f_n = 1;$$

$$p_n(x) = \sum_{k=0}^{n} \frac{1}{k!} x^k, \qquad x \in R$$
$$\lim_n p_n = \exp;$$

$$F_n(A) = \sum_{k=0}^{n} A^k, \qquad |A| < 1$$
$$\lim_n F_n(A) = (I - A)^{-1}.$$

What is a vector $Y = (y_1, \ldots, y_m)$ in R^m? It is a real-valued function on the set $\{1, 2, \ldots, m\}$:
$$Y(k) = y_k.$$

So, a sequence of vectors Y_n in R^m is a sequence of functions $Y_n = (y_{n1}, \ldots, y_{nm})$ with common domain
$$D = \{1, 2, \ldots, m\}$$
and, these functions converge pointwise if and only if the sequence of vectors $\{Y_n\}$ converges in R^m.

Pointwise convergence of functions seems so natural that one might wonder why we bother to formalize it. We do so not only because it is a fundamental idea but also because we wish to emphasize its limitations. We would like to be able to transfer to $F = \lim_n F_n$ properties of the functions F_n; for instance, we would like to know that F is continuous if each F_n is continuous. And we would like to establish the continuity of some operations (interchange the order of two limits), for example,

$$\int F = \lim_n \int F_n.$$

Pointwise convergence usually is not strong enough to give positive results in these directions. Here is a stronger form of convergence.

Definition. *The sequence of functions* $\{F_n\}$ **converges uniformly to (the function)** F *if, for each* $\epsilon > 0$, *there exists a positive integer* N *such that*

$$|F(X) - F_n(X)| < \epsilon, \qquad n > N, X \in D.$$

In other words, F_n converges uniformly to F, provided (for large n) $F_n(X)$ is near $F(X)$, where "near" means uniformly near for all X in the domain. To rephrase, pointwise convergence means that, given $\epsilon > 0$ and given X, there exists a positive integer $N_{X,\epsilon}$ such that

$$|F(X) - F_n(X)| < \epsilon, \qquad n > N_{X,\epsilon}.$$

The convergence is uniform if and only if the integers $N_{x,\epsilon}$ can be chosen so as not to depend on X, i.e., if and only if $\{N_{x,\epsilon}; X \in D\}$ is bounded (for each $\epsilon > 0$).

A convenient notation for uniform convergence is provided by the **sup norm**:

(5.2)
$$D \xrightarrow{F} R^m$$
$$\|F\|_\infty = \sup\{|F(X)|; X \in D\}.$$

We will explain in Chapter 6 the reason for the use of the subscript "∞" in this notation. At times we need to discuss the supremum of $|F|$ on a subset K of D and we will write

$$\sup_K |F| = \sup\{|F(X); X \in K\}.$$

In this notation

$$\|F\|_\infty = \sup_D |F|.$$

If $\{F_n\}$ is a sequence with a common domain of definition, we say the sequence is **bounded** if the sequence of numbers $\{\|F_n\|_\infty\}$ is bounded.

Plainly, F_n converges uniformly to F if and only if

$$\lim_n \|F - F_n\|_\infty = 0.$$

In order to understand uniform convergence, one should have clearly in mind some picture of the ϵ-ball about F which is defined by the sup norm: $\{G; \|F - G\|_\infty < \epsilon\}$. Suppose f is a real-valued function on (a subset of) the real line. The "uniform ϵ-ball about f" consists of the functions g for which the graph stays in a band about the graph of f. (See Figure 18.)

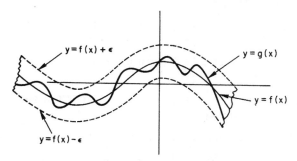

$y = f(x) + \epsilon$

$y = g(x)$

$y = f(x)$

$y = f(x) - \epsilon$

FIGURE 18

EXAMPLE 1. This is an elementary but important example. Suppose f_n is defined on the real line by

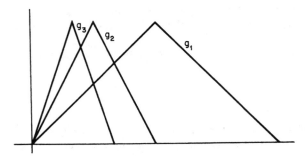

FIGURE 19

$$f_n(x) = \begin{cases} n^2x, & 0 \le x \le \dfrac{1}{n} \\ n - n^2\left(x - \dfrac{1}{n}\right), & \dfrac{1}{n} \le x \le \dfrac{2}{n} \\ 0, & \text{otherwise.} \end{cases}$$

In other words f_n is a tent function which rises linearly from 0 to n on the interval $[0, 1/n]$, falls linearly from n to 0 on $[1/n, 2/n]$, and is 0 elsewhere. The sequence $\{f_n\}$ converges pointwise to 0; however, the convergence is not uniform because

$$\|0 - F_n\|_\infty = \|f_n\|_\infty$$

does not even stay bounded, much less go to 0. But, the lack of boundedness is not the essential element in the lack of uniformity. The tent functions $g_n = (1/n)f_n$ constitute a bounded sequence which converges pointwise to 0; and yet $\|g_n\|_\infty = 1$ for all n. See Figure 19.

EXAMPLE 2. Let h_n be the real-valued function on the real line

$$h_n(x) = \begin{cases} 0, & x \le n \\ x - n, & n \le x \le n + 1 \\ 1, & x \ge n + 1. \end{cases}$$

Then $\{h_n\}$ is bounded; $h_1 \ge h_2 \ge h_3 \ge \cdots$; h_n converges pointwise to 0. But, the convergence is not uniform.

EXAMPLE 3. Let's look at the convergence of the polynomials

$$p_n(x) = \sum_{k=0}^{n} \frac{1}{k!}x^k, \qquad x \in R$$

to the exponential function. Is that convergence uniform? Well,

(5.3) $$|e^x - p_n(x)| \le \sum_{n+1}^{\infty} \frac{1}{k!}|x|^k.$$

On any interval $[-b, b]$, the convergence is uniform because

$$\sup_{[-b, b]} |\exp - p_n| \le \sum_{n+1}^{\infty} \frac{1}{k!} b^k.$$

On R, the convergence is *not* uniform, because (for each n)

$$\lim_{x \to \infty} |e^x - p_n(x)| = \infty.$$

For many things we wish to do, uniform convergence on compact subsets will suffice.

EXAMPLE 4. Let F be a continuous function from the interval $[0, 1]$ into R^m. If n is a positive integer, we partition $[0, 1]$ into n subintervals of equal length using the points

$$0, \frac{k}{n}, \frac{2k}{n}, \ldots, \frac{n-1}{n}, 1.$$

Define

$$F_n(t) = \begin{cases} F\left(\dfrac{k}{n}\right), & \dfrac{k-1}{n} < t \le \dfrac{k}{n} \\ F\left(\dfrac{1}{n}\right), & t = 0. \end{cases}$$

Then F_n is a "step function" which has the constant value $F(1/n)$ on the interval $[0, 1/n]$, the constant value $F(2/n)$ on the interval $(1/n, 2/n]$, and so on. We assert that the sequence $\{F_n\}$ converges uniformly to F. Let $\epsilon > 0$. Since F is continuous on $[0, 1]$, F is uniformly continuous. Hence there exists $\delta > 0$ such that

$$|F(x) - F(t)| < \epsilon, \qquad |x - t| < \delta.$$

Let N be any positive integer such that $1/N \le \delta$. Then

$$\|F - F_n\|_\infty \le \epsilon, \qquad n \ge N.$$

Why? Let t be any point of $[0, 1]$ and n be any integer greater than N. If $t \ne 0$, then t belongs to one of the intervals $[(k-1)/n, k/n]$ and $F_n(t) = F(k/n)$. Thus

$$|F(t) - F_n(t)| = \left| F(t) - F\left(\frac{k}{n}\right) \right|$$

and $|F(t) - F(k/n)| < \epsilon$, because $|t - (k/n)| < 1/n < \delta$. So we have

$$|F(t) - F_n(t)| < \epsilon, \qquad \text{all } n \ge N \qquad \text{and all } t \in [0, 1].$$

Lemma. *Let $\{F_n\}$ be a sequence of functions from the set* D *into* R^m.

(i) $\{F_n\}$ *converges pointwise to some function* F *from* D *into* R^m *if and only if* $\{F_n(X)\}$ *is a Cauchy sequence for each* X *in* D.

(ii) $\{F_n\}$ *converges uniformly to some function* F *from* D *into* R^m *if and only if*

$$\lim_{k, n} \|F_k - F_n\|_\infty = 0.$$

Proof. In each case it is the "if" half of the result which requires comment.

(i) Let $X \in D$. If $\{F_n(X)\}$ is a Cauchy sequence, the completeness of R^m tells us that $\{F_n(X)\}$ converges to a vector in R^m. Call this vector $F(X)$. If the Cauchy condition holds for each X, then F associates with each X in D a vector

$$F(X) = \lim_n F_n(X)$$

in R^m.

(ii) The condition

$$\lim_{k,n} \| F_k - F_n \|_\infty = 0$$

states that, given $\epsilon > 0$, there is a positive integer N_ϵ such that

$$|F_n(X) - F_k(X)| < \epsilon, \qquad \text{for all } k, n \geq N_\epsilon \text{ and all } X \in D.$$

In other words $\{F_n(X)\}$ is Cauchy and converges at a rate which is independent of X. If we let $F(X) = \lim_n F_n(X)$, then

$$|F(X) - F_n(X)| \leq \epsilon, \qquad \text{for all } k \geq N_\epsilon \text{ and all } X \in D.$$

That is,

$$\| F - F_k \|_\infty \leq \epsilon, \qquad k \geq N_\epsilon.$$

EXAMPLE 5. Weierstrass noted long ago a basic (but special) case in which an infinite series of functions

$$\sum_{n=1}^\infty F_n$$

converges uniformly. Suppose that

$$\sum_{n=1}^\infty \| F_n \|_\infty < \infty.$$

Then look at the partial sums

$$S_n = \sum_{k=1}^n F_k.$$

We have

$$\| S_n - S_k \|_\infty \leq \sum_{k+1}^\infty \| F_j \|_\infty, \qquad k \leq n$$

and by the last lemma $\{S_n\}$ converges uniformly to S:

$$S(X) = \sum_{n=1}^\infty F_n(X).$$

There is one important set of circumstances in which pointwise convergence implies uniform convergence.

Theorem 1 (Dini). *Let* D *be a compact set in* R^k, *and let* $\{f_n\}$ *be a sequence of continuous real-valued functions on* D *which is monotone-decreasing:*

$$f_1 \geq f_2 \geq f_3 \geq \cdots.$$

If f_n *converges pointwise to* 0, *then* f_n *converges uniformly to* 0.

Proof. Let $\epsilon > 0$. Let

$$U_n = \{X \in D; f_n(X) < \epsilon\}.$$

Since f_n is continuous, U_n is open relative to D. Note also that, since $f_n \geq f_{n+1}$, we have

$$U_1 \subset U_2 \subset U_3 \subset \cdots.$$

Now, $\{U_n\}$ is a cover of D. Why? If X is in D, then $f_n(X)$ converges to 0; in particular, $f_n(X) < \epsilon$ for some n, i.e., X is in U_n. Since D is compact and the U_n's are increasing, some U_N covers D, i.e., for some N, $U_N = D$. We then have

$$0 \leq f_n < \epsilon, \qquad n \geq N.$$

Corollary. *Let* $\{f_n\}$ *be a sequence of non-negative continuous functions on a compact set* D. *If*

$$\sum_n f_n(X) < \infty$$

for each $X \in D$ *and if the function*

(5.4) $$f = \sum_n f_n$$

is continuous, then the series (5.4) *converges uniformly.*

It would be difficult to emphasize too often how crucial the hypotheses of *monotone* convergence and *compact* domain are in Dini's theorem. In Example 1, the continuous functions f_n converge pointwise to 0 as do the continuous functions g_n, but in neither case is the convergence uniform. In Example 2 the convergence to 0 is monotone but the domain is not compact and the convergence is not uniform. In the corollary, the monotonicity of the sequence of partial sums $s_n = \sum_{k=1}^{n} f_k$ is ensured by the condition $f_k \geq 0$. The *a priori* knowledge that $f = \lim_n s_n$ is continuous then guarantees that Dini's theorem applies to $\{f - s_n\}$.

EXAMPLE 9. Let

$$f(x) = x^{1/2}, \qquad 0 \leq x \leq 1.$$

Here is one way to construct a sequence of polynomials which converges uniformly to f. Define

(5.5)
$$f_0 = 0$$
$$f_{n+1}(x) = f_n(x) + \tfrac{1}{2}[x - f_n(x)^2], \qquad n \geq 0.$$

Now we verify that

  (i) f_n is a polynomial function;
  (ii) $0 \leq f_n \leq f$.

Obviously, f_{n+1} is a polynomial function if f_n is. We verify (ii) by mathematical induction. Certainly (ii) holds for $n = 0$. Suppose it holds for $n = k$. It is then clear from (5.5) that

$$f_{k+1} \geq f_k \geq 0.$$

Now

$$\begin{aligned}
f_{k+1} &= f_k + \tfrac{1}{2}(f^2 - f_k^2) \\
&= f_k + \tfrac{1}{2}(f + f_k)(f - f_k) \\
&\leq f_k + f \cdot (f - f_k)
\end{aligned}$$

and (since we are on the compact interval $[0, 1]$) $f \leq 1$. Thus,

$$\begin{aligned}
f_{k+1} &\leq f_k + (f - f_k) \\
&= f.
\end{aligned}$$

We conclude that $\{f_n\}$ is a monotone increasing sequence of continuous functions on $[0, 1]$, bounded above by f:

$$0 \leq f_n \leq f_{n+1} \leq f.$$

Thus $\{f_n\}$ converges pointwise. If g is the limit function, then (5.5) shows us that

$$g = g + \tfrac{1}{2}(f^2 - g^2).$$

Thus $g = f$. *Since f is continuous,* $\{f - f_n\}$ is a sequence of continuous functions on $[0, 1]$ which converges pointwise monotonely down to 0. By Dini's theorem, the convergence is uniform.

Exercises

1. Let D be a non-empty set and let f be a complex-valued function on D such that

$$|f(X)| < 1, \qquad \text{for each } X \in D.$$

Show that the sequence of powers $\{f_n\}$ converges pointwise to the zero function on D. Prove that f^n converges uniformly to 0 if and only if

$$\sup_D |f| < 1.$$

2. Let D be the set of positive integers and let $\{f_n\}$ be the sequence of functions on D defined by

$$f_n(k) = \frac{n-k}{n+k}.$$

Show that $\{f_n\}$ converges pointwise to (the constant function) 1. Is this convergence uniform?

3. Construct a sequence $\{f_n\}$ of continuous functions on the real line R such that

(i) $\{f_n\}$ converges pointwise to 0;
(ii) none of the functions f_n is bounded.

4. True or false? If $\{F_n\}$ is a sequence of functions which converges uniformly, then $\{F_n\}$ is a bounded sequence.

5. For which $x \in R$ does $\lim_n \cos nx$ exist?

6. Let

$$f_n(x) = \int_0^x e^{int} \, dt, \qquad x \in R.$$

The sequence $\{f_n\}$ converges uniformly to 0 on the real line.

7. Let f be the complex-valued function defined on the open unit disk $D = \{z \in C; |z| < 1\}$ by

$$f(z) = (1 - z)^{-1}, \qquad z \in D.$$

Let

$$f_n(z) = \sum_{k=0}^{n} z^k.$$

Then (as we know) f_n converges pointwise to f. Show that this convergence is *not* uniform on D. Then prove that it is uniform on each compact subset of D, i.e., if $K \subset D$ is compact,

$$\lim_n \sup_K |f - f_n| = 0.$$

8. Let $\{T_n\}$ be a sequence of *linear* transformations from R^k into R^m. If $\{T_n\}$ converges pointwise to T, then

(i) T is a linear transformation;
(ii) the convergence is uniform on each compact subset of R^k.

9. Prove Dini's theorem for functions on closed-bounded sets, using the Bolzano-Weierstrass theorem rather than the Heine-Borel theorem.

10. Let F_n, F be functions of bounded variation on the interval $[a, b]$. Suppose that $V_a^b(F - F_n)$ converges to 0 and $F_n(a)$ converges to $F(a)$. Prove that $\{F_n\}$ converges uniformly to F. Why did we have to make the assumption about $F_n(a)$ and $F(a)$?

11. Find a sequence of polynomials f_n which converges uniformly to 0 on $[a, b]$ and for which

$$\lim_n V_a^b(f_n) = \infty.$$

12. Let $\{p_n\}$ be a sequence of polynomial functions on the real line. Suppose that

(i) $\{p_n\}$ converges pointwise to a function f;
(ii) the degrees $\deg(p_n)$ are bounded.

Prove that f is a polynomial function. *Hint*: Try to show that the coefficients of the various powers of x involved in p_n are converging.

13. (Abel's lemma) Let $\sum V_n$ be a series of vectors in R^m such that the partial sums of the series lie in a convex set K. Let $c_1 \geq c_2 \geq \cdots \geq 0$ be a decreasing sequence of non-negative real numbers. Show that the partial sums of the series $\sum c_n V_n$ lie in the set $c_1 K$.

14. (Abel's theorem) Let $\{F_n\}$ be a sequence of functions from a set D into R^m, and let $p_1 \geq p_2 \geq \cdots \geq 0$ be a decreasing sequence of non-negative functions on D. Suppose that the series $\sum F_n$ converges uniformly and that the sequence $\{p_n\}$ is bounded (i.e., p_1 is bounded). Use Abel's lemma (Exercise 13) to show that the series $\sum p_n F_n$ converges uniformly.

15. (Dirichlet-Hardy) Let $\{F_n\}$ be a sequence of functions from a set D into R^m, and let $p_1 \geq p_2 \geq \cdots \geq 0$ be a decreasing sequence of non-negative functions on D. If the partial sums of the series $\sum F_n$ are bounded (in sup norm) and if the sequence $\{p_n\}$ converges uniformly to 0, then the series $\sum p_n F_n$ converges uniformly.

5.2. *Calculus and Convergence*

Now let us take up the questions which we cited as a motivation for the definition of uniform convergence.

Theorem 2. *If $\{F_n\}$ is a sequence of continuous functions*

$$D \xrightarrow{F_n} R^m, \qquad D \subset R^k$$

and if F_n converges uniformly to F, then F is continuous.

Proof. The idea of the proof is this. We have

$$(5.6) \qquad F(X) - F(X_0) = F(X) - F_n(X) + F_n(X) - F_n(X_0)$$
$$+ F_n(X_0) - F(X_0)$$

for each n. For large n, $F_n(X)$ is near $F(X)$, $F_n(X_0)$ is near $F(X_0)$, and (since F_n is continuous) $F_n(X)$ will be near $F_n(X_0)$ if $|X - X_0|$ is small. We have to watch our "nears".

Let $X_0 \in D$. Let $\epsilon > 0$. Choose one particular positive integer N so that

$$\|F - F_N\|_\infty < \frac{\epsilon}{3}.$$

Then (5.6) gives us

$$(5.7) \qquad |F(X) - F(X_0)| < \frac{2\epsilon}{3} + |F_N(X) - F_N(X_0)|.$$

Since F_N is continuous at X_0 there exists $\delta > 0$ such that

$$|F_N(X) - F_N(X_0)| < \frac{\epsilon}{3}, \qquad |X - X_0| < \delta.$$

(This δ depends upon ϵ, X_0, and N.) Thus from (5.7)

$$|F(X) - F(X_0)| < \epsilon, \qquad |X - X_0| < \delta.$$

Corollary. *Let*

$$D \xrightarrow{F_n} R^m, \qquad D \subset R^k$$

and let X_0 *be a cluster point of* D. *Suppose*

(i) *each* F_n *has a limit at* X_0;

(ii) F_n *converges uniformly to* F.

Then F *has a limit at* X_0 *and*

$$\lim_{X \to X_0} F(X) = \lim_n \lim_{X \to X_0} F_n(X).$$

Proof. The reader should be able to see that this is a corollary of the previous theorem. It may prove more profitable to go through the proof of the theorem again, observing that the corollary has substantially the same proof: If L_n is the limit of F_n at X_0, then $|L_n - L_k| \leq ||F_n - F_k||_\infty$; thus L_n converges to a point L in R^m. Then

$$|L - F(X)| \leq |L - L_n| + |L_n - F_n(X)| + |F_n(X) - F(X)|$$

and argue as before.

The corollary states that uniform convergence allows us to interchange the order of two limits (in certain cases). In this section we repeat that fact again and again. The various results are illustrations of this one principle.

As a first application of the principle, suppose we have a **double sequence** of points in R^m:

$$X_{nk}, \qquad n, k = 1, 2, 3, \ldots.$$

For descriptive language, think of $\{X_{nk}\}$ as an infinite matrix

$$
\begin{matrix}
X_{11} & X_{12} & X_{13} & \cdots \\
X_{21} & X_{22} & X_{23} & \cdots \\
X_{31} & X_{32} & X_{33} & \cdots \\
\cdot & \cdot & \cdot & \\
\cdot & \cdot & \cdot & \\
\cdot & \cdot & \cdot & \cdot
\end{matrix}
$$

One problem in the interchange of order of limits is this. Suppose each row converges and each column converges:

$$A_n = \lim_k X_{nk} \quad \text{exists for each } n$$

$$B_k = \lim_n X_{nk} \quad \text{exists for each } k$$

When can we conclude that $\lim_n A_n$ and $\lim_k B_k$ exist and are equal? Evidently, that conclusion is false in general as the following triangular matrix shows:

$$
\begin{matrix}
1 & 0 & 0 & 0 & \cdots \\
1 & 1 & 0 & 0 & \cdots \\
1 & 1 & 1 & 0 & \cdots \\
1 & 1 & 1 & 1 & \cdots \\
\cdot & \cdot & \cdot & \cdot \\
\cdot & \cdot & \cdot & \cdot \\
\cdot & \cdot & \cdot & \cdot
\end{matrix}
$$

But, the conclusion is valid if either the row or column convergence is uniform (over the columns or rows).

Theorem 3 (*Moore-Osgood Double Limit Theorem*). *Let* $\{X_{nk}\}$ *be a double sequence of vectors in* $\mathbf{R}^m$. *Suppose that*

(i) *for each* n *the* nth *row converges*

$$A_n = \lim_k X_{nk}$$

(ii) *for each* k *the* kth *column converges*

$$B_k = \lim_n X_{nk}$$

(iii) *either the row or column convergence is uniform* (*in* n *or* k).

Then $\lim_n A_n$ *exists,* $\lim_k B_k$ *exists and these limits are equal. Furthermore, if* X *is their common value, the double sequence* $\{X_{nk}\}$ *converges to* X:

$$\lim_{n,k} |X - X_{nk}| = 0.$$

Proof. Suppose that row limits A_n exist and that the column limits B_k exist, uniformly in k. The latter condition means that, for each $\epsilon > 0$, there exists an N such that

$$|B_k - X_{nk}| < \epsilon, \qquad \text{for all } n \geq N \text{ and for all } k.$$

The functions

$$Z_+ \xrightarrow{\;F_n\;} R^m$$

$$F_n(k) = X_{nk}$$

converge uniformly to the function F, $F(k) = B_k$. And each F_n has a limit at ∞

$$\lim_{k \to \infty} F_n(k) = A_n.$$

Thus we are in the situation of the last corollary, except that the "cluster" point is the point at ∞. The reader may wish to repeat the proof in the present context, to be certain that the fact that the limits are at ∞ changes nothing. We conclude that

$$\lim_n A_n = \lim_k B_k.$$

If X is the common value of these two limits, then

$$|X - X_{nk}| \leq |X - B_k| + |B_k - X_{nk}|.$$

Given $\epsilon > 0$, choose N so that $|B_k - X_{nk}| < \epsilon/2$ for all k, provided $n \geq N$. Then choose K so that $|X - B_k| < \epsilon/2$ provided $k \geq K$. If $M = \max(K, N)$, then

$$|X - X_{nk}| < \epsilon, \qquad n, k \geq M.$$

There are some further remarks we should make about the double limit theorem. It can be restated in terms of the Cauchy criterion for convergence; for example, if $\{X_{nk}\}$ is a Cauchy sequence for each fixed n and as $n \to \infty$ is Cauchy uniformly in k, then

$$\lim_n \lim_k X_{nk} = \lim_k \lim_n X_{nk} = \lim_{n,k} X_{nk}.$$

A result may also be restated for double infinite series $\sum X_{nk}$. We have already dealt with the most important special case—absolute convergence. We proved that if

$$\sum_{n,k} |X_{nk}| < \infty$$

then

$$\sum_n \sum_k X_{nk} = \sum_k \sum_n X_{nk}$$

and hence that we can talk unambiguously about

$$\sum_{n,k} X_{nk}.$$

The reader should verify that this follows from the double limit theorem applied to the sequence

$$S_{nk} = \sum_{i=1}^{n} \sum_{j=1}^{k} X_{ij}$$

Absolute convergence ensures uniformity in the convergence of both the row and column sums of the matrix $\{S_{nk}\}$.

Let us continue with our applications of uniform convergence and the interchange of limit operations.

Theorem 4. *Let $\{F_n\}$ be a sequence of continuous functions on a closed box* B *in* R^k. *If* F_n *converges uniformly to* F, *then*

$$\lim_n \int_B F_n = \int_B F.$$

Proof. Observe that

$$\left| \int_B F - \int_B F_n \right| \leq \int_B |F - F_n|$$
$$\leq \|F - F_n\|_\infty m(B).$$

Thus the result seems trivial. It is not, however, because we need Theorem 2 to ensure that F is continuous and therefore integrable.

Theorem 5. *Let* I *be an interval on the real line, and let* $\{F_n\}$ *be a sequence of differentiable functions from* I *into* $\mathbb{R}^m$. *Suppose that*

(i) F_n *converges to* F;

(ii) *the derivatives* F'_n *converge uniformly.*

Then F *is differentiable and* $F' = \lim\limits_n F'_n$.

Proof. Fix $x \in I$. What we want to prove is that

(5.8) $$\lim_{t \to x} \lim_n \frac{F_n(t) - F_n(x)}{t - x} = \lim_n \lim_{t \to x} \frac{F_n(t) - F_n(x)}{t - x}.$$

This will follow, by the same reasoning we have been using if we can show that the limit

$$\lim_n \frac{F_n(t) - F_n(x)}{t - x} = \frac{F(t) - F(x)}{t - x}$$

exists uniformly in t. This is why we have assumed that the derivatives F'_n converge uniformly. We want

(5.9) $$\frac{F_n(t) - F_n(x)}{t - x} - \frac{F_k(t) - F_k(x)}{t - x}$$

to be small, uniformly for all $t \neq x$, provided k, n are sufficiently large. The argument is cleaner if we use coordinates. We need only verify the smallness of (5.9) for real-valued functions. In the real-valued case, the mean value theorem tells us that (5.9) is

$$(f_n - f_k)'(c)$$

where c is between t and x. Thus

$$\left| \frac{f_n(t) - f_n(x)}{t - x} - \frac{f_k(t) - f_k(x)}{t - x} \right| \leq \| f'_n - f'_k \|_\infty.$$

It should now be clear that (5.8) holds.

Of course, the hypotheses of Theorem 5 imply that F_n converges uniformly to F. In fact, if one knows that $\{F'_n\}$ converges uniformly and that the values $\{F_n(x)\}$ at any one point converge, then it follows easily that $\{F_n\}$ converges uniformly.

Theorem 5 tells us that we may differentiate under integral signs in some cases:

Corollary. *Let* I *be a closed interval and let* B *be a closed box in* $\mathbb{R}^k$. *Let* F *be a continuous function on* $I \times B$ *such that the partial derivative*

$$(D_1 F)(x, X) = \lim_{t \to x} \frac{F(t, X) - F(x, X)}{t - x}$$

exists and is continuous on $I \times B$. *Then*

$$G(t) = \int_B F(t, X) \, dX$$

is a differentiable function on I *and*

$$G'(t) = \int_B (D_1 F)(t, X) \, dX.$$

Proof. The result follows from Theorem 5 and basic properties of integration. The point is that the difference quotient converges uniformly to $D_1 F$ as t tends to 0. Some people prefer to write the conclusion as

$$\frac{dG}{dt} = \int_B \frac{\partial F}{\partial t} \, dX$$

which is fine, if one knows what it means.

Exercises

1. True or false? If $\{f_n\}$ is a sequence of uniformly continuous functions and if f_n converges uniformly to f, then f is uniformly continuous.

2. True or false? If F_n converges uniformly to F and G_n converges uniformly to G, then $\langle F_n, G_n \rangle$ converges uniformly to $\langle F, G \rangle$.

3. Let f be a non-negative continuous function on the interval $[0, 1]$. Suppose that the sequence $f^{1/n}$ converges uniformly. How many zeros does f have?

4. If f and g are continuous functions on the real line, let

$$(f * g)(x) = \int_0^1 f(x - t) g(t) \, dt.$$

True or false? If f is of class C^∞, then $f * g$ is of class C^∞.

5. Let

$$f(x, y) = \frac{x - y}{1 - xy}, \qquad x > 1, y > 1.$$

True or false?

$$\lim_{x \to 1} f(x, y) = 1$$

uniformly in y. True or false?

$$\lim_{x \to \infty} f(x, y) = -\frac{1}{y}$$

uniformly in y.

6. Prove that, if $\{a_{kn}\}$ is *any* double sequence of *non-negative* numbers and if ∞ is allowed, then

$$\sum_n \left(\sum_k a_{kn} \right) = \sum_k \left(\sum_n a_{kn} \right).$$

7. Let $\{X_{nk}\}$ be the double sequence of complex numbers

$$X_{nk} = \exp\left(-n + \frac{1}{k}i\right).$$

(a) Is the Moore-Osgood double limit theorem applicable to $\{X_{nk}\}$?
(b) Is the double *series* $\sum X_{nk}$ absolutely convergent?

8. If $\{g_n\}$ is a sequence of Riemann-integrable functions, if

$$f_n(x) = \int_a^x g_n(t)\, dt$$

and $\{g_n\}$ converges uniformly on $[a, b]$, then $\{f_n\}$ converges uniformly on $[a, b]$.

9. True or false? If f is continuous on D and $\| f \|_\infty < 1$, then the series

$$\sum_{n=0}^\infty f^n \qquad (f^0 = 1)$$

converges uniformly to $1/(1 - f)$.

10. Let F be a function of class C^1 on the real line. Prove that, as t tends to 0, the difference quotient

$$\frac{F(x + t) - F(x)}{t}$$

converges to $F'(x)$, uniformly on compact subsets of R.

*11. Suppose that, for each $t \in [0, 1]$, we have a continuous function f_t on D. Assume that the map from t to f_t is continuous, in the sense that, if t_n converges to t, then f_{t_n} converges uniformly to f_t. Define and discuss

$$\int_0^1 f_t\, dt.$$

5.3. Real Power Series

Beyond polynomials, the nicest real-valued functions on the real line are those which can be expanded in a power series:

$$f(x) = \sum_{n=0}^\infty a_n x^n.$$

We have been working with some such functions:

$$e^x = \sum_{n=0}^\infty \frac{1}{n!} x^n, \qquad\qquad x \in R$$

$$\sin x = \sum_{n=0}^\infty \frac{(-1)^n}{(2n + 1)!} x^{2n+1}, \qquad x \in R$$

$$\cos x = \sum_{n=0}^\infty \frac{(-1)^n}{(2n)!} x^{2n}, \qquad x \in R$$

$$\frac{1}{1 - x} = \sum_{n=0}^\infty x^n, \qquad\qquad |x| < 1.$$

Definition. Let f *be a complex-valued function on a subset of the real line. We call* f *a* **real-analytic function** *if*

(i) *the domain of definition of* f *is an open set* U;
(ii) *if* $x_0 \in U$, *there is some power series about* x_0

$$\sum_{n=0}^{\infty} a_n(x - x_0)^n$$

which converges to f *in a neighborhood of* x_0.

In other words, a real-analytic function is one which is locally the sum of a convergent power series. If we want to know about the local behavior of such functions, we need only study the behavior of a single convergent power series; and, we may as well center that series at the point $x_0 = 0$.

So, suppose we have a sequence of numbers $a_0, a_1, a_2, \ldots$. They may be complex. Consider the formal power series

(5.10)　　　　　　　　　$\sum_{n=0}^{\infty} a_n x^n.$

Does it converge for any x? Obviously it does for $x = 0$, but that's not inspiring. Suppose it does converge for some $t \neq 0$. Then it is easy to see that it converges in the interval $|x| < |t|$, for this reason. Since the series converges at $x = t$, we have

$$\lim_n a_n t^n = 0$$

so that there exists an N such that

$$|a_n t^n| \leq 1, \quad n > N.$$

Then, if $|x| < |t|$, we have

$$\sum_n a_n x^n = \sum_n a_n \left(\frac{x}{t}\right)^n t^n$$

(5.11)　　　$\sum_n |a_n x^n| = \sum_{n=1}^{N} |a_n x^n| + \sum_{N+1}^{\infty} |a_n t^n| \left|\frac{x}{t}\right|^n$

$$\leq \sum_{1}^{N} |a_n x^n| + \sum_{N+1}^{\infty} \left|\frac{x}{t}\right|^n.$$

Since $|x/t| < 1$, the series converges absolutely at each point of the open interval $|x| < |t|$; and, the convergence is uniform on each compact subset of that interval, that is, on each set $\{x; |x| \leq c|t|\}$, where $0 \leq c < 1$. The uniformity is clear from (5.11).

Now we see that attached to each formal power series (5.10) is a **radius of convergence**

$$r = \sup \{|t|; \sum a_n t^n \text{ converges}\}$$

and that radius of convergence possesses and is determined by these properties:

　　(i) $0 \leq r \leq \infty$;
　　(ii) the series converges uniformly and absolutely on each compact subset of the interval $|x| < r$;
　　(iii) the series does not converge for any x with $|x| > r$.

Theorem 6. *If* r *is the radius of convergence of the power series*

$$\sum_{n=0}^{\infty} a_n x^n,$$

then

(5.12) $$\frac{1}{r} = \limsup_n |a_n|^{1/n}.$$

Proof. If the series converges at $x = t$, we have

$$|a_n t^n| \leq 1, \qquad n > N.$$

Thus

$$|a_n|^{1/n} \leq \frac{1}{|t|}, \qquad n > N.$$

Therefore if

$$L = \limsup_n |a_n|^{1/n}$$

we have

$$L \leq \frac{1}{r}.$$

Now we want to prove the reverse inequality, $r \geq 1/L$. We show that if $0 \leq t < 1/L$ then $r \geq t$. That will do it. Given such a number t, choose c so that $t < c < 1/L$. By the definition of lim sup, there is an N such that

$$|a_n|^{1/n} \leq \frac{1}{c}, \qquad n > N.$$

Thus

$$|a_n t^n| \leq \left(\frac{t}{c}\right)^n, \qquad n > N$$

and since $t/c < 1$, that shows that the series converges at t. Hence $r \geq t$. This completes the proof of the theorem.

Suppose we have a power series with radius of convergence $r > 0$. Let us look at the function f which is its sum:

(5.13) $$f(x) = \sum_{n=0}^{\infty} a_n x^n, \qquad |x| < r.$$

Certainly f is continuous, because the polynomials (partial sums)

(5.14) $$p_n(x) = \sum_{k=0}^{n} a_k x^k$$

converge uniformly to f on each compact subset of $|x| < r$. In fact f is a function of class C^∞, for the following reason. Formally differentiate the series to obtain a new power series

(5.15) $$\sum_{n=0}^{\infty} n a_n x^{n-1}.$$

Since

$$\lim_n n^{1/n} = 1$$

the series (5.15) has the same radius of convergence as does the series for f. Therefore, the partial sums of the differentiated series converge uniformly on compact subsets of $|x| < r$. Those partial sums are

$$p'_n(x) = \sum_{k=0}^{n} k a_k x^{k-1}.$$

On each closed and bounded subinterval, p_n converges to f and p'_n converges uniformly. By Theorem 5, f is differentiable and

$$f'(x) = \lim_n p'_n(x) = \sum_{n=0}^{\infty} n a_n x^{n-1}.$$

By the same argument, f' is differentiable, etc.

Theorem 7. *Each real-analytic function* f *is of class* C^∞. *If*

$$f(x) = \sum_{n=0}^{\infty} a_n (x - x_0)^n$$

in a neighborhood of X_0, *then (in that same neighborhood) the* nth *derivative of* f *is the sum of the power series obtained by* n *times (term-by-term) differentiating the power series for* f; *in particular*

$$a_n = \frac{1}{n!} f^{(n)}(x_0).$$

This seems a good point at which to clear up an irritating technical question. As we defined real-analytic function, it is required that there exist a convergent power series expansion about each point in the (open) domain of definition. It is not immediately apparent that a function which is defined as the sum of a single convergent power series has that property. We need to verify that, if

$$f(x) = \sum_{n=0}^{\infty} a_n x^n, \qquad |x| < r$$

and if $t \in (-r, r)$, then near t there is a power series in $(x - t)$ which converges to $f(x)$. That series expansion will (of course) be

$$f(x) = \sum_{n=0}^{\infty} \frac{1}{n!} f^{(n)}(t)(x - t)^n.$$

Why is it valid? Well, repeated differentiation yields

$$f^{(n)}(t) = \sum_{k=n}^{\infty} k(k - 1) \cdots (k - n + 1) a_k t^{k-n}$$

$$\frac{1}{n!} f^{(n)}(t) = \sum_{k=n}^{\infty} \binom{k}{n} a_k t^{k-n}.$$

Formally, we have then

$$\sum_{n=0}^{\infty} \frac{1}{n!} f^{(n)}(t)(x-t)^n = \sum_{n=0}^{\infty} \sum_{k=n}^{\infty} \binom{k}{n} a_k t^{k-n}(x-t)^n$$

$$= \sum_{k=0}^{\infty} a_k \sum_{n=0}^{k} \binom{k}{n} t^{k-n}(x-t)^n$$

$$= \sum_{k=0}^{\infty} a_k [t + (x-t)]^k$$

$$= f(x).$$

In the presence of absolute convergence of the double series, this calculation will be correct. Replace a_k by $|a_k|$, replace t by $|t|$, and replace $(x-t)$ by $|x-t|$. Then make the corresponding calculation. A finite sum is obtained, provided

$$|t| + |x-t| < r.$$

Thus, the series expansion about t is valid on the interval $|x-t| < r - |t|$.

Conclusion: *Let* f *be a complex-valued function on an open set* U. *Then* f *is real-analytic if and only if the domain* U *can be covered by open sets, on each of which* f *is the sum of a convergent power series.*

Now, suppose we are given a function f on a neighborhood of 0 and we want to know whether or not f can be represented in some neighborhood of 0 as the sum of a convergent power series:

$$f(x) = \sum_{n=0}^{\infty} a_n x^n.$$

We know two things. First, we must start with a function of class C^∞. Second, the only possible choice of coefficients is

$$a_n = \frac{1}{n!} f^{(n)}(0).$$

So, given a C^∞-function f, we write down the formal series

(5.16) $$\sum_{n=0}^{\infty} \frac{1}{n!} f^{(n)}(0) x^n.$$

This is usually called the Taylor series for the function f. We ask two questions:

1. Does the series (5.16) converge for any $x \neq 0$ (i.e., does it converge on a neighborhood of 0)?

2. If the series does converge on a neighborhood of 0, does it converge to f?

One answer to question (1) is provided by Theorem 6: The series (5.16) converges for some non-zero x if and only if

$$\limsup_n \left| \frac{1}{n!} f^{(n)}(0) \right|^{1/n} < \infty.$$

Suppose that the radius of convergence of the series is positive. The series still need not converge to f. The standard example of this phenomenon is:

$$f(x) = \exp\left(-\frac{1}{x^2}\right), \qquad x \neq 0$$

$$f(0) = 0.$$

This f is of class C^∞ on the real line, and

$$f^{(n)}(0) = 0, \qquad n = 0, 1, 2, \ldots.$$

The associated series expansion about $x = 0$ certainly converges—but not to f. (Incidentally this f is real-analytic on the intervals $x > 0$ and $x < 0$.)

Taylor's theorem gives an expression for the error involved in approximating a real-valued function f by the partial sums of its Taylor series. The result states that the nth remainder

$$R_n(x) = f(x) - \sum_{k=0}^{n} \frac{1}{k!} f^{(k)}(0) x^k$$

is given by

$$R_n(x) = \frac{1}{(n+1)!} f^{(n+1)}(\theta x) x^{n+1}$$

where $0 \leq \theta \leq 1$. This can be obtained from the mean value theorem. We shall use a different form of the remainder, one which is not restricted to real-valued functions.

Theorem 8. *Let f be a complex-valued function of class C^{n+1} on a neighborhood of 0, and let*

$$R_n(x) = f(x) - \sum_{k=0}^{n} \frac{1}{k!} f^{(k)}(0) x^k.$$

Then

$$R_n(x) = \frac{x^{n+1}}{n!} \int_0^1 f^{(n+1)}(tx)(1-t)^n \, dt.$$

Proof. Let

$$g_k(x) = \frac{x^{k+1}}{k!} \int_0^1 f^{(k+1)}(tx)(1-t)^k \, dt, \qquad k = 0, \ldots, n.$$

Integrate by parts and you will obtain

$$g_k(x) = -\frac{x^k}{k!} f^{(k)}(0) + g_{k-1}(x)$$

$$g_0(x) = f(x) - f(0).$$

Hence, $g_n(x) = R_n(x)$.

EXAMPLE 7. We should become acquainted with power series and the function

$$f(x) = \log x, \quad x > 0.$$

This is a real-analytic function. If $x_0 > 0$, we would not expect the power series about x_0:

$$\sum_{0}^{\infty} \frac{1}{n!} f^{(n)}(x_0)(x - x_0)^n$$

to converge outside the interval $(0, 2x_0)$, because f is unbounded near $x = 0$. Since $f'(x) = 1/x$ (Example 1 of Chapter 4),

$$f^{(n)}(x) = \frac{(-1)^{n-1}(n-1)!}{x^n}, \quad n \geq 1.$$

The formal series about x_0 is therefore

$$\log x_0 + \sum_{n=1}^{\infty} \frac{(-1)^{n-1}}{n} \left(\frac{x - x_0}{x_0}\right)^n.$$

The radius of convergence of this series is plainly x_0. Why does the series converge to $\log x$ on $(0, 2x_0)$? Differentiate the series term-by-term and you get $1/x$.

Constant use is made of the series expansion of log about $x = 1$. It is more convenient to translate by 1 and write

$$(5.17) \qquad \log(1 + x) = \sum_{n=1}^{\infty} \frac{(-1)^{n-1}}{n} x^n, \quad |x| < 1.$$

This series can be obtained immediately from the series

$$\frac{1}{1 + x} = \sum_{n=0}^{\infty} (-x)^n, \quad |x| < 1$$

by term-by-term integration. That is a legitimate procedure since the series for $(1 + x)^{-1}$ converges uniformly on compact subsets of $(-1, 1)$.

EXAMPLE 8. Once upon a time, there may have been a mystery about the function

$$f(x) = \frac{1}{1 + x^2}, \quad x \in R.$$

This is a real-analytic function on the entire real line. If we expand in a power series about $x = 0$, it is clear what we must get because

$$(5.18) \qquad \frac{1}{1 + x^2} = \sum_{n=0}^{\infty} (-x^2)^n, \quad |x^2| < 1$$

$$\frac{1}{1 + x^2} = \sum_{n=0}^{\infty} (-1)^n x^{2n}, \quad |x| < 1.$$

Furthermore, the series does not converge either at $x = 1$ or $x = -1$ (because the terms don't go to 0). Why should that be the case? This function is not badly behaved at one of the end points as is the function log $(1 + x)$. Yet it appears that the power series thinks that the function is badly behaved at the boundary of the interval. This is because the series doesn't know that we are only using real values of x. We can just as well obtain (5.18) for complex x, and the series recognizes that $(1 + x^2)^{-1}$ is not at all well-behaved at the points $x = \pm i$ in the complex plane.

Exercises

1. Expand $(1 + x^2)^{-1}$ in a power series about $x = 1$.

2. What is the radius of convergence of the power series

$$\sum_{n=0}^{\infty} n^2 x^n ?$$

3. What is the radius of convergence of the power series

$$\sum_{n=0}^{\infty} n! \, x^n ?$$

4. log $(1 + x) \leq x$.

5. Let
$$f(x) = x^{1/3}, \qquad -1 < x < 1.$$
Is f of class C^{∞}? Is f real-analytic?

6. Give an example of a sequence of real-analytic functions which converges uniformly to a function which is not real-analytic.

7. If f is real-analytic on a neighborhood of 0,
$$f(x) = \sum_{n=0}^{\infty} a_n x^n$$
define
$$||| f ||| = \sum_{n=0}^{\infty} |a_n|.$$
If $||| f_n - f_k ||| \longrightarrow 0$, does $\{f_n\}$ converge uniformly on a neighborhood of 0?

8. If f is a function of class C^{∞} on the real line and if $f(0) = 0$, then
$$g(x) = \begin{cases} \dfrac{f(x)}{x}, & x \neq 0 \\ f'(0), & x = 0 \end{cases}$$
is of class C^{∞}.

9. Extend Exercise 8 as follows. If f is of class C^{∞}, then either
$$\lim_{x \to 0} \frac{f(x)}{x^n} = 0$$
for every n or we can write
$$f(x) = x^n g(x)$$
where g is of class C^{∞} and $g(0) \neq 0$. If f is real-analytic and $f \neq 0$, the first possibility cannot occur.

10. Let f be real-analytic on an interval about the origin. Suppose that there exists a sequence of points $x_n \neq 0$ such that

$$f(x_n) = 0, \qquad n = 1, 2, 3, \ldots$$
$$\lim_n x_n = 0.$$

Prove that $f = 0$.

11. Show that the product of two real-analytic functions is real-analytic. *Hint:* If

$$f(x) = \sum_{n=0}^{\infty} a_n x^n, \qquad g(x) = \sum_{n=0}^{\infty} b_n x^n$$

on the interval $|x| < r$, show that

$$\sum_{n=0}^{\infty} \left(\sum_{k=0}^{n} |a_k||b_{n-k}| \right) |x|^n < \infty$$

for each x in the interval $(-r, r)$.

12. Show that the quotient of two real-analytic functions is real-analytic off the set where the denominator is 0. *Hint:* If $g(0) = 1$, then

$$\frac{1}{g} = \frac{1}{1 - h}$$

where $h(x)$ is small near $x = 0$.

13. Show that the composition of real-analytic functions is real-analytic.

5.4. *Multiple Power Series*

This is a concise discussion of power series in n variables and of the functions which can be represented by such series. The situation does not require new results or techniques, provided one adopts a tractable notation.

A power series (about the origin) in n variables has the form

$$(5.19) \qquad \sum_{k_1, \ldots, k_n = 0}^{\infty} a_{k_1, \ldots, k_n} x_1^{k_1} \cdots x_n^{k_n}$$

where each $a_{k_1, \ldots, k_n}$ is a complex number. The multiple subscripts are cumbersome; hence, we rewrite (5.19) as

$$(5.20) \qquad \sum_k a_k X^k$$

where $k = (k_1, \ldots, k_n)$ is a vector index (an n-tuple of non-negative integers) and where

$$X^k = x_1^{k_1} \cdots x_n^{k_n}.$$

There is a potential ambiguity in the meaning of the series (5.20), because we did not stipulate the order in which the terms are to be summed. That problem needn't worry us too much, since we will be concerned with multiple power series only on sets in R^n where they converge absolutely:

(5.21) $$\sum_k |a_k X^k| < \infty.$$

The meaning of (5.21) is not ambiguous and has nothing to do with the order in which we sum. It means that there is a number $M < \infty$ such that

$$\sum_{k \in K} |a_k X^k| \leq M$$

for each *finite* set of (vector) indices K. In the presence of absolute convergence, the terms in (5.20) may be summed in any order—the result will be the same. The order which many people regard as most natural is this one.

Let

(5.22) $$P_N(X) = \sum_{|k|=N} a_k X^k$$

so that $P_N(X)$ is the sum of all the terms $a_k X^k$ for which the degree $|k| = k_1 + \cdots + k_n$ is equal to N. Then we rewrite the series:

(5.23) $$\sum_k a_k X^k = \sum_{N=0}^\infty P_N(X).$$

The latter series has a well-defined meaning. It has partial sums

(5.24) $$\begin{aligned} S_N(X) &= P_0(X) + \cdots + P_N(X) \\ &= \sum_{|k| \leq N} a_k X^k \end{aligned}$$

and it specifies one order in which to sum the terms of the power series, if it converges absolutely.

EXAMPLE 9. The power series

$$\sum_k X^k = \sum_{k_1, \ldots, k_n} x_1^{k_1} \cdots x_n^{k_n}$$

converges absolutely if $|x_i| < 1$ for each i. Clearly

(5.25) $$\sum_{|k| \leq N} |x_1|^{k_1} \cdots |x_n|^{k_n} \leq \prod_{j=1}^n (1 + |x_j| + \cdots + |x_j|^N)$$

so that

$$\sum_k |X^k| \leq (1 - |x_1|)^{-1} \cdots (1 - |x_n|)^{-1}.$$

A bit more thought along the lines of (5.25) shows that

$$\sum_k X^k = (1 - x_1)^{-1} \cdots (1 - x_n)^{-1}.$$

Suppose that the series (5.20) converges absolutely at the point $T = (t_1, \ldots, t_n)$. If $|x_i| \leq |t_i|$ for each index i, then

$$\begin{aligned} \sum_k |a_k X^k| &= \sum_k |a_k| |x_1|^{k_1} \cdots |x_n|^{k_n} \\ &\leq \sum_k |a_k| |t_1|^{k_1} \cdots |t_n|^{k_n} \\ &< \infty. \end{aligned}$$

Accordingly, the series converges uniformly and absolutely on the box $\{X; |x_i| \le |t_i|, i = 1, \ldots, n\}$.

The **radius of convergence** of the series is the largest r, $0 \le r \le \infty$, such that the series converges at each point in the symmetric box $\{X; |x_i| < r, i = 1, \ldots, n\}$. One might expect a vector radius of convergence $r = (r_1, \ldots, r_n)$ to be definable, extending as far out as possible in each coordinate direction. Unfortunately, such a concept does not make sense. For instance, a power series $\sum a_{jk}x^j y^k$ in two variables may converge in the rectangle $|x| < 1$, $|y| < \frac{1}{2}$ and in the rectangle $|x| < \frac{1}{2}$, $|y| < 1$ but not converge (everywhere) in the rectangle $|x| < 1$, $|y| < 1$. For such a series the radius of convergence is $\frac{1}{2}$. This should explain why we used symmetric boxes.

Definition. *Let* f *be a complex-valued function on a subset of* R^n. *We call* f *a* **real-analytic function** *if*

   (i) *the domain of* f *is an open set* U *in* R^n;
   (ii) *for each* $X_0 \in U$ *there is a power series*

$$\sum_k a_k(X - X_0)^k$$

which (*is absolutely convergent and*) *converges to* f *in a neighborhood of* X_0.

Theorem 9. *Each real-analytic function* f *is of class* C^∞. *If*

(5.26) $$f(X) = \sum_k a_k(X - X_0)^k$$

then the partial derivatives of f *are obtained from term-by-term differentiation of the power series; in particular,*

(5.27) $$a_k = \frac{1}{k!}(D^k f)(X_0).$$

Proof. If (5.26) holds in a neighborhood of X_0, then the radius of convergence of the power series is some number $r > 0$. If we fix values $x_i = t_i$ for $i \ne j$, then $U(t_1, \ldots, t_{j-1}, x_j, t_{j+1}, \ldots, t_n)$ is a real-analytic function of x_j. Theorem 7 makes it clear that all partial derivatives of f exist and are obtained from term-by-term differentiation. The formula (5.27) for the coefficients is then immediate. We remind the reader what the notation is:

$$k! = k_1! \cdots k_n!$$
$$D^k f = D_1^{k_1} \cdots D_n^{k_n} f$$
$$= \frac{\partial^{|k|} f}{\partial x_1^{k_1} \cdots \partial x_n^{k_n}}.$$

The attempt to expand a given function in a power series follows the pattern of the one variable case. There is a Taylor's formula for the remainder after the Nth degree approximation of a smooth function by its Taylor series. Again, we confine ourselves to the integral expression for the remainder.

Theorem 10. *Let* f *be a function of class* C^{N+1} *on* U, *an open convex neighborhood of the origin in* R^n. *If*

$$R_N(X) = f(X) - \sum_{|k| \leq N} \frac{1}{k!} (D^k f)(0) X^k$$

then

$$R_N(X) = (N + 1) \sum_{|k| = N+1} \frac{X^k}{k!} \int_0^1 (D^k f)(tX)(1 - t)^N \, dt.$$

Proof. Fix an X in the given neighborhood of 0. Let

$$g(t) = f(tX), \qquad 0 \leq t \leq 1.$$

Then g is a function of class C^{N+1} on the interval $[0, 1]$. We have

$$g'(t) = \sum_{i=1}^{n} x_i (D_i f)(tX)$$

$$g''(t) = \sum_{i,j} x_i x_j (D_i D_j f)(tX)$$

$$\vdots \qquad \vdots$$

$$g^{(M)}(t) = \sum_{i_1, \ldots, i_M} x_{i_1} \cdots x_{i_M} (D_{i_1} \cdots D_{i_M} f)(tX).$$

This is valid for $M = 0, 1, \ldots, N + 1$. Each monomial

$$x_{i_1} \cdots x_{i_M}$$

is of the form X^k with $|k| = M$. Given $k_1, \ldots, k_n$ with $k_1 + \cdots + k_n = M$ there are $M!/k!$ different M-tuples $(i_1, \ldots, i_M)$ in which 1 occurs k_1 times, 2 occurs k_2 times, etc. Thus,

$$g^{(M)}(t) = \sum_{|k|=M} \frac{M!}{k!} X^k (D^k f)(tX), \qquad 0 \leq M \leq N + 1.$$

We now apply Theorem 8 to the function g:

$$g(1) = \sum_{M=0}^{N} \frac{1}{M!} g^{(M)}(0) + \frac{1}{N!} \int_0^1 g^{(N+1)}(t)(1 - t)^N \, dt.$$

Simply observe that $g(1) = f(X)$ and

$$f(X) - \sum_{M=0}^{N} \frac{1}{M!} g^{(M)}(0) = R_N(X)$$

$$= \frac{1}{N!} \int_0^1 g^{(N+1)}(t)(1 - t)^N \, dt$$

$$= (N + 1) \sum_{|k|=N+1} \frac{X^k}{k!} \int_0^1 (D^k f)(tX)(1 - t)^N \, dt.$$

Corollary. *If* f *is a function of class* C^{N+1} *on a neighborhood of the origin in* R^n *and if* $f(0) = 0$, *then*

$$f(X) = x_1 f_1(X) + \cdots + x_n f_n(X)$$

where $f_1, \ldots, f_n$ *are functions of class* C^N *on a* (*possibly smaller*) *neighborhood of* 0. *If* f *is of class* C^∞, *the functions* $f_1, \ldots, f_n$ *may be chosen to be of class* C^∞.

Proof. Apply the theorem in the case $N = 0$:

$$f(X) - f(0) = \sum_{j=1}^{n} x_j \int_0^1 (D_j f)(tX) \, dt.$$

Since $D_1 f, \ldots, D_n f$ are of class C^N, each function

$$f_j(X) = \int_0^1 (D_j f)(tX) \, dt$$

is of class C^N. If f is of class C^∞, each D_j (hence, each f_j) is also.

Exercises

1. If f is real-analytic on U and g is real-analytic on V, then

$$h(X; Y) = f(X)g(Y)$$

is real-analytic on $U \times V$.

2. Let f be of class C^{N+1} on a neighborhood of the origin. Show that there exists a *unique* polynomial P of degree at most N such that

$$\lim_{X \to 0} \frac{f(X) - P(X)}{|X|^N} = 0.$$

3. If

$$h(X) = \sum_{i_1, \ldots, i_M} x_{i_1} \cdots x_{i_M}, \qquad X \in R^n, \ |X| < 1$$

show that (if $|k| = M$)

$$(D^k h)(0) = M!$$

4. Let f be a real-analytic function on a neighborhood of the origin. If $f(0, \ldots, 0) = 0$, then

$$f(X) = x_1 f_1(X) + \cdots + x_n f_n(X)$$

where $f_1, \ldots, f_n$ are real-analytic on a neighborhood of the origin.

5.5. Complex Power Series

A power series

$$\sum_0^\infty a_n x^n$$

is a formal thing. It is really nothing more than a sequence of coefficients a_n, together with an indicated intent to plug various objects x into the series and ask whether the resulting series converges. Throughout this book, we have been substituting complex matrices A in such series, in order to work with e^A, $(I + A)^{-1}$, etc. There is nothing to prevent us from using the obvious series to define $\sin A$, $\cos A$, $\log (I + A)$ for $|A| < 1$, etc. Matrices do not represent the ultimate generality in which such substitution is interesting or important. We could also allow the coefficients themselves to be matrices and all sorts of other mathematical objects. Obviously, we cannot discuss all possibilities. But, we are obligated to say something about (what is indisputably) the most important case.

We consider a power series

$$(5.28) \qquad \sum_{n=0}^{\infty} a_n z^n$$

where the a_n's are complex numbers, and we discuss the behavior of the resulting function of the complex variable z. If the power series (5.28) converges for a particular value $z = w$, then it converges uniformly and absolutely on compact subsets of the open disk $|z| < |w|$. The proof is precisely as in (5.11). Thus the series has a radius of convergence r, given by

$$\frac{1}{r} = \lim \sup |a_n|^{1/n}.$$

The series converges uniformly and absolutely on compact subsets of the disk $D_r = \{z; |z| < r\}$; and it converges at no point z for which $|z| > r$.

Let's consider a power series with a positive radius of convergence and study the function which is the sum of the series:

$$(5.29) \qquad f(z) = \sum_{n=0}^{\infty} a_n z^n, \qquad z \in D_r.$$

We expect f to be a very well-behaved function. In what sense does f have a derivative? It is easy to see that f has partial derivatives of all orders; but, let's delay the discussion of that and proceed by a formal analogy with what we did earlier for real power series. Let $z_0 \in D_r$, then

$$(5.30) \qquad \lim_{z \to z_0} \frac{f(z) - f(z_0)}{z - z_0} = \sum_{n=0}^{\infty} n a_n z_0^n.$$

One can show this the same way as we did in the real case. It can also be done directly, as follows. Let

$$p_n(z) = \sum_{k=0}^{n} a_k z^k.$$

Obviously,

$$\lim_{z \to z_0} \frac{p_n(z) - p_n(z_0)}{z - z_0} = \sum_{k=1}^{n} k a_k z_0^{k-1}.$$

Also,

$$(5.31) \qquad \begin{aligned} \frac{f(z) - f(z_0)}{z - z_0} - \frac{p_n(z) - p_n(z_0)}{z - z_0} &= (z - z_0)^{-1} \sum_{k=n+1}^{\infty} a_k(z^k - z_0^k) \\ &= \sum_{k=n+1}^{\infty} a_k \frac{z^k - z_0^k}{z - z_0}. \end{aligned}$$

Now
$$\left| \frac{z^k - z_0^k}{z - z_0} \right| = |z^{k-1} + z^{k-2}z_0 + \cdots + z_0^{k-1}|$$
$$\leq kc^{k-1}$$

where $c = \max [|z|, |z_0|]$. From (5.31) we then see that

$$\lim_n \frac{p_n(z) - p_n(z_0)}{z - z_0} = \frac{f(z) - f(z_0)}{z - z_0}$$

and the convergence is uniform on a neighborhood of z_0. Therefore we may interchange the two limits and obtain (5.30). Now we are ready for two definitions and a theorem.

Definition. *Let* f *be a complex-valued function on a subset of* **C**. *Then* f *is a* **complex-analytic** (*or* **holomorphic**) **function** *if*

(i) *the domain of* f *is an open set* **D** *in* **C**;

(ii) *for each point* $z_0 \in$ **D**, *there is a power series about* z_0,

$$\sum_{n=0}^{\infty} a_n(z - z_0)^n$$

which converges to f(z) *in a neighborhood of* z_0.

Definition. *Let* f *be a complex-valued function on an open set* **D** *in* **C**. *If* $z_0 \in$ **D**, *we say that* f *is* **complex-differentiable at** z_0 (*or,* **has a complex derivative at** z_0) *if the limit*

(5.32)
$$f'(z_0) = \lim_{z \to z_0} \frac{f(z) - f(z_0)}{z - z_0}$$

exists.

Theorem 11. *A complex-analytic function has complex derivatives of all orders. If*

$$f(z) = \sum_{n=0}^{\infty} a_n(z - z_0)^n$$

in a neighborhood of z_0, *then* (*in that same neighborhood*) $f^{(n)}$, *the* nth *complex derivative of* f, *is the sum of the power series obtained by* n *times formally differentiating the power series for* f; *in particular,*

$$a_n = \frac{1}{n!} f^{(n)}(z_0).$$

The same formal argument which we used for real-analytic functions shows that a function f is complex-analytic if and only if its (open) domain can be covered by open sets, on each of which f is the sum of a convergent power series. In other words, if

$$f(z) = \sum_{n=0}^{\infty} a_n z^n, \qquad |z| < r$$

and if $|z_0| < r$, then

$$f(z) = \sum_{n=0}^{\infty} \frac{1}{n!} f^{(n)}(z_0)(z - z_0)^n, \qquad |z - z_0| < r - |z_0|.$$

Thus, we know some complex-analytic functions:

(i) polynomials:
$$p(z) = a_0 + a_1 z + \cdots + a_n z^n$$

(ii) the exponential function:
$$e^z = \sum_{n=0}^{\infty} \frac{1}{n!} z^n$$

(iii) the sine function:
$$\sin z = \sum_{n=0}^{\infty} \frac{(-1)^n}{(2n+1)!} z^{2n+1}$$

(iv) the cosine function:
$$\cos z = \sum_{n=0}^{\infty} \frac{(-1)^n}{(2n)!} z^{2n}.$$

Their complex derivatives may be computed using term-by-term differentiation of the power series. Hence if

$$f(z) = e^z$$
$$g(z) = \sin z$$
$$h(z) = \cos z$$

we have $f' = f$, $g' = h$, $h' = -g$.

Soon, we shall list many ways in which known analytic functions can be combined to yield new ones.

Theorem 12 (*Identity Theorem*). *Let* f *be a complex-analytic function on a connected open set* D *in the plane. Suppose that there is a sequence of distinct points* $z_n \in$ D *such that*

(i) f(z_n) = 0;
(ii) $\{z_n\}$ *converges to a point* z_0 *in* D.

Then f = 0.

Proof. For convenience, assume that $z_0 = 0$. We have

$$f(z) = \sum_{n=0}^{\infty} a_n z^n, \qquad |z| < r.$$

Since $f(z_n) = 0$ and $\lim_n z_n = 0$, the constant $a_0 = f(0)$ is 0. Let

$$g(z) = \sum_{n=1}^{\infty} a_n z^{n-1}, \qquad |z| < r$$

so that $f(z) = zg(z)$. Now $0 = z_n g(z_n)$, and since $z_1, z_2, z_3, \ldots$ are distinct points, not more than one of them is 0. *Conclusion:* $g(0) = 0$, i.e., $a_1 = 0$. Now divide $g(z)$ by z and repeat the argument. By induction, we obtain $a_n = 0$ for all n.

What we just showed was this: If $z_0 \in D$ and if there is a sequence of distinct points z_n with $f(z_n) = 0$ and $z_0 = \lim_n z_n$, then $f(z) = 0$ on a neighborhood of z_0. Let N be the set of all points $z_0 \in D$ for which such a sequence $\{z_n\}$ exists. We just showed that N is open relative to D. Obviously, N is closed relative to D. By hypothesis, N is non-empty. Since D is connected, $N = D$ (which tells us that $f = 0$).

To the reader who understood what we did with real power series, the things we have had to say thus far about complex power series are cause for little excitement. Although we did not state it earlier, the identity theorem is obviously valid for real-analytic functions on an interval. We indicated that the study of complex-analytic functions is important. It is also a fascinating subject which has quite a different flavor from the sort of analysis which we have been discussing. At this point, we begin to see why.

If we were to proceed as in Section 5.3, it would seem natural for us now to prove a complex version of Taylor's theorem. There is no need for such a theorem. One of the remarkable facts in mathematics is this: If f is a complex-differentiable function, then f is complex-analytic. That is, the existence of just the first complex derivative implies that (f has derivatives of all orders and) f is locally the sum of a convergent power series. This result, which we shall prove later when the derivative is known to be continuous, contrasts sharply with the situation on the real line, and it suggests that perhaps we should take a close look at the connection between real and complex derivatives.

Suppose that f is complex-analytic near the origin in the plane:

$$f(z) = \sum_n a_n z^n, \qquad z \in D_r.$$

Identify C with R^2 in the usual way and regard f as a complex-valued function on R^2, i.e., regard $f(x + iy)$ as a function of the two real variables x, y. Certainly f is a smooth function on R^2. In fact it is a real-analytic function on R^2, because we can expand $f(x + iy)$ in a double power series in x and y. Use the binomial theorem on each term z^n:

$$f(x + iy) = \sum_n a_n (x + iy)^n$$

(5.33)
$$= \sum_n a_n \sum_{k=0}^{n} \binom{n}{k} x^k (iy)^{n-k}$$

$$= \sum_{k,p} c_{kp} x^k y^p$$

where

$$c_{kp} = a_{k+p} \binom{k+p}{k} i^p.$$

The last equality in (5.33) is legitimate if the series

$$\sum_{k,p} c_{kp} x^k y^p$$

is absolutely convergent, for then, we may regroup its terms in the form $\sum a_n(x + iy)^n$. Now

$$\sum_{k,p} |c_{kp}| |x|^k |y|^p = \sum_n |a_n| (|x| + |y|)^n.$$

(Regrouping is always legitimate for a series with non-negative terms.) Thus the double series is absolutely convergent if $|x| + |y| < r$, and therefore f is a real-analytic function of x, y on a neighborhood of the origin.

EXAMPLE 10. Not every real-analytic function of x, y is complex-analytic. There are so many ways to see this fact that it is difficult to choose one. Perhaps the most obvious one is to let

$$g(x + iy) = x.$$

It should be clear that $x = \sum a_n(x + iy)^n$ is a bit difficult to arrange through any choice of the constants a_n. We could also show that g fails to have a complex derivative:

$$\lim_{z \to 0} \frac{g(z) - g(0)}{z} = \lim_{z \to 0} \frac{x}{z}.$$

If the origin is approached along the imaginary axis, xz^{-1} tends to, in fact is, 0; but, as the origin is approached along the real axis, xz^{-1} tends to 1.

If f is complex-analytic, the fact that $f(x + iy)$ is (locally) the sum of a power series in x and y tells us in particular that

$$R^2 \xrightarrow{\ f\ } C$$

is of class C^∞. There is another way to show that f has partial derivatives of all orders, and it provides a description of one extra condition which a smooth function on R^2 must satisfy in order to be complex-analytic. In presenting this, we warn the reader not to confuse the complex derivative f' with the gradient of f.

If the complex derivative

$$f'(z_0) = \lim_{z \to z_0} \frac{f(z) - f(z_0)}{z - z_0}$$

exists, then the limit exists as z approaches z_0 along any line through z_0. The derivative of f at z_0 in the direction of the vector (complex number) w is

$$(D_w f)(z_0) = \lim_{t \to 0} \frac{f(z_0 + tw) - f(z_0)}{t}$$

$$= w \lim_{t \to 0} \frac{f(z_0 + tw) - f(z_0)}{tw}$$

$$= w f'(z_0).$$

In particular,

(5.34)

$$\frac{\partial f}{\partial x} = f'$$

$$\frac{\partial f}{\partial y} = if'.$$

If f is complex-analytic, then f' exists and is itself complex-analytic. Repeated application of (5.34) shows that f is of class C^∞. We have also picked up an important piece of information about complex-differentiability.

Theorem 13. *Let* f *be a complex-valued function of class* C^1 *on an open set* U *in the plane. The following are equivalent.*

(i) f *is complex-differentiable on* U.

(ii) $\dfrac{\partial f}{\partial x} + i\dfrac{\partial f}{\partial y} = 0$ *(throughout* U*).*

(iii) *If* u $=$ Re f, v $=$ Im f, *then*

$$\frac{\partial u}{\partial x} = \frac{\partial v}{\partial y} \quad and \quad \frac{\partial u}{\partial y} = -\frac{\partial v}{\partial x} \qquad (throughout\ U).$$

Proof. In (5.34) we just showed that (i) implies (ii). Replace f by $u + iv$ in (ii) and remember that a complex number is 0 if and only if its real and imaginary parts are 0. You will see immediately that (ii) and (iii) are equivalent. Suppose that (ii) holds. We shall show that f is complex-differentiable at $z_0 \in U$. For convenience, assume that $z_0 = 0$. Since f is of class C^1, we have

$$f(x + iy) - f(0) = x \int_0^1 (D_1 f)(tz)\, dt + y \int_0^1 (D_2 f)(tz)\, dt$$

for all $z = x + iy$ near the origin. This is a special case of Theorem 5.9; or, it may be viewed as an application of Theorem 5.7. But, all it amounts to is the fundamental theorem of calculus applied to the function $g(t) = f(tz)$ on the interval $0 \le t \le 1$. From the differential equation (ii), which relates $D_1 f$ and $D_2 f$, we have

$$f(z) - f(0) = z \int_0^1 (D_2 f)(tz)\, dt.$$

Thus

$$\lim_{z \to 0} \frac{f(z) - f(0)}{z} = \lim_{z \to 0} \int_0^1 (D_1 f)(tz)\, dt$$

and since $D_1 f$ is continuous, the last limit is $(D_1 f)(0)$. This shows that f is complex-differentiable and (in the process) that (5.34) holds.

The equations in condition (iii) of the theorem are known as the **Cauchy-Riemann equations** and (ii) is known as the **Cauchy-Riemann con-**

dition. This condition

$$\frac{\partial f}{\partial x} + i\frac{\partial f}{\partial y} = 0$$

is a simple way of phrasing complex-differentiability for a function of class C^1. If f is of class C^2 (which is automatic but we don't know that yet), we can differentiate the Cauchy-Riemann equations:

$$\frac{\partial^2 f}{\partial x^2} + i\frac{\partial^2 f}{\partial x\,\partial y} = 0$$

$$\frac{\partial^2 f}{\partial y\,\partial x} + i\frac{\partial^2 f}{\partial y^2} = 0$$

and, since the mixed partial derivatives are equal, we obtain

(5.35) $$\frac{\partial^2 f}{\partial x^2} + \frac{\partial^2 f}{\partial y^2} = 0.$$

The differential equation (5.35) is called **Laplace's equation** and its solutions are called **harmonic functions**. If f is complex-differentiable and of class C^2, then the real and imaginary parts of f (as well as f) are harmonic functions.

Exercises

1. If f is complex-analytic on a neighborhood of w, then

$$g(z) = \frac{f(z) - f(w)}{z - w}$$

is complex-analytic on that same neighborhood.

2. If f is complex-analytic and non-constant on a connected open set D and if $w \in D$, then for some n

$$f(z) - f(w) = (z - w)^n g(z)$$

where g is complex-analytic on D and $g(w) \neq 0$.

3. Let g be a piecewise-continuous (or an integrable) function on $[0, 1]$. Then

$$f(z) = \int_0^1 \frac{g(x)}{x - z}\,dx$$

is complex-analytic off the interval $[0, 1]$.

4. Suppose f is complex-analytic on a disk D. If f is real-valued, then f is constant.

5. Let f be a complex-analytic function on a connected open set D. If $f' = 0$, show that f is constant.

6. The function $f(z) = e^z$ is (of course) complex-analytic on the entire plane. Verify by direct calculation that the real part of f is harmonic.

7. Introduce polar coordinates in the plane by $x = r \cos \theta$, $y = r \sin \theta$. Verify that the Cauchy-Riemann condition can be expressed this way:

$$r\frac{\partial f}{\partial r} + i\frac{\partial f}{\partial \theta} = 0.$$

8. Let f be a complex-valued function of class C^2 on an open set in the plane. Prove that f is complex-differentiable if and only if $f(z)$ is harmonic and $zf(z)$ is harmonic.

9. Let u be a real-valued harmonic function. Show that, if u^2 is also harmonic, then u is constant.

10. Use a power series to define

$$\log(1 + z), \qquad |z| < 1.$$

Of course, you want $\exp(\log(1 + z)) = 1 + z$. Refer to Example 7 and be sure to verify that $\exp(\log(1 + z)) = 1 + z$.

11. Show that it is not possible to extract the square root of z analytically on a neighborhood of the origin. In other words, prove that no complex-analytic function f satisfies $f(z)^2 = z$ on a neighborhood of the origin.

*12. Show that it is not possible to extract the square root of z continuously on a neighborhood of the origin.

13. Show that it is possible to extract the square root of z analytically in a neighborhood of $z = 1$. (*Hint:* See Exercise 10.)

14. Use a power series to define (for complex matrices A)

$$\log(I + A), \qquad |A| < 1.$$

Show that $\exp(\log(I + A)) = I + A$.

15. If N is a $k \times k$ matrix which is nilpotent ($N^k = 0$), then $I + N = e^B$ for some complex matrix B.

16. If you know what the Jordan canonical form for a complex matrix is, use it to prove that the exponential map

$$A \longrightarrow e^A$$

maps the $k \times k$ complex matrices *onto* the invertible $k \times k$ matrices.

5.6. Fundamental Results
on Complex-Analytic Functions

Now that we understand a bit about complex differentiation, we'll begin to sample some of the elegant methods and results in the study of complex-analytic functions. A number of special properties of complex-analytic functions can be derived from an elementary one, known as the mean value property.

Lemma. *If* $\{z; |z - z_0| \leq \rho\}$ *is a closed disk contained in the domain of the complex-analytic function* f, *then*

$$(5.36) \qquad f(z_0) = \frac{1}{2\pi} \int_{-\pi}^{\pi} f(z_0 + \rho e^{i\theta}) \, d\theta.$$

Proof. The hypothesis is that f is complex-analytic both inside and on C_ρ, the circle of radius ρ about the point z_0. We are asked to show that the average (or mean) value of f over C_ρ is equal to the value of f at the center of C_ρ. It's not difficult to see how someone guessed that such is the case. For simplicity, take $z_0 = 0$. Then

$$f(z) = \sum_{n=0}^{\infty} a_n z^n, \qquad |z| < \epsilon$$

or

$$f(re^{i\theta}) = \sum_{n=0}^{\infty} a_n r^n e^{in\theta}, \qquad 0 \leq r < \epsilon.$$

Now

$$\frac{1}{2\pi} \int_{-\pi}^{\pi} e^{in\theta} \, d\theta = \begin{cases} 0, & n \neq 0 \\ 1, & n = 0. \end{cases}$$

If $r < \epsilon$, then (by the uniform convergence of the series)

$$\frac{1}{2\pi} \int_{-\pi}^{\pi} f(re^{i\theta}) \, d\theta = \sum_{n=0}^{\infty} a_n r^n \frac{1}{2\pi} \int_{-\pi}^{\pi} e^{in\theta} \, d\theta$$

$$= a_0$$

$$= f(0).$$

It appears that we are finished; however, we are not quite. The hypothesis is that f is analytic for $|z| \leq \rho$. We have shown that

$$f(0) = \frac{1}{2\pi} \int_{-\pi}^{\pi} f(re^{i\theta}) \, d\theta, \qquad r < \epsilon$$

where ϵ is the radius of convergence of the series expansion of f about the origin. It is a theorem that, since f is complex-analytic on $|z| \leq \rho$, the series expansion for f about 0 must converge on $|z| \leq \rho$; but, we don't know this yet. We have to face the possibility that $\epsilon < \rho$.

Define

$$(5.37) \qquad I(r) = \frac{1}{2\pi} \int_{-\pi}^{\pi} f(re^{i\theta}) \, d\theta, \qquad 0 < r \leq \rho.$$

We'll verify that $I(r)$ is constant, by showing that

$$\frac{\partial I}{\partial r} = 0.$$

Now f is complex-differentiable and therefore has radial and angular partial derivatives:

$$\frac{\partial f}{\partial r}(z_0) = \lim_{t \to r_0} \frac{f(te^{i\theta_0}) - f(r_0 e^{i\theta_0})}{t - r_0}$$

$$= e^{i\theta_0} f'(z_0)$$

$$\frac{\partial f}{\partial \theta}(z_0) = iz_0 f'(z_0).$$

Thus the complex-differentiability implies that

$$r\frac{\partial f}{\partial r} + i\frac{\partial f}{\partial \theta} = 0.$$

So

$$\frac{\partial I}{\partial r} = \frac{1}{2\pi} \int_{-\pi}^{\pi} \frac{\partial f}{\partial r}(re^{i\theta})\, d\theta$$

$$= -\frac{i}{r}\frac{1}{2\pi} \int_{-\pi}^{\pi} \frac{\partial f}{\partial \theta}(re^{i\theta})\, d\theta$$

$$= \frac{-i}{2\pi r}[f(re^{i\pi}) - f(re^{-i\pi})]$$

$$= 0.$$

Now we know that $I(r)$ is constant. Furthermore, $I(r) = f(0)$ for small r. This proves the mean value property.

Many people prefer to write the integral involved in the mean value property as a Riemann-Stieltjes integral with respect to the function $z(\theta) = \rho e^{i\theta}$ on the interval $[-\pi, \pi]$. We have

$$dz(\theta) = \frac{dz}{d\theta}\, d\theta$$

$$= i\rho e^{i\theta}\, d\theta$$

$$= iz\, d\theta$$

so that (as in Example 9) we can also write the mean value property this way:

$$f(0) = \frac{1}{2\pi i} \int_{-\pi}^{\pi} \frac{f(z)}{z}\, dz(\theta).$$

This is a very useful way to write the mean value property and it is in many respects a more "natural" way when dealing with analytic functions. As it stands, the notation has a defect in that it does not indicate which circle $|z| = \rho$ is involved. In other words, in order to regard z as a function of θ we had to choose some ρ and fix $|z| = \rho$; the notation for the integral should reflect this choice. Accordingly, if g is any continuous complex-valued function on the circle $|z| = \rho$, we define to be the integral

$$\int_{|z|=\rho} g(z)\,dz$$

$$\int_{-\pi}^{\pi} g(z)\,dz(\theta), \qquad z = \rho e^{i\theta}.$$

The reader will probably recognize

$$\int_{|z|=\rho} g(z)\,dz$$

as the line integral over the (counterclockwise oriented) circle $|z| = \rho$ of the differential 1-form

$$g(z)\,dz = g(x, y)(dx + i\,dy).$$

At the moment it is merely a piece of shorthand.

Theorem 14 (*Cauchy*). *If f is complex-analytic on (an open set which contains) the closed disk of radius ρ about the origin, then*

$$(5.38) \qquad f(w) = \frac{1}{2\pi i} \int_{|z|=\rho} \frac{f(z)}{z - w}\,dz, \qquad |w| < \rho.$$

Proof. Fix w. Let

$$g(z) = z \cdot \frac{f(z) - f(w)}{z - w}.$$

Of course, we define $g(w) = wf'(w)$. Then g is complex-analytic where f is. (Exercise 1 of the last section.) Apply the mean value theorem to g:

$$0 = g(0)$$

$$= \frac{1}{2\pi i} \int_{|z|=\rho} \frac{g(z)}{z}\,dz$$

$$= \frac{1}{2\pi i} \int_{|z|=\rho} \frac{f(z)}{z - w}\,dz - f(w) \cdot \frac{1}{2\pi i} \int_{|z|=\rho} \frac{dz}{z - w}.$$

Simply calculate that

$$\frac{1}{2\pi i} \int_{|z|=\rho} \frac{1}{z - w}\,dz = 1, \qquad |w| < \rho$$

and Cauchy's formula follows. The power series argument in the proof of the next corollary should make it clear how to carry out this calculation.

Corollary. *Let f be a complex-analytic function on an open set D, and let z_0 be a point in D. The power series expansion for f converges to f in the largest disk which is centered at z_0 and contained in D.*

Proof. We may as well assume that $z_0 = 0$. Suppose that the closed disk $|z| = \rho$ is contained in D. We want to show that the power series for f about the origin converges to f on the disk $|z| \le \rho$. Look at Cauchy's formula for $f(w)$, given in (5.38). If $|z| \leqq \rho$ and $|w| < \rho$, then

$$\frac{1}{z - w} = \frac{1}{z} \cdot \frac{1}{1 - z^{-1}w}$$

$$= \frac{1}{z}\left(1 + \frac{w}{z} + \left(\frac{w}{z}\right)^2 + \cdots\right)$$

$$= \sum_{n=0}^{\infty} \frac{w^n}{z^{n+1}}$$

and so

$$f(w) = \frac{1}{2\pi i} \int_{|z|=\rho} \frac{f(z)}{z - w} \, dz$$

$$= \sum_{n=0}^{\infty} a_n w^n$$

where

(5.39) $$a_n = \frac{1}{2\pi i} \int_{|z|=\rho} \frac{f(z)}{z^{n+1}} \, dz.$$

The last corollary removes an irritating detail which bothered us before. It is not, however, a minor detail. To understand this, the reader should think very carefully about what the corollary says, starting from our definition of complex-analytic function.

Theorem 15. *Every (continuously) complex-differentiable function is complex-analytic.*

Proof. The sum, product, and quotient of complex-differentiable functions are complex-differentiable (where defined). That follows readily from the definition of complex-differentiable, by the same arguments one uses for differentiation on the real line.

In this theorem, we are given f, a function of class C^1 on an open set in the plane which is complex-differentiable, i.e., which satisfies the Cauchy-Riemann condition. We want to show that f is (locally) the sum of a convergent power series. We may assume that the domain of f contains the closed disk of radius ρ about the origin and content ourselves with showing that f is analytic on that disk. Let w be a point with $|w| < \rho$. Define

$$g(z) = z \cdot \frac{f(z) - f(w)}{z - w}, \qquad z \neq w$$

$$g(w) = wf'(w).$$

Then g is (obviously) continuously complex-differentiable, except at w; and g is continuous at w. We claim that

$$g(0) = \frac{1}{2\pi} \int_{-\pi}^{\pi} g(re^{i\theta}) \, d\theta, \qquad r \leq \rho.$$

To prove this, let $I(r)$ be the integral on the right. Then $I(r)$ is continuous on $0 \leq r \leq \rho$ and differentiable except possibly at $r = |w|$. Since g is complex-differentiable except (possibly) at w,

$$r\frac{\partial g}{\partial r} + i\frac{\partial g}{\partial \theta} = 0, \qquad r \neq |w|.$$

Thus,

$$\frac{dI}{dr} = 0 \qquad (r \neq |w|)$$

as in our proof of the mean value property for analytic functions. So $I(r)$ is constant. Obviously,

$$\lim_{r \to 0} I(r) = g(0).$$

Consequently, we have proved the mean value property for g:

$$g(0) = \frac{1}{2\pi} \int_{-\pi}^{\pi} g(\rho e^{i\theta})\, d\theta$$

$$= \frac{1}{2\pi i} \int_{|z|=\rho} \frac{g(z)}{z}\, dz.$$

This says that

$$0 = g(0)$$

$$= \frac{1}{2\pi i} \int_{|z|=\rho} \frac{f(z)}{z-w}\, dz - f(w) \cdot \frac{1}{2\pi i} \int_{|z|=\rho} \frac{dz}{z-w}$$

$$= \frac{1}{2\pi i} \int_{|z|=\rho} \frac{f(z)}{z-w}\, dz - f(w).$$

So Cauchy's formula is valid for f. That integral formula shows us immediately how to expand f in a power series about the origin.

We remarked earlier that Theorem 15 is valid without the assumption that f' is continuous, i.e., just the existence of the complex derivative (on an open set) implies analyticity. The sort of proof we have given can be refined to cover that case; however, the refinement tends to cloud the simplicity of the arguments, so we have omitted it.

Here is another basic corollary of the mean value property (or Cauchy's formula).

Theorem 16. *Let* D *be an open set in the plane and let* $\{f_n\}$ *be a sequence of complex-analytic functions on* D. *If* f_n *converges uniformly to* f, *then* f *is complex-analytic.*

Proof. Analyticity is a local property; hence, we need only prove the theorem when D is a disk centered at the origin. The Cauchy formula states that

$$f_n(w) = \frac{1}{2\pi i} \int_{|z|=r} \frac{f_n(z)}{z - w} dz$$

$$= \frac{1}{2\pi} \int_{-\pi}^{\pi} f_n(re^{i\theta}) \frac{re^{i\theta}}{re^{i\theta} - w} d\theta$$

if $|w| < r$ and r is less than the radius of D. Fix an r and a w. Since $f_n(re^{i\theta})$ converges to $f(re^{i\theta})$, uniformly in θ, and since $f_n(w)$ converges to $f(w)$, we have

$$f(w) = \frac{1}{2\pi i} \int_{|z|=r} \frac{f(z)}{z - w} dz.$$

Therefore, f is complex-analytic on the disk $|z| < r$:

$$f(w) = \sum_n a_n w^n$$

where

$$a_n = \frac{1}{2\pi i} \int_{|z|=r} \frac{f(z)}{z^{n+1}} dz.$$

(See the proof of the previous corollary.)

Corollary. *Let* f, g *be complex-analytic functions. Then (where defined) the sum* f + g, *the product* fg, *the quotient* f/g, *and the composition* f ∘ g *are complex-analytic functions.*

Proof. This result follows rather easily from Theorem 15 and the corresponding result about sums, products, etc. for complex-differentiable functions. It is informative to think of it this way. Theorem 16 tells us that a function is complex-analytic if and only if it is locally the uniform limit of a sequence of polynomials $p_n(z)$. Since sums, products, and compositions of polynomials are polynomials, it follows that $f + g$, fg, and $f \circ g$ are analytic if f and g are. The function $1/z$ is analytic for $z \neq 0$:

$$\frac{1}{z} = \frac{1}{z_0 + (z - z_0)}$$

$$= \frac{z_0^{-1}}{1 + z_0^{-1}(z - z_0)}$$

$$= \sum_{n=1}^{\infty} (-1)^n z_0^{-(n+1)}(z - z_0)^n, \qquad |z - z_0| < |z_0|$$

Thus (by composing with $1/z$) f/g is analytic off the zero set of g.

It follows that the class of real-analytic functions is closed under sum, product, quotient, and composition (where defined). That is because the power series for such functions about x_0 can be used to extend them to complex-analytic functions in a neighborhood of x_0 in the complex plane. But note that the theorem on uniform convergence breaks down completely. It is a famous theorem (which we shall soon prove) that any continuous function on a closed interval can be uniformly approximated by real-analytic functions. Thus, for power series, there is an enormous difference between uniform convergence on an interval and uniform convergence on a disk.

Let us summarize briefly what we know about complex-analytic functions. We have proved that, for a complex-valued function f on an open set D in the plane, the following conditions are equivalent

1. f is complex-analytic; i.e., at each point $z_0 \in D$ there is a power series in $(z - z_0)$ which converges to $f(z)$ in a neighborhood of z_0.

2. f is complex-differentiable.

3. If $f = u + iv$, where u, v are real-valued, the map $(x, y) \rightarrow (u, v)$ from D into R^2 is of class C^1 and satisfies the Cauchy-Riemann equations:

$$\frac{\partial u}{\partial x} = \frac{\partial v}{\partial y}, \qquad \frac{\partial u}{\partial y} = -\frac{\partial v}{\partial x}.$$

4. If the closed disk $|z - z_0| \leq \rho$ is contained in D, then

$$f(w) = \frac{1}{2\pi i} \int_{|z-z_0|=\rho} \frac{f(z)}{z - w} \, dz, \qquad |w - z_0| < \rho$$

That is,

$$f(w) = \sum_{n=0}^{\infty} a_n (w - z_0)^n, \qquad |w - z_0| < \rho$$

where

$$a_n = \frac{1}{2\pi i} \int_{|z-z_0|=\rho} \frac{f(z)}{(z - z_0)^{n+1}} \, dz.$$

5. If z_0 is any point of D, there is a sequence of polynomials $p_n(z)$ which converges to $f(z)$, uniformly on some neighborhood of z_0.

We only verified that condition 2 implies analyticity under the assumption that f' is continuous.

We also showed that each complex-analytic function has the mean value property on circles. It is not in the list above because it does not characterize complex-analytic functions. (In fact, it characterizes harmonic functions.) There is, however, a stronger form of the mean value property which does characterize analytic functions. If the disk $|z - z_0| \leq \rho$ is contained in the domain of the analytic function f, then $g(z) = (z - z_0)f(z)$ is complex-analytic on $|z - z_0| \leq \rho$ and satisfies $g(z_0) = 0$. By the Cauchy formula

$$0 = g(z_0)$$

$$= \frac{1}{2\pi i} \int_{|z-z_0|=\rho} \frac{g(z)}{z - z_0} \, dz$$

$$= \frac{1}{2\pi i} \int_{|z-z_0|=\rho} f(z) \, dz.$$

In other words the line integral of the differential 1-form $f(z)\,dz$ over any circle $\gamma = \{z; |z - z_0| = \rho\}$ contained in D is 0:

$$\int_\gamma f(z)\,dz = 0.$$

We can add to the list of characterizations 1–5:

6. If γ is the circular boundary of any closed disk in the domain D, then

$$\int_\gamma f(z)\,dz = 0.$$

Why does condition 6 imply analyticity? We shall deal with this in the case where f is known to be of class C^1. We borrow the result from the calculus of functions of two variables known as Green's theorem: If P and Q are real or complex-valued functions of class C^1 on (a neighborhood of) the closed disk Δ, and *if* γ is the (circular) boundary of Δ, oriented counterclockwise, then

$$\int_\gamma (P\,dx + Q\,dy) = \int_\Delta \left(\frac{\partial Q}{\partial x} - \frac{\partial P}{\partial y}\right) dx\,dy.$$

Apply this with $P\,dx + Q\,dy = f\,dz = f\,dx + if\,dy$ and one has

$$\int_\gamma f(z)\,dz = \int_\Delta \left(i\frac{\partial f}{\partial x} - \frac{\partial f}{\partial y}\right) dx\,dy$$

Thus condition 6 says that the continuous function $i(\partial f/\partial x) - (\partial f/\partial y)$ has the property that

$$\int_\Delta \left(i\frac{\partial f}{\partial x} - \frac{\partial f}{\partial y}\right) dx\,dy = 0$$

for every closed disk Δ in the domain D. This implies that

$$i\frac{\partial f}{\partial x} - \frac{\partial f}{\partial y} = 0$$

which is the Cauchy-Riemann condition for complex-differentiability.

At this point we should remark that conditions 4 and 6 are valid more generally. If f is analytic and γ is any simple, closed rectifiable path in D, then

(Cauchy's formula) $f(w) = \dfrac{1}{2\pi i} \displaystyle\int_\gamma \dfrac{f(z)}{z - w}\,dz$ (w inside γ)

(Cauchy integral theorem) $\displaystyle\int_\gamma f(z)\,dz = 0.$

Loosely speaking, a simple, closed rectifiable path is one which winds around each point inside it exactly once. It would take us too far afield to discuss these matters precisely at this point.

Exercises

1. If f is a complex-analytic function on a disk D, then $f = g'$, where g is complex-analytic on D.

2. Let f be a function which is complex-analytic and has no zeros on the disk D. Show that $f = e^g$, where g is complex-analytic on D. *Hint:* What differential equation must g satisfy?

3. If f is complex-analytic and bounded in the unit disk $|z| < 1$, then

$$\sum_{n=0}^{\infty} \left| \frac{1}{n!} f^{(n)}(0) \right|^2 < \infty.$$

Hint: What is

$$\frac{1}{2\pi} \int_{-\pi}^{\pi} |f(re^{i\theta})|^2 \, d\theta \,?$$

4. If f is complex-analytic on $|z| \leq \rho$, then the complex derivatives of f are given by

$$f^{(n)}(w) = \frac{n!}{2\pi i} \int_{|z| = \rho} \frac{f(z)}{(z - w)^{n+1}} \, dz, \qquad |w| < \rho.$$

5. (Liouville's theorem) If f is bounded and complex-analytic on the entire plane, then f is constant. *Hint:* Use Exercise 4 to look at $f'(w)$. Think big (circles).

6. Let f be a function which is complex-analytic on the punctured disk $0 < |z| < r$. If f is bounded, show that f can be extended to a complex-analytic function on the disk $|z| < r$. *Hint:* First show that $z^2 f(z)$ is analytic in $|z| < r$.

7. Use Exercises 5 and 6 to show that, if f is complex-analytic on the entire plane and if there is a constant K such that

$$|f(z)| \leq K |z|^n, \qquad \text{for large } |z|$$

then f is a polynomial of degree not more than n.

8. Let $\{f_n\}$ be a sequence of complex-analytic functions on D. If f_n converges uniformly to f, then f_n' converges to f'.

9. Let f be complex-analytic on the annulus $r_1 < |z| < r_2$. Prove that $f = g + h$, where g is complex-analytic on the disk $|z| < r_2$ and h is complex-analytic outside the disk $|z| \leq r_1$. *Hint:* What integral would you think would define g?

*10. Let

$$f(z) = \sum_{n=0}^{\infty} z^{n!}, \qquad |z| < 1.$$

If $|\alpha| = 1$, show that f cannot be extended so as to be analytic in a neighborhood of α.

*11. Let

$$f(z) = \sum_{n=0}^{\infty} \frac{z^{2^n}}{n^2}, \qquad |z| \leq 1.$$

Show that f is continuous on $|z| \leq 1$, analytic on $|z| < 1$, but f cannot be extended analytically to any point on the unit circle $|z| = 1$.

5.7. Complex-Analytic Maps

One of the most important aspects of the theory of analytic functions is concerned with how an analytic mapping

$$D \xrightarrow{\ f\ } C$$

behaves geometrically. We do not have the time nor the tools to go into that question in depth; however, a few basic facts are within easy reach.

Theorem 17 (*Weak Maximum Principle*). *Let* f *be complex-analytic on an open set* D, *and let* V *be an open subset of* D *such that* $\bar{V}$ *is compact. There exists a point* w *on the boundary of* V *such that*

$$|f(w)| = \sup_V |f(z)|.$$

Proof. Since f is continuous, there exists a point $w \in \bar{V}$ such that

$$|f(w)| = \max_{\bar{V}} |f(z)|$$
$$= \sup_V |f(z)|.$$

Among such points w, choose one which is nearest to the boundary of V. (We can do that because the boundary of V is closed.) This w cannot be in V, for the following reason.

If $w \in V$, choose $r > 0$ so that the closed disk of radius r about w is contained in V. Then $f(w)$ is the mean value of f over C_r, the circle of radius r about w:

$$f(w) = \frac{1}{2\pi} \int_{-\pi}^{\pi} f(w + re^{i\theta})\, d\theta.$$

Therefore,

$$|f(w)| \leq \frac{1}{2\pi} \int_{-\pi}^{\pi} |f(w + re^{i\theta})|\, d\theta$$
$$\leq \sup_{C_r} |f| \frac{1}{2\pi} \int_{-\pi}^{\pi} d\theta$$
$$= \sup_{C_r} |f|$$
$$\leq \sup_V |f|.$$

If $|f(w)| = \sup_V |f|$, then equality must hold at every step in the chain of inequalities just derived. That can only happen if $f(z)$ is constantly equal to $f(w)$ on the circle C_r, because $|f(w)| - |f(z)|$ is a non-negative continuous function with integral 0. Some point of C_r is nearer to the boundary of V than w is; hence $\sup_V |f|$ is attained at some point closer to the boundary of V than w is. This contradicts the choice of w.

Corollary (Strong Maximum Principle). *Let* f *be a non-constant complex-analytic function on a connected open set* V. *Then*

$$|f(w)| < \sup_V |f(z)|, \qquad w \in V;$$

in other words, $|f|$ *connot attain a maximum at any point in* V.

Proof. It suffices to prove the result when $\bar{V}$ is compact. (Why?) From the proof of Theorem 17, we see that, if $w \in V$ and

$$|f(w)| = \sup_V |f(z)|$$

then f is constant on a neighborhood of w. Since V is connected, that cannot happen unless f is constant.

With a little information about the behavior of the modulus of an analytic function, we can prove one of the most basic theorems in mathematics.

Theorem 18 (Fundamental Theorem of Algebra). *Let*

$$p(z) = a_0 + a_1 z + \cdots + a_n z^n$$

be a non-constant polynomial with complex coefficients. There exists a complex number z *such that* $p(z) = 0$.

Proof. We may assume that $n > 1$ because the case $n = 1$ is trivial. We may also assume that $a_n = 1$:

$$p(z) = a_0 + a_1 z + \cdots + a_{n-1} z^{n-1} + z^n.$$

We claim that $|p(z)|$ is big outside big sets, i.e., that

$$\lim_{|z| \to \infty} |p(z)| = \infty.$$

We have

$$|p(z)| \geq |z|^n - |\sum_0^{n-1} a_k z^k|$$

$$\geq |z|^n - \sum_0^{n-1} |a_k| |z|^k.$$

If $|z| > 1$, then $|z|^k \leq |z|^{n-1}, k = 1, \ldots, n-1$; hence

$$|p(z)| \geq |z|^n - c|z|^{n-1}, \qquad |z| > 1$$

where $c = |a_0| + \cdots + |a_{n-1}|$. Thus

$$|p(z)| \geq |z|^{n-1}, \qquad |z| \geq 1 + c.$$

The last inequality shows that $|p(z)| \to \infty$. In particular, we can choose $r > 0$ so that

(5.40) $|p(z)| > 1 + |a_0|, \qquad |z| \geq r.$

Any zero of p must lie in the disk $\bar{D}_r = \{|z| \leq r\}$. Does p have a zero on $\bar{D}_r$? There exists a point $w \in \bar{D}_r$ such that

(5.41) $|p(w)| = \inf\{|p(z)|; z \in \bar{D}_r\} \leq |a_0|$

because p is continuous and $\bar{D}_r$ is compact. By (5.40) and (5.41) the point w is in the open disk D_r. If $p(w) \neq 0$, then $1/p$ is analytic on D_r and $|1/p|$ has a local maximum at w. By the maximum principle, $1/p$ (and hence p) is constant. We cannot have that; hence $p(w) = 0$.

Corollary. *If* p *is a polynomial with complex coefficients*

$$p(z) = a_0 + a_1 z + \cdots + a_n z^n, \qquad a_n \neq 0,$$

there exists complex numbers $z_1, \ldots, z_n$ *such that*

$$p(z) = a_n(z - z_1)(z - z_2) \cdots (z - z_n).$$

Proof. This results from repeated application of the theorem plus the fact that if $p(w) = 0$, then $p(z) = (z - w)q(z)$, where q is a polynomial.

We now establish a result about the zeros of the general analytic function. It is an integral formula for counting the zeros. When we count zeros, we shall always include multiplicities. That means the following. If f is analytic near z_0 and if $f \neq 0$, we can write

$$f(z) = (z - z_0)^n g(z)$$

where g is analytic near z_0 and $g(z_0) \neq 0$. We call n either the **order** of the zero at z_0 or the **multiplicity** of z_0 as a zero of f. We then count z_0 as n zeros of f.

Theorem 19. *Let* f *be a function which is complex-analytic on the closed disk* $|z - z_0| \leq \rho$ *with no zeros on* $|z - z_0| = \rho$. *Then*

$$\frac{1}{2\pi i} \int_{|z - z_0| = \rho} \frac{f'(z)}{f(z)} \, dz$$

is equal to the number of zeros which f *has in the disk* $|z - z_0| < \rho$.

Proof. We may assume that $z_0 = 0$. If f has no zeros on $|z| \leq \rho$, then f'/f is analytic on that disk; consequently (by the mean value property),

$$\frac{1}{2\pi i} \int_{|z| = \rho} \frac{f'(z)}{f(z)} dz = \frac{z f'(z)}{f(z)} \Big|_{z=0}$$

$$= 0.$$

So, the formula is correct when f is zero-free.

Now, take a general f. The identity theorem (Theorem 12) tells us that

f, being analytic on an open set which contains $|z| \leq \rho$ and not identically zero, can vanish at only a finite number of points in that disk. Let $z_1, \ldots, z_n$ be the distinct points z such that $|z| < \rho$ and $f(z) = 0$. Let k_j be the order of the zero at z_j. Then

$$f(z) = (z - z_1)^{k_1} \cdots (z - z_n)^{k_n} g(z)$$

where g is analytic and zero-free on $|z| \leq \rho$. Compute that

$$\frac{f'(z)}{f(z)} = \frac{g'(z)}{g(z)} + \sum_{j=1}^{n} \frac{k_j}{z - z_j}.$$

By the result from the zero-free case

$$\frac{1}{2\pi i} \int_{|z|=\rho} \frac{f'(z)}{f(z)} dz = \sum_{j=1}^{n} k_j \cdot \frac{1}{2\pi i} \int_{|z|=\rho} \frac{dz}{z - z_j}$$

$$= \sum_{j=1}^{n} k_j.$$

Lemma. *Let* f *be complex-analytic and non-constant on a neighborhood of the point* z_0. *Let* n *be the order of the zero of* f $-$ f(z_0) *at* z_0. *For every sufficiently small number* $\rho > 0$, *there exists* $\epsilon > 0$ *such that, if* $0 < |\alpha - f(z_0)| < \epsilon$, *then there are precisely* n *points* z *with* $|z - z_0| < \rho$ *and* f$(z) = \alpha$.

Proof. Let

$$D_\rho = \{z; |z - z_0| < \rho\}$$
$$C_\rho = \{z; |z - z_0| = \rho\}, \qquad \rho > 0.$$

We choose ρ sufficiently small that

    (i) f is complex-analytic on $\bar{D}_\rho$;
    (ii) $f(z) \neq f(z_0)$, if $z \in \bar{D}_\rho$ and $z \neq z_0$;
    (iii) $f'(z) \neq 0$, if $z \in D_\rho$ and $z \neq z_0$.

The identity theorem (Theorem 12) tells us that, if (ii) were not satisfied for small ρ, then f would be constant near z_0. Similarly, if (iii) were not satisfied we would have $f' = 0$ near z_0 and hence f constant. Note that $f'(z_0)$ may or may not be 0.

Condition (ii) and Theorem 19 tell us that

$$\frac{1}{2\pi i} \int_{C_\rho} \frac{f'(z)}{f(z) - f(z_0)} dz = n$$

where n is the order of the zero which $f - f(z_0)$ has at the point z_0. Similarly, if $\alpha \notin f(C_\rho)$, then

$$N(\alpha) = \frac{1}{2\pi i} \int_{C_\rho} \frac{f'(z)}{f(z) - \alpha} dz$$

is the number of zeros of $f - \alpha$ inside C_ρ, i.e., the number of times f assumes the value α on the disk D_ρ. Since $N(\alpha)$ is an integer, we will have

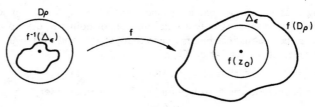

$N(\alpha) = n$ provided that $|N(\alpha) - n| < 1$. Look at the integrals which give us n and $N(\alpha)$. It should be clear that those integrals are less than 1 apart for all α such that $|\alpha - f(z_0)| < \epsilon$, i.e., for all α sufficiently near $f(z_0)$.

We now have a picture like that in Figure 20 for small ρ, ϵ. Of course, $\Delta_\epsilon = \{\alpha; |\alpha - f(z_0)| < \epsilon\}$. If $\alpha \in \Delta_\epsilon$, then $f(z) = \alpha$ for exactly n points $z \in D_\rho$. Presumably, that counts multiplicities. For instance, if $\alpha = f(z_0)$, then $f(z) = \alpha$ for n points of D_ρ only because of the convention to count z_0 as an n times repeated zero of $f(z) - f(z_0)$. But, for points $\alpha \ne f(z_0)$ multiplicities do not enter because $f^{-1}(\alpha)$ consists of n distinct points. Why? If $f(w) = \alpha$, then $f'(w) \ne 0$ by condition (iii) on ρ; so, $f - \alpha$ has a first order zero at w.

The conclusion of the lemma can be roughly summarized this way. If f is analytic and non-constant at z_0, then near z_0

$$f(z) = f(z_0) + a_n(z - z_0)^n + a_{n+1}(z - z_0)^{n+1} + \cdots$$

where $a_n \ne 0$, and f behaves there much the way that z^n behaves near the origin. Two basic facts are easily derived from the lemma.

Theorem 20 (*Open Mapping Theorem*). *Let* f *be a non-constant complex-analytic function on a connected open set* D. *Then* f *is an open mapping; that is, if* V *is an open subset of* D, *then* f(V) *is an open set.*

Proof. Let z_0 be a point in V, an open subset of D. We must show that $f(V)$ is a neighborhood of $f(z_0)$. Since D is connected and f is not constant, $f - f(z_0)$ has a zero at z_0 of some well-defined order n, $1 \le n$. The last corollary shows us that, for small ρ, the image under f of $D_\rho = \{z; |z - z_0| < \rho\}$ contains the disk about $f(z_0)$ of some radius $\epsilon > 0$. Since $D_\rho \subset V$ for small ρ, the result follows.

Theorem 21. *Let* f *be a complex-analytic map*

$$D \xrightarrow{\ f\ } C$$

on an open set D *in the plane.*

(i) *If* $z_0 \in$ D, *then* f *is* $1:1$ *on a neighborhood of* z_0 *if and only if* f'(z_0) $\ne 0$.

(ii) *If* f *is* $1:1$ *on* D, *then* f^{-1}:

$$\overset{f^{-1}}{D \longleftarrow f(D)}$$

is complex-analytic.

Proof. Statement (i) is a special case of the lemma prior to Theorem 20. To prove (ii), observe that f is an open mapping. That tells us that $f(D)$ is an open set. It also tells us that f^{-1} is continuous. Since f is $1:1$, $f'(z) \neq 0$ throughout D. Now we can see that f^{-1} is complex-differentiable at each point $w = f(z)$ in $f(D)$:

$$\lim_{\alpha \to w} \frac{f^{-1}(\alpha) - f^{-1}(w)}{\alpha - w} = \lim_{\alpha \to w} \frac{f^{-1}(\alpha) - z}{\alpha - f(z)}$$

$$= \frac{1}{f'(z)}.$$

We also see that $(f^{-1})'$ is continuous:

$$(f^{-1})'(w) = [f'(f^{-1}(w))]^{-1}.$$

So f^{-1} is complex-analytic, by Theorem 15.

EXAMPLE 11. Let f be the exponential function, $f(z) = e^z$. Then f is analytic on the entire plane. Since $f' = f$, the derivative has no zeros. Thus f is locally $1:1$. The function f is not $1:1$ on the plane, because

$$f(z + 2n\pi i) = f(z).$$

Evidently, the equation

$$w = e^z = e^x e^{iy}$$

can be solved for z if $w \neq 0$; hence, the image of f is the punctured plane $\{w \in C; w \neq 0\}$. In fact, any horizontal strip

$$-\infty < x < \infty$$

$$y_0 < y \leq y_0 + 2\pi$$

is mapped $1:1$ onto the punctured plane. Fix $w_0 \neq 0$ and select any z_0 such that $w_0 = e^{z_0}$. Theorem 21 tells us (as is obvious in this case) that there is a neighborhood U of z_0 which is mapped by f in a $1:1$ manner onto V, a neighborhood of w_0. Also, the inverse map on V is analytic. What does all this say? It says, start with any $w_0 \neq 0$. Given any logarithm z_0 of w_0, we can analytically extract logarithms on a neighborhood of w_0, in such a way that $\log w_0 = z_0$.

Exercises

1. (Schwarz lemma) Suppose that f is complex-analytic and bounded by M on the unit disk:

$$|f(z)| \leq M, \qquad |z| < 1.$$

If $f(0) = 0$, then $|f(z)| \leq M|z|$ on the unit disk.

2. Let f be a complex-analytic function on the unit disk such that
$$|f(z)| \leq 1, \qquad |z| < 1$$
$$f'(0) = 1.$$

Prove that $f(z) = z$. *Hint:* Express $f'(0)$ as an integral over $|z| = 1$.

3. If $|\alpha| < 1$, let
$$L(z) = \frac{z - \alpha}{1 - \alpha^* z}, \qquad |z| < 1.$$

Show that L is a $1:1$ complex-analytic map of the unit disk *onto* the unit disk.

4. Let f be complex-analytic on the unit disk. Let $|\alpha| < 1$ and suppose that $f(\alpha) = 0$. Then
$$f(z) = \frac{z - \alpha}{1 - \alpha^* z} g(z)$$

where g is complex-analytic on the unit disk and
$$\sup_{|z|<1} |f(z)| = \sup_{|z|<1} |g(z)|.$$

5. Suppose that f is analytic on the unit disk and $|f(z)| \leq M$. If $f(\alpha) = 0$, then
$$|f(z)| \leq M \left| \frac{z - \alpha}{1 - \alpha^* z} \right|, \qquad |z| < 1.$$

If equality holds at any point z, then f is a constant multiple of the function L from Exercise 3.

6. Let f be a complex-analytic map from the unit disk into the unit disk. Prove that
$$\left| \frac{f(\beta) - f(\alpha)}{1 - f(\alpha)^* f(\beta)} \right| \leq \left| \frac{\beta - \alpha}{1 - \alpha^* \beta} \right|.$$

Hint: Use Exercise 5 and the function
$$g(z) = \frac{f(z) - f(\alpha)}{1 - f(\alpha)^* f(z)}.$$

7. Let f be a $1:1$ analytic map of the unit disk onto the unit disk. Show that
$$f(z) = c \cdot \frac{z - \alpha}{1 - \bar{\alpha} z}$$

where $|c| = 1$ and $|\alpha| < 1$.

8. Let $\{f_n\}$ be a sequence of complex-analytic functions on D which converges uniformly to f. If f has a zero somewhere on D, there is an N such that f_n has a zero on D for all $n \geq N$.

9. Is the result in Exercise 8 valid if we change "has a zero" to "has no zero" in the two places where it occurs?

10. Let $p(z)$ be a polynomial of degree $n \geq 1$. Show that
$$\lim_{\rho \to \infty} \int_{|z|=\rho} \left[\frac{p'(z)}{p(z)} - \frac{n}{z} \right] dz = 0.$$

In view of Theorem 19 what does that tell you about the number of zeros which p has inside large disks?

11. Show that

$$f(z) = \frac{1+z}{1-z}$$

maps the unit disk 1:1 onto the right half-plane Re $z > 0$. What is the image under f of the upper half of the unit disk.

12. Find explicitly a 1 : 1 analytic map of half of the unit disk onto the full unit disk. *Hint:* Compose maps as in Figure 21.

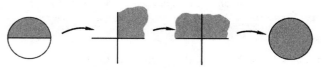

FIGURE 21

13. Prove the open mapping theorem directly from the strong maximum principle and the fact that the reciprocal of a non-vanishing analytic function is analytic.

*14. (Rouche's theorem) Let f and g be complex-analytic on the closed disk $|z| \leq \rho$ and suppose that $|g(z)| < |f(z)|$ on the circle $|z| = \rho$. Prove that f and $f + g$ have the same number of zeros on the disk $|z| < \rho$.

5.8. *Fourier Series*

On an interval of the real line, the attempt to expand functions in power series fails even for some very smooth functions. Of course, many of the functions with which we work regularly are real-analytic and their series expansions are important in a variety of problems. But, the class of real-analytic functions is extremely special, and that limits the applicability of power series. Now, we are going to talk a bit about an equally important type of series expansion—the expansion into a trigonometric (or Fourier) series. Such expansions can be obtained for every smooth function on an interval. Furthermore, expansions into trigonometric series are natural and important in many mathematical problems (and in many applications of mathematics).

Suppose that f is a (complex-valued) function on the interval $[-\pi, \pi]$. We ask if it is possible to expand f in a series

(5.42) $$f(z) = \sum_{n=-\infty}^{\infty} c_n e^{inx}.$$

Such an expansion would be equivalent to

$$f(x) = c_0 + \sum_{n=1}^{\infty} (a_n \cos nx + b_n \sin nx)$$

where

$$a_n = c_n + c_{-n}$$
$$b_n = i(c_n - c_{-n}).$$

We shall continue to work with the exponential functions.

In the presence of absolute convergence, there can be no ambiguity about the meaning of a Fourier expansion (5.42); however, to allow for more general situations, we should stipulate the order in which the terms are to be summed. As one might guess, what we have in mind is to form the **partial sums** this way:

$$s_n(x) = \sum_{k=-n}^{n} c_k e^{ikx}.$$

The meaning of (5.42) is then that $f(x) = \lim_n s_n(x)$.

If any Fourier expansion is to be valid, it is quite easy to see which coefficients c_n we should use. If, for instance, (5.42) were uniformly convergent, we could legitimately compute

$$\frac{1}{2\pi} \int_{-\pi}^{\pi} e^{-ikx} f(x)\, dx = \sum_{n=-\infty}^{\infty} c_n \cdot \frac{1}{2\pi} \int_{-\pi}^{\pi} e^{i(n-k)x}\, dx$$
$$= c_n$$

because

$$\frac{1}{2\pi} \int_{-\pi}^{\pi} e^{i(n-k)x}\, dx = \begin{cases} 0, & n \neq k \\ 1, & n = k. \end{cases}$$

Therefore, if we start with a Riemann-integrable function f, we define the **Fourier coefficients** of f to be

$$(5.43) \qquad c_n = \frac{1}{2\pi} \int_{-\pi}^{\pi} f(x) e^{-inx}\, dx, \qquad n = 0, \pm 1, \pm 2, \ldots$$

and we call the formal series

$$(5.44) \qquad\qquad \sum_{n=-\infty}^{\infty} c_n e^{inx}$$

the **Fourier series** for f. Then we ask:

(i) Does the Fourier series converge?

(ii) If it converges, does it converge to f?

We are going to limit the discussion to piecewise-continuous functions although the results are valid more generally.

There are some elementary observations we should make. We are attempting to expand f in a series of functions each periodic with period 2π. Hence, if we want to have the series converge to f at each point, we had better assume that $f(-\pi) = f(\pi)$. Of course, we would not expect the Fourier series for a piecewise-continuous function to converge to $f(x)$ at each point, since the values of f at finitely many points could be changed without altering the Fourier coefficients c_n. Thus, question (ii) above should be reworded along the lines of: If the series converges, does it converge to f at most points?

Before we discuss the two basic questions (i), (ii) for the general continuous (or piecewise-continuous) function, let's talk about smooth functions f. For a function f of class C^1, which satisfies $f(-\pi) = f(\pi)$ it is not too difficult to see that the Fourier series converges uniformly to f. In order to prove that, we first show that the Fourier coefficients tend to 0.

Theorem 22 (*Riemann-Lebesgue*). *Let* f *be a piecewise-continuous function on the interval* [a, b]. *Then*

$$\lim_{|t|\to\infty} \int_a^b f(x)e^{itx}\,dx = 0.$$

Proof. The function f is essentially the sum of a finite number of functions, each of which is uniformly continuous on an interval and 0 outside that interval. So, it should be clear that we need only prove the theorem for continuous functions f. Any continuous function f can be uniformly approximated by step functions, that is, functions which are piecewise constant. Why? Let $\epsilon > 0$. By the uniform continuity of f, there exists $\delta > 0$ such that

$$|f(x_1) - f(x_2)| < \epsilon, \qquad |x_1 - x_2| < \delta.$$

Choose a partition $a = x_0 < x_1 < \cdots < x_n = b$ of $[a, b]$ such that $x_k - x_{k-1} < \delta, k = 1, \ldots, n$. Define

$$s(x) = f(x_{k-1}), \qquad x_{k-1} \le x < x_k.$$

(See Figure 22.) Then s is piecewise-constant and is uniformly within ϵ of f:

$$|f(x) - s(x)| < \epsilon, \qquad a \le x \le b.$$

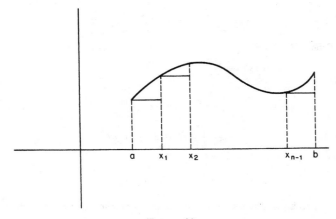

FIGURE 22

Therefore, it suffices to prove the theorem for step functions. We have left the proof of that sufficiency to the exercises.

But now the proof is easy, because each step function is a finite sum of functions which are constant on an interval and 0 outside that interval. Thus, all we need to do is to prove the result for constant functions, i.e., to show that

$$\lim_{|t| \to \infty} \int_a^b e^{itx}\, dx = 0.$$

That is easy to see, since the integral can be calculated explicitly:

$$\int_a^b e^{itx}\, dx = \frac{1}{it} e^{itx} \Big|_a^b$$

$$\left| \int_a^b e^{itx}\, dx \right| \le \frac{2}{|t|}.$$

Corollary. *If* f *is piecewise-continuous, then*

$$\lim_{|t| \to \infty} \int_a^b f(x) \cos tx\, dx = \lim_{|t| \to \infty} \int_a^b f(x) \sin tx\, dx = 0.$$

The Riemann-Lebesgue theorem is valid for many more functions than piecewise-continuous ones. It is valid for Riemann-integrable functions as well as ones which are "integrable" in the sense of Lebesgue. (Chapter 7). What interests us now is that the Riemann-Lebesgue theorem shows that the question of whether or not the Fourier series for f converges at the point x depends only upon the behavior of f near x.

Lemma. *Let* f *be a piecewise-continuous function on* $[-\pi, \pi]$*. The* nth *partial sum of the Fourier series for* f *is*

$$(5.45) \qquad s_n(x) = \frac{1}{2\pi} \int_{-\pi}^{\pi} f(x - t) \frac{\sin (n + \frac{1}{2})t}{\sin \frac{1}{2}t}\, dt.$$

Proof. We have

$$s_n(x) = \sum_{k=-n}^{n} c_k e^{ikx}$$

$$= \sum_{k=-n}^{n} e^{ikx} \cdot \frac{1}{2\pi} \int_{-\pi}^{\pi} f(t) e^{-ikt}\, dt$$

$$= \frac{1}{2\pi} \int_{-\pi}^{\pi} f(t) D_n(x - t)\, dt$$

where

$$D_n(x) = \sum_{k=-n}^{n} e^{ikx}.$$

Extend f to a piecewise-continuous function on the real line in such a way that $f(x + 2\pi) = f(x)$. (Forget about the values at π and $-\pi$. They don't affect the integrals we're considering.) The substitution $u = x - t$ yields

$$s_n(x) = \frac{1}{2\pi} \int_{x-\pi}^{x+\pi} f(x - u)D_n(u)\, du$$

and since both f and D_n are periodic with period 2π

$$s_n(x) = \frac{1}{2\pi} \int_{-\pi}^{\pi} f(x - u)D_n(u)\, du.$$

The rest of the lemma is a calculation:

$$
\begin{aligned}
D_n(x) &= \sum_{k=-n}^{n} e^{ikx} \\
&= \sum_{k=0}^{n} + \sum_{k=-1}^{-n} \\
&= \frac{e^{i(n+1)x} - 1}{e^{ix} - 1} + \frac{e^{-i(n+1)x} - 1}{e^{-ix} - 1} - 1 \\
&= \frac{\cos nx - \cos(n+1)x}{1 - \cos x} \\
&= \frac{\sin(n + \frac{1}{2})x}{\sin \frac{1}{2}x}.
\end{aligned}
$$

The sequence $\{D_n\}$ is called **Dirichlet's kernel.**

   Theorem 23. *Let* f *be a piecewise-continuous function of period 2π on the real line. At any point* x *where* f *is differentiable, the Fourier series for* f *converges to* f(x).

   Proof. The result is stated for periodic functions on the real line in order to be clear about what is meant here by differentiability at the end points of $[-\pi, \pi]$. If f is a function on $[-\pi, \pi]$ and if we extend f periodically by $f(x + 2\pi) = f(x)$, then the extended function is differentiable at π exactly when $f(\pi) = f(-\pi)$ and $f'(\pi) = f'(-\pi)$. This assumes more than differentiability of the corresponding function on the interval $[-\pi, \pi]$.

   Fix a point x where f is differentiable. We have

(5.46) $$s_n(x) = \frac{1}{2\pi} \int_{-\pi}^{\pi} f(x - t)D_n(t)\, dt$$

where

$$D_n(t) = \frac{\sin(n + \frac{1}{2})t}{\sin \frac{1}{2}t}.$$

If we apply (5.46) to the constant function 1, we have

(5.47) $$1 = \frac{1}{2\pi} \int_{-\pi}^{\pi} D_n(t)\, dt.$$

Multiply (5.47) by the constant $f(x)$ and subtract from (5.46):

$$s_n(x) - f(x) = \frac{1}{2\pi} \int_{-\pi}^{\pi} [f(x - t) - f(x)] D_n(t)\, dt$$

(5.48)

$$= \frac{1}{2\pi} \int_{-\pi}^{\pi} \frac{f(x - t) - f(x)}{t} t D_n(t)\, dt.$$

Since f is differentiable at x, the difference quotient

$$\frac{f(x - t) - f(x)}{t}$$

is piecewise-continuous for $-\pi \leq t \leq \pi$ (with a suitable definition of its value at 0). Now

$$t D_n(t) = \frac{t}{\sin \frac{1}{2} t} \sin (n + \tfrac{1}{2}) t.$$

The first factor on the right is piecewise-continuous because it has the limit 2 at $t = 0$. Thus,

$$s_n(x) - f(x) = \frac{1}{2\pi} \int_{-\pi}^{\pi} h(t) \sin \left(n + \frac{1}{2} \right) t\, dt$$

where h is piecewise-continuous. By the (Corollary of the) Riemann-Lebesgue theorem,

$$\lim_n [s_n(x) - f(x)] = 0.$$

It is not too difficult to strengthen the proof of the last theorem to show that, on any closed interval where f is of class C^1, the partial sums s_n converge uniformly to f. In Exercises 10–13, we have indicated one way to do that. We shall obtain the result later as a simple consequence of a more general convergence theorem.

The hypothesis in Theorem 23 can be weakened somewhat, e.g., if f is continuous at x and of bounded variation near x, then $s_n(x)$ converges to $f(x)$; however, the smoothness condition on f cannot be weakened too much. If f is merely continuous at x, the Fourier series need not converge at x. That is disappointing but not a disaster. One of the oldest problems in the study of Fourier series was solved only a few years ago when the Swedish mathematician Lennart Carleson proved that the Fourier series for a continuous function (or even an integrable function) converges to the function "almost everywhere". The proof is not too far beyond us conceptually; but, it is much too long and difficult for us to present.

Prior to Carleson's result, and presumably this will continue to be the case, the attitude had been this: If the Fourier series of a continuous function need not converge to the function at all points, then let us look for other ways to recapture the function, given its Fourier series. One method of doing that is the following, known as the **Cesaro summability**

method. Form the averages

(5.49) $$\sigma_n = \frac{1}{n}(s_0 + \cdots + s_{n-1}).$$

These functions are called the **Cesaro means** of the Fourier series for f. It is a theorem that, if f is continuous and $f(-\pi) = f(\pi)$, then the Cesaro means σ_n converge uniformly to f. We shall not stop to prove this, because in the next section we shall prove the analogous result for a summation method known as Abel-Poisson summation. It is a much more illuminating method.

Exercises

1. What is the Fourier series for

$$f(x) = \sin x \cos 2x?$$

(Think. Don't compute any integrals.)

2. Let f be a function of class C^2 on $[-\pi, \pi]$ with $f(-\pi) = f(\pi)$. Use integration by parts to show that

$$\sum_n |c_n| < \infty.$$

3. If f is of class C^2 and $f(-\pi) = f(\pi)$, then the Fourier series for f converges uniformly to f.

4. True or false? If f is of class C^1 and $f(-\pi) = f(\pi)$, the Fourier series for f' is obtained from term-by-term differentiation of the Fourier series for f.

5. If f is an even function, $f(-x) = f(x)$, then its Fourier coefficients satisfy $c_{-n} = c_n$ and so the Fourier series is really a "cosine series". What about odd functions, $f(-x) = -f(x)$?

6. Let $\{f_n\}$ be a sequence of piecewise-continuous functions, for each of which the Riemann-Lebesgue theorem is known to be valid. If f_n converges uniformly to f, show that the Riemann-Lebesgue theorem is valid for f.

7. Show that in Exercise 6 it would be enough to know that

$$\lim_n \int_{-\pi}^{\pi} |f_n - f| = 0.$$

8. Use the result of Exercise 7 to (define the objects in and) prove the Riemann-Lebesgue theorem for functions which are piecewise-continuous in this sense: There is a partition of $[-\pi, \pi]$ such that f is continuous and bounded on each open interval defined by the partition.

9. Use Exercise 8 to show that, if f is piecewise-continuous on $[-\pi, \pi]$ and satisfies a Lipschitz condition at x:

$$|f(x) - f(y)| \le M|x - y|, \qquad \text{all } y$$

then $s_n(x)$ converges to $f(x)$.

10. If f is a function on the real line and if $t \in R$, the t-translate of f is the function f_t defined by
$$f_t(x) = f(x + t).$$
If f is piecewise-continuous, show that translation of f is continuous in this sense:
$$\lim_{t \to 0} \int_a^b |f(x) - f_t(x)| \, dx = 0.$$

Hint: It follows from uniform continuity in case f is continuous; and, if f is piecewise-continuous, there is a continuous g such that $\int |f - g| < \epsilon$.

11. Let f be piecewise-continuous and let $0 < \delta \leq \pi$. Show that
$$\lim_n \int_\delta^\pi f(t) D_n(t) \, dt = 0.$$

12. Let f be piecewise-continuous on the real line and let $0 < \delta \leq \pi$. Show that
$$\lim_n \int_\delta^\pi f(x - t) D_n(t) \, dt = 0$$
uniformly for $x \in [-\pi, \pi]$. *Hint:* Use Exercises 10 and 11.

13. If f is of class C^1, show that the Fourier series for f converges uniformly to f. *Suggestion:* You need to show that
$$\lim_n \int_{-\pi}^\pi \frac{f(x - t) - f(x)}{t} \sin\left(n + \frac{1}{2}\right) t \, dt = 0$$
uniformly in x. Choose δ, $0 < \delta \leq \pi$, so that the difference quotient is within ϵ of $f'(x)$ for $|t| \leq \delta$ (and for all x). Break up the integral into the integral over $[-\delta, \delta]$ plus what's left over. Apply Exercise 12 to the left-overs.

*14. The series
$$\sum_{n \neq 0} \frac{1}{n} e^{inx}$$
is the Fourier series of a piecewise-continuous function. Find that function.

*15. Prove that the series
$$\sum_{n=1}^\infty \frac{1}{n} e^{inx}$$
is *not* the Fourier series of a piecewise-continuous function.

5.9. *Abel-Poisson Summation*

Suppose that f is a complex-valued function on the interval $[-\pi, \pi]$ and $f(-\pi) = f(\pi)$. Then f may be regarded as a function on the unit circle in the complex plane:
$$(5.50) \qquad\qquad f(e^{i\theta}) = f(\theta), \qquad -\pi \leq \theta \leq \pi.$$
If f is [piecewise] continuous on $[-\pi, \pi]$, then the corresponding function on the circle is [piecewise] continuous. There is nothing much to that;

however, there is an important point here. When we think of f as a function on the unit circle, it suggests some things to do with f that otherwise we might have missed. In particular, it suggests a strong connection between Fourier series and complex power series. That connection will lead us to the Abel-Poisson summation method for Fourier series.

Let's begin with a very special situation. Suppose that f is a complex-analytic function in a disk of radius greater than 1 about the origin:

$$(5.51) \qquad f(z) = \sum_{n=0}^{\infty} a_n z^n, \qquad |z| < 1 + \epsilon.$$

Now

$$f(re^{i\theta}) = \sum_{n=0}^{\infty} a_n r^n e^{in\theta}, \qquad 0 \le r < 1 + \epsilon.$$

In particular, the restriction of f to the unit circle is a continuous function on that circle, and the Fourier series for that function is staring us in the face:

$$(5.52) \qquad f(e^{i\theta}) = \sum_{n=0}^{\infty} a_n e^{in\theta}.$$

In fact, for each r we have a function f_r on the unit circle

$$(5.53) \qquad f_r(e^{i\theta}) = f(re^{i\theta})$$

and the Fourier coefficients for f_r are

$$\frac{1}{2\pi} \int_{-\pi}^{\pi} e^{-in\theta} f_r(e^{i\theta}) \, d\theta = \begin{cases} a_n r^n, & n \ge 0 \\ 0, & n < 0. \end{cases}$$

Since f is continuous on $|z| < 1 + \epsilon$,

$$\lim_{r \to 1} f_r = f$$

and the convergence is uniform on the unit circle.

Here is the idea of Abel-Poisson summability. Start with any piecewise-continuous complex-valued function f on the unit circle. Let c_n be the nth Fourier coefficient of f. *Define*

$$(5.54) \qquad f_r(e^{i\theta}) = \sum_{n=-\infty}^{\infty} c_n r^{|n|} e^{in\theta}, \qquad 0 \le r < 1.$$

Why does this make sense? The sequence $\{c_n\}$ is bounded:

$$|c_n| \le \sup \{|f(z)|; |z| = 1\}.$$

Clearly then, if $r < 1$, the series in (5.54) is uniformly absolutely convergent because

$$\sum_{n=-\infty}^{\infty} r^{|n|} = 2(1 - r)^{-1} - 1, \qquad 0 \le r < 1.$$

Now, we're going to prove that (as $r \to 1$) f_r converges to f (at most points). First, we obtain an integral representation for f_r:

$$f_r(e^{i\theta}) = \sum_{n=-\infty}^{\infty} r^{|n|}e^{in\theta} \cdot \frac{1}{2\pi} \int_{-\pi}^{\pi} f(e^{it})e^{-int}\, dt$$

$$= \frac{1}{2\pi} \int_{-\pi}^{\pi} f(e^{it})[\sum_{n} r^{|n|}e^{in(\theta-t)}]\, dt.$$

Thus we have

(5.55)
$$f_r(e^{i\theta}) = \frac{1}{2\pi} \int_{-\pi}^{\pi} f(e^{it})P_r(\theta - t)\, dt$$

where

(5.56)
$$\begin{aligned}
P_r(\theta) &= \sum_{n=-\infty}^{\infty} r^{|n|}e^{in\theta} \\
&= \sum_{n=0}^{\infty} r^n e^{in\theta} + \sum_{n=1}^{\infty} r^n e^{-in\theta} \\
&= \frac{1}{1 - re^{i\theta}} + \frac{1}{1 - re^{-i\theta}} - 1 \\
&= \frac{1 - r^2}{1 - 2r\cos\theta + r^2}.
\end{aligned}$$

The family of functions $\{P_r\}$ is known as the **Poisson kernel**. That family of functions has these properties:

(i) $\dfrac{1}{2\pi} \displaystyle\int_{-\pi}^{\pi} P_r(\theta)\, d\theta = 1$;

(ii) $P_r \geq 0$; $P_r(-\theta) = P_r(\theta)$;

(iii) if $0 < \delta < \pi$,

then
$$\lim_{r \to 1} \sup \{P_r(\theta); \delta \leq |\theta| \leq \pi\} = 0.$$

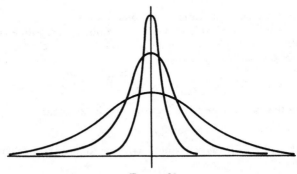

FIGURE 23

Property (ii) is obvious. No calculation is necessary in order to verify property (i). It results from (5.55) applied to the constant function $f = 1$. What does (iii) say? It says, fix a small neighborhood of 0. On that part of $[-\pi, \pi]$ which is outside that neighborhood, P_r goes uniformly to 0 as r

tends to 1. On the other hand, $P_r(0)$ tends to ∞ as r tends to 1, and so the picture is something like that indicated in Figure 23.

In order to carefully verify (iii), observe that

$$P_r(\theta) \leq \frac{1 - r^2}{1 - 2r \cos \delta + r^2}, \qquad \delta \leq |\theta| \leq \pi.$$

Theorem 24. *Let f be a piecewise-continuous function on the unit circle with Fourier coefficients $\{c_n\}$. At each point $e^{i\theta}$ where f is continuous, the Abel-Poisson means*

$$f_r(e^{i\theta}) = \sum_{n=-\infty}^{\infty} c_n r^{|n|} e^{in\theta}$$

$$= \frac{1}{2\pi} \int_{-\pi}^{\pi} f(e^{it}) P_r(\theta - t) \, dt$$

converge to $f(e^{i\theta})$ as r tends to 1. The convergence is uniform on each closed arc on which f is continuous.

Proof. It is technically convenient to (again) regard f as a function on the real line with period 2π:

$$f(t) = f(e^{it}).$$

Then

$$f_r(\theta) = \frac{1}{2\pi} \int_{-\pi}^{\pi} f(t) P_r(\theta - t) \, dt$$

$$= \frac{1}{2\pi} \int_{-\pi}^{\pi} f(\theta - t) P_r(t) \, dt.$$

Thus

$$f(\theta) - f_r(\theta) = \frac{1}{2\pi} \int_{-\pi}^{\pi} [f(\theta) - f(\theta - t)] P_r(t) \, dt$$

$$|f(\theta) - f_r(\theta)| \leq \frac{1}{2\pi} \int_{-\pi}^{\pi} |f(\theta) - f(\theta - t)| \, P_r(t) \, dt$$

(We just used property (i) and property (ii) of P_r.)

Let $\epsilon > 0$. If f is continuous at θ, we can choose $\delta > 0$ such that

(5.57) $$|f(\theta) - f(t)| < \frac{\epsilon}{2}, \qquad |\theta - t| < \delta.$$

Then

$$|f(\theta) - f_r(\theta)| \leq \frac{1}{2\pi} \int_{-\delta}^{\delta} |f(\theta) - f(\theta - t)| \, P_r(t) \, dt + \frac{1}{2\pi} \int_{|t| \geq \delta}$$

$$< \frac{\epsilon}{2} + \frac{1}{2\pi} \int_{|t| \geq \delta} |f(\theta) - f(\theta - t)| \, P_r(t) \, dt$$

$$< \frac{\epsilon}{2} + K \sup \{P_r(t); \delta \leq |t| \leq \pi\}$$

where

$$K = 2 \cdot \frac{1}{2\pi} \int_{-\pi}^{\pi} |f(t)| \, dt.$$

From property (iii) of P_r we have

$$|f(\theta) - f_r(\theta)| < \epsilon, \qquad r \text{ near } 1.$$

If f is continuous on the arc $a \leq \theta \leq b$, then uniform continuity allows us to choose δ so that (5.57) holds uniformly in θ. Our inequalities then show that f_r converges uniformly to f on that arc.

Corollary. *If* f *is continuous on the unit circle and*

$$\frac{1}{2\pi} \int_{-\pi}^{\pi} e^{-in\theta} f(e^{i\theta}) \, d\theta = 0, \qquad n = 0, \pm 1, \pm 2, \ldots,$$

then f $= 0$.

Corollary. *If* f *is continuous on the unit circle, then* f *can be uniformly approximated by trigonometric polynomials*

$$p(e^{i\theta}) = \sum_{n=-N}^{N} a_n e^{in\theta}.$$

Proof. Choose r so that f_r is uniformly near f. Then take a partial sum of the series for f_r:

$$\sum_{n=-N}^{N} c_n r^{|n|} e^{in\theta}.$$

Corollary (*The Weierstrass Approximation Theorem*). *Let* f *be a continuous complex-valued function on a closed interval* [a, b] *of the real line. Then* f *can be uniformly approximated by polynomials.*

Proof. Evidently we may assume that the interval is $[a, \pi]$, where $-\pi < a < \pi$. Extend f to a continuous function on $[-\pi, \pi]$ such that $f(-\pi) = f(\pi)$. (Just use a straight line.) If $\epsilon > 0$, there is a trigonometric polynomial

$$g(x) = \sum_{n=-N}^{N} c_n e^{inx}$$

with $\|f - g\|_\infty < \epsilon$. Each e^{inx} can be expanded in a power series in x. Thus, there is a polynomial

$$p_n(x) = \sum_{k=0}^{Nn} a_{nk} x^k$$

with complex coefficients so that $|e^{inx} - p_n(x)|$ is uniformly small on $[-\pi, \pi]$. The polynomial

$$p(x) = \sum_{n=-N}^{N} c_n p_n(x)$$

will be uniformly close to f.

Theorem 25. *Let* f *be a piecewise-continuous function on the interval* $[-\pi, \pi]$. *Let* c_n *be the* nth *Fourier coefficient of* f *and let* s_n *be the* nth *partial sum of the Fourier series for* f. *Then*

(i) $\displaystyle\sum_{n=-\infty}^{\infty} |c_n|^2 = \frac{1}{2\pi} \int_{-\pi}^{\pi} |f(x)|^2 \, dx$

(ii) $\displaystyle\lim_n \int_{-\pi}^{\pi} |f(x) - s_n(x)|^2 \, dx = 0.$

Proof. Let $r < 1$. From the uniform convergence of the series

$$f_r(e^{i\theta}) = \sum_{n=-\infty}^{\infty} c_n r^{|n|} e^{in\theta}$$

it is easy to compute that

$$\frac{1}{2\pi} \int_{-\pi}^{\pi} |f_r(e^{i\theta})|^2 \, d\theta = \frac{1}{2\pi} \int_{-\pi}^{\pi} f_r(e^{i\theta}) f_r(e^{i\theta})^* \, d\theta$$

$$= \sum_{n=-\infty}^{\infty} |c_n|^2 r^{2|n|}.$$

Now, let r tend to 1. If f is continuous, then f_r converges uniformly to f and so we obtain statement (i) for f. But that argument works equally well for a function f which is continuous on $a \leq \theta \leq b$ and is 0 off that interval. Since a piecewise-continuous function is the sum of a finite number of such functions (disregarding the values at a finite number of points), we have the relation (i) for each piecewise continuous f.

To prove (ii), just calculate that

$$\frac{1}{2\pi} \int_{-\pi}^{\pi} |f(x) - s_n(x)|^2 \, dx = \frac{1}{2\pi} \int_{-\pi}^{\pi} |f(x)|^2 \, dx - \sum_{k=-n}^{n} |c_k|^2.$$

Then apply (i).

Corollary. *If* f *is a function of class* C^1 *on the unit circle, the Fourier series for* f *converges uniformly to* f.

Proof. We have

$$f(x) = f(0) + \int_0^x f'(t) \, dt.$$

Then integration by parts shows that the nth Fourier coefficient of f' is

$$\frac{1}{2\pi} \int_{-\pi}^{\pi} f'(x) e^{-inx} \, dx = \frac{in}{2\pi} \int_{-\pi}^{\pi} f(x) e^{-inx} \, dx$$

$$= inc_n.$$

Therefore, the nth partial sum of the Fourier series for f' is s_n', the derivative of the nth partial sum of the Fourier series for f. So

$$f(x) - s_n(x) = f(0) - s_n(0) + \int_0^x [f'(t) - s_n'(t)] \, dt$$

$$|f(x) - s_n(x)| \leq |f(0) - s_n(0)| + \int_0^{2\pi} |f'(t) - s_n'(t)| \, dt.$$

We know that $s_n(0)$ converges to $f(0)$, by Theorem 23. We also know that

$$\lim_n \int_0^{2\pi} |f'(t) - s_n'(t)|^2 \, dt = 0.$$

The result now follows from the Schwartz inequality:

$$\frac{1}{2\pi} \int_{-\pi}^{\pi} |g| \leq \left(\frac{1}{2\pi} \int_{-\pi}^{\pi} |g|^2 \right)^{1/2}$$

We have left that inequality for the exercises.

Exercises

1. Let f be a piecewise-continuous function on the unit circle. Suppose that all of the Fourier coefficients of f are non-negative (real numbers). Prove that the Fourier series for f converges uniformly to f.

2. For piecewise-continuous f and g, define

$$\langle f, g \rangle = \frac{1}{2\pi} \int_{-\pi}^{\pi} f(x) g(x)^* \, dx.$$

(a) Show that $\langle ., . \rangle$ is a pseudo inner product on the space of piecewise-continuous functions. That means that it satisfies the conditions

  (i) $\langle cf + g, h \rangle = c\langle f, h \rangle + \langle g, h \rangle$;
  (ii) $\langle g, f \rangle = \langle g, f \rangle^*$;
  (iii) $\langle f, f \rangle \geq 0$.

(b) Show that the Cauchy-Schwartz inequality holds:

$$|\langle f, g \rangle|^2 \leq \langle f, f \rangle \langle g, g \rangle.$$

(Consult the proof for Euclidean space.)

3. Use Theorem 25 to show that, if f has Fourier coefficients a_n and g has Fourier coefficients b_n, then

$$\langle f, g \rangle = \sum_{n=-\infty}^{\infty} a_n b_n^*.$$

(See Exercise 2 for the notation $\langle f, g \rangle$.)

4. If f and g are piecewise-continuous functions on the unit circle, define the convolution $f * g$ by

$$(f * g)(x) = \frac{1}{2\pi} \int_{-\pi}^{\pi} f(x - t) g(t) \, dt.$$

Show that

(a) $f * g = g * f$;
(b) $f * (g + h) = f * g + f * h$;
(c) if f has Fourier coefficients a_n and g has Fourier coefficients b_n, then $f * g$ has Fourier coefficients $a_n b_n$.

5. If f and g are piecewise-continuous functions, then the Fourier series for the convolution $f * g$ converges uniformly to $f * g$. (*Hint:* Use Theorem 25 and Exercise 4(c).)

6. Refer to the convolution multiplication defined in Exercise 4. If there were an identity element for that multiplication ($e * f = f$ for all f), what would the Fourier series for e have to be? (Use Exercise 4(c).) Is that the Fourier series of a function? (See Exercise 1 or Theorem 25.)

7. If f is a function on the real line, you know what the t-translate of f is:

$$f_t(x) = f(x + t).$$

What is the relation between the Fourier coefficients of f and those of the function f_t?

8. If f has Fourier coefficients c_n, translate the coefficients by k:

$$a_n = c_{k+n}.$$

Which function has $\{a_n\}$ for its sequence of Fourier coefficients?

9. Use the result of Exercise 8 to show the following. If some (sort of a) function f had the Fourier series

$$\sum_{n=-\infty}^{\infty} e^{inx}$$

then that function would have to satisfy

(i) $f(x) = 0$, $x \neq 0$;

(ii) $\dfrac{1}{2\pi} \displaystyle\int_{-\pi}^{\pi} f(x)\, dx = 1$.

Dirac called that function the "delta function". Obviously no such function exists in the sense of our definition of function. But, find a function of bounded variation F such that

$$\frac{1}{2\pi} \int_{-\pi}^{\pi} e^{-inx}\, dF(x) = 1, \qquad \text{all } n.$$

10. Look at Figure 23. Does it look like the Poisson kernel P_r converges (as r goes to 1) to the mythical delta function of Exercise 9? What is the convolution of f with the non-existent delta function? In view of those things, does it seem plausible that $f * P_r$ converges to f as r tends to 1?

11. Let f be a function which is continuous on the closed disk $|z| \leq 1$ and is complex-analytic on the open disk $|z| < 1$. Then f defines a function on the unit circle $|z| = 1$ and that function satisfies

$$\frac{1}{2\pi} \int_{-\pi}^{\pi} f(e^{i\theta}) e^{in\theta}\, d\theta = 0, \qquad n = 1, 2, 3, \ldots .$$

Prove that, if we are given a continuous function on the circle which satisfies that condition, it can be extended to a function which is continuous on the disk $|z| \leq 1$ and analytic on $|z| < 1$.

12. Which continuous functions on the closed unit disk $|z| \leq 1$ can be uniformly approximated by polynomials $p(z)$? Compare with the Weierstrass approximation theorem.

13. We mentioned the Cesaro means of the Fourier series for f:

$$\sigma_n = \frac{1}{n}(s_0 + s_1 + \cdots + s_{n-1}), \qquad n = 1, 2, 3, \ldots.$$

Show that

$$\sigma_n(x) = \frac{1}{2\pi}\int_{-\pi}^{\pi} f(t)K_n(x - t)\, dt$$

where K_n is the nth Cesaro mean of the formal series

$$\sum_{n=-\infty}^{\infty} e^{inx}.$$

Verify that the sequence $\{K_n\}$ satisfies the conditions analogous to properties (i), (ii), and (iii) of the Poisson kernel (positive, integral 1, etc.). From those properties show that the Cesaro means for a continuous f on the circle converge uniformly to f.

5.10. The Dirichlet Problem

Recall that a (complex-valued) function u on an open set in the plane is called harmonic if it is of class C^2 and satisfies Laplace's equation:

$$\frac{\partial^2 u}{\partial x^2} + \frac{\partial^2 u}{\partial y^2} = 0.$$

We came across such functions in our work with complex-analyticity. We did not say much about them, except to note that an analytic function is harmonic and hence, that its real and imaginary parts are harmonic as well. Harmonic functions are important in their own right, for a variety of reasons. Thus, we shall devote this little section to a few of their fundamental properties. The discussion will also help to round out our brief treatment of power series and Fourier series.

Theorem 26 (Mean Value Property). *Let* u *be a harmonic function on an open set* D *in the plane. If the closed disk* $\{z;\, |z - z_0| \leq \rho\}$ *is contained in* D, *then*

$$u(z_0) = \frac{1}{2\pi}\int_{-\pi}^{\pi} u(z_0 + \rho e^{i\theta})\, d\theta.$$

Proof. As usual, we assume that $z_0 = 0$, for convenience. Since u is of class C^2, it has continuous radial and angular derivatives of the second order (except at the origin). A simple calculation shows that Laplace's equation in terms of such derivatives is:

(5.58)
$$r\frac{\partial}{\partial r}\left(r\frac{\partial u}{\partial r}\right) + \frac{\partial^2 u}{\partial\theta^2} = 0.$$

Define

$$I(r) = r\int_{-\pi}^{\pi}\frac{\partial u}{\partial r}\, d\theta, \qquad 0 < r \leq \rho.$$

By Laplace's equation (5.58),

$$\frac{\partial I}{\partial r} = -\frac{1}{r} \int_{-\pi}^{\pi} \frac{\partial^2 u}{\partial \theta^2} \, d\theta$$

$$= -\frac{1}{r} \frac{\partial u}{\partial \theta}\Big|_{-\pi}^{\pi}$$

$$= 0.$$

Thus $I(r)$ is constant, and since obviously

$$\lim_{r \to 0} I(r) = 0$$

we have $I(r) = 0$. In other words,

(5.59) $$\int_{-\pi}^{\pi} \frac{\partial u}{\partial r} \, d\theta = 0, \qquad 0 < r \leq \rho.$$

Now use that fact to compute that

$$\frac{1}{2\pi} \int_{-\pi}^{\pi} u(\rho e^{i\theta}) \, d\theta = u(0) + \frac{1}{2\pi} \int_{-\pi}^{\pi} \left[\int_{0}^{\rho} \frac{\partial u}{\partial r} \, dr \right] d\theta$$

$$= u(0) + \frac{1}{2\pi} \int_{0}^{\rho} \left[\int_{-\pi}^{\pi} \frac{\partial u}{\partial r} \, d\theta \right] dr$$

$$= u(0).$$

Corollary (*Weak Maximum Principle*). *Let* V *be a bounded open set in the plane. Let* u *be a function which is continuous on* $\bar{V}$ *and harmonic on* V. *The maximum of* $|u(z)|$ *over* $\bar{V}$ *occurs on the boundary of* V. *In case* u *is real-valued, both the maximum and the minimum of* $u(z)$ *over* $\bar{V}$ *are attained on the boundary of* V.

Proof. This follows from the mean value property, just as it did for analytic functions.

A particular consequence of the maximum principle is that a harmonic function on V which vanishes on the boundary of V is 0. Thus, each harmonic function is determined by its values on the boundary. The **Dirichlet problem** for an open set V is this. Suppose that we are given a (continuous) function on the boundary of V. Can we find a harmonic function on V which takes on those values on the boundary? The question should be made a little bit more precise. It turns out to be somewhat complicated for general open sets V; but, when V is a disk, we can handle it nicely. In fact, we did that in the last section without saying so. Let us restate the result in the present context.

Theorem 27. *Let* f *be a continuous function on the unit circle. There exists a function* F *which is continuous on the closed unit disk, harmonic on the open unit disk, and which coincides with* f *on the unit circle. That function*

F *is unique, and it is given by the Poisson integral formula:*

$$F(re^{i\theta}) = \frac{1}{2\pi} \int_{-\pi}^{\pi} f(e^{it}) P_r(\theta - t)\, dt.$$

Proof. As we just remarked, if there is such a function F, there is only one. Define F on the open disk by the Poisson formula and define $F(e^{i\theta}) = f(e^{i\theta})$. According to Theorem 5.23,

$$\lim_{r \to 1} F(re^{i\theta}) = f(e^{i\theta})$$

uniformly in θ. From that it is clear that F is continuous on the closed unit disk.

Why is F harmonic on the open unit disk? There are several ways to see that. One is to compute that F satisfies Laplace's equation by differentiating under the integral sign. The harmonicity derives from the fact that the function $P(r, \theta) = P_r(\theta)$ is harmonic. But, let's observe that F is harmonic by this method. We have

$$P_r(\theta - t) = \mathrm{Re}\left[\frac{e^{it} + re^{i\theta}}{e^{it} - re^{i\theta}}\right].$$

Suppose that f is real-valued. Then

$$F(z) = \mathrm{Re}\, h(z)$$

where

$$h(z) = \frac{1}{2\pi} \int_{-\pi}^{\pi} f(e^{it})\frac{e^{it} + z}{e^{it} - z}\, dt.$$

Obviously h is complex-analytic on $|z| < 1$. So F is the real part of an analytic function and is therefore harmonic.

Corollary. *A real-valued function is harmonic if and only if it is locally the real part of an analytic function.*

Proof. If u is a real-valued harmonic function on an open set V and if $z_0 \in V$, let D be any disk which contains z_0 and lies in V. By the proof of the theorem, there is a function h, analytic on D, such that $u(z) = \mathrm{Re}\, h(z)$, $z \in D$.

Corollary. *Let* V *be a connected open set in the plane and let* u *be a harmonic function on* V. *If* u *vanishes on a non-empty open subset of* V, *then* u $= 0$.

Proof. We may assume that u is real-valued. Let D be the union of all open sets $U \subset V$ such that $u(z) = 0$, $z \in U$. Then D is a non-empty open subset of V. We shall show that D is closed relative to V. Since V is connected, that will prove the result.

Suppose w is a point in V which is in $\bar{D}$. Choose an open disk W so that $w \in W \subset V$. From the definition of D and the fact that $w \in \bar{D}$, it follows that u vanishes on a non-empty open subset of W. But, on W we have $u = \text{Re } h$ where h is analytic. So h is pure imaginary (and therefore constant) on a non-empty open subset of W. Thus h is an imaginary constant. So u is zero on W. *Conclusion*: $w \in D$.

Corollary (Strong Maximum Principle). *Let* u *be a real-valued harmonic function on a connected open set* V. *Then* u *cannot attain either a maximum or a minimum value on* V *unless* u *is constant.*

Proof. The mean value property shows that u is constant in a neighborhood of any point in V where it attains a maximum or minimum. By the previous corollary, we are done.

Corollary. *Let* u *be a continuous complex-valued function on any open set* V *in the plane. Then* u *is harmonic if and only if* u *has the mean value property.*

Proof. It is a local theorem; i.e., we need only prove it on a disk. So, let us say that u is continuous on $|z| \leq 1$ and has the mean value property

$$u(z_0) = \frac{1}{2\pi} \int_{-\pi}^{\pi} u(z_0 + \rho e^{i\theta})\, d\theta$$

on each disk $|z - z_0| \leq \rho$ contained in the unit disk. Let v be the harmonic function on $|z| < 1$ which (is continuous on $|z| \leq 1$ and) agrees with u on the unit circle. Then, both u and v have the mean value property; consequently, $u - v$ has that property. Thus the maximum of $|u - v|$ is attained on the unit circle, where $u - v = 0$. *Conclusion*: $u = v$, so u is harmonic.

Exercises

1. If f is complex-analytic and u is harmonic, then the composition $u \circ f$ is harmonic.

2. True or false? If u is harmonic on the unit disk and if there exists a sequence $\{z_n\}$ such that $z_n \longrightarrow 0$ and $u(z_n) = 0$, then $u = 0$.

3. Show that a function u on the unit disk is harmonic if and only if there exists a sequence of numbers c_n such that

$$u(re^{i\theta}) = \sum_{n=-\infty}^{\infty} c_n r^{|n|} e^{in\theta}$$

on the disk.

4. Let u be a harmonic function on the entire plane. Show that $u = \text{Re } f$, where f is analytic on the plane.

5. (Liouville) Let u be a harmonic function on the plane. If u is bounded above or bounded below, then u is constant.

6. Let
$$u(re^{i\theta}) = \log r, \qquad r \neq 0.$$
Show that u is harmonic on the punctured plane. Does there exist f, analytic on the punctured plane, so that $u = Re\, f$?

7. Let $p(x, y)$ be a polynomial in two real variables with real coefficients. Some such polynomials are harmonic functions on the plane and some are not. True or false? If $p(x, y)$ is harmonic, then $p(x, y) = Re\, f(x + iy)$ where $f(z)$ is a complex polynomial.

8. (Harnack) Let u be a *non-negative* harmonic function in the unit disk. Show that
$$\frac{1-r}{1+r} u(0) \leq u(re^{i\theta}) \leq \frac{1+r}{1-r} u(0).$$

9. Any uniform limit of harmonic functions is harmonic.

10. (Harnack's principle) Let $\{u_n\}$ be a sequence of real-valued harmonic functions on the connected open set D. Suppose that the sequence is increasing: $u_1 \leq u_2 \leq \cdots$. Let
$$u(z) = \lim_n u_n(z).$$
Then either $u(z) = \infty$ for every $z \in D$ or u is a harmonic function on D. *Hint:* See Exercises 8 and 9.

6. Normed Linear Spaces

6.1. Linear Spaces and Norms

We are going to consider a type of "space" which is more general than Euclidean space. There are three reasons for doing so.

1. The discussion will sharpen our understanding of what we have done up to this point.

2. The terminology will enable us to formulate clearly and naturally some results which we have not yet dealt with.

3. The more general concept will pave the way to a thorough discussion of integration in the next chapter.

Definition. A **linear space** *is a vector space over the field of real numbers or the field of complex numbers.*

Thus, a linear space consists of a set L, together with a vector addition on L and a scalar multiplication of vectors by numbers. The vector addition is required to satisfy:

1. Addition is commutative,

$$X + Y = Y + X.$$

2. Addition is associative,

$$(X + Y) + Z = X + (Y + Z).$$

3. There is a unique vector 0 (zero) in L such that

$$0 + X = X$$

for all X in L.

4. To each X in L there corresponds a unique vector $-X$ in L such that

$$X + (-X) = 0.$$

The scalar multiplication is required to satisfy

5. $\qquad\qquad (ab)X = a(bX)$

6. $\qquad\qquad 1X = X$

and to be related to vector addition by the two distributive laws:

7. $\qquad\qquad c(X + Y) = cX + cY$

8. $\qquad\qquad (a + b)X = aX + bX.$

In the case of a **real linear space**, we multiply vectors X by real numbers c; and, in the case of a **complex linear space**, we use complex numbers c. When arguments are being presented which are applicable to both cases, we shall refer to the field of numbers as the field of **scalars**.

If we start with a linear space L, each subset of which is "closed" under the linear operations in L provides us with an example of another linear space: A **(linear) subspace** of L is a subset M such that

(i) if $X, Y \in M$, then $(X + Y) \in M$;
(ii) if $X \in M$, then $(cX) \in M$ for all scalars c.

Nearly all of the linear spaces we meet will be spaces of functions—subspaces of the space of functions on a set. The richness of this class of spaces is one indicator of the fact that they form the natural settings for many questions in analysis.

EXAMPLE 1. Let S be any set. The collection of all real- [complex-] valued functions on S forms a real [complex] linear space, with the (usual) operations

$$(f + g)(s) = f(s) + g(s)$$
$$(cf)(s) = cf(s).$$

In the particular case $S = R$ the basic space under discussion is the space of all real- [complex-] valued functions on the real line. Here is a decreasing sequence of subspaces: those which consist of the continuous functions, the differentiable functions, the functions of class $C^1, \ldots$, the functions of class $C^k, \ldots$, the functions of class C^∞, the real-analytic functions, the polynomials, $\ldots$, the polynomials of degree at most $n, \ldots$, the affine functions, the linear functions, the zero function.

In case S is the interval $[0, 1]$, another little decreasing chain of spaces of functions consists of the spaces of bounded functions, functions of bounded variation, functions which are continuously differentiable, functions which satisfy the homogeneous differential equation $f'' + f = 0$.

If S consists of the first n positive integers, $S = \{1, \ldots, n\}$, a real-valued function on S is an ordered n-tuple $X = (x_1, \ldots, x_n)$. Therefore the space of real-valued functions on S is the basic space R^n. The subspaces of R^n were described in Chapter 1. Each has a dimension k, $0 \leq k \leq n$, and can be defined by imposing $(n - k)$ homogeneous linear conditions on the coordinates $x_1, \ldots, x_n$. When $n = 3$, the subspaces other than R^3 and $\{0\}$ are the planes through the origin and the straight lines through the origin.

If S is the set of positive integers, the space of functions on S is the space of sequences of real [complex] numbers, $X = (x_1, x_2, x_3, \ldots)$. We shall mention in subsequent examples some important subspaces of the space of sequences.

Much of analysis deals with problems of approximation and convergence in linear spaces. Frequently the appropriate measure of approximation is provided by the following type of function on vectors, which generalizes the idea of length.

Definition. *If* L *is a linear space, a* **semi-norm** *on* L *is a real-valued function* $\| \ldots \|$ *on* L *with these properties:*

(i) $\|X\| \geq 0$;
(ii) $\|X + Y\| \leq \|X\| + \|Y\|$ *(triangle inequality);*
(iii) $\|cX\| = |c| \|X\|$.

A semi-norm is called a **norm** *if it also satisfies*

(i)′ *if* $\|X\| = 0$, *then* $X = 0$.

A [**semi-**] **normed linear space** *is a linear space* L, *together with a specified* [*semi-*] *norm on* L.

EXAMPLE 2. *Euclidean space* of dimension n is the normed linear space which consists of the linear space R^n, together with the norm

$$|X| = (x_1^2 + \cdots + x_n^2)^{1/2}.$$

This special norm $|\cdots|$ is referred to as **length**. There are other norms on R^n which are important and useful. Two of these are

(6.1)
$$\|X\|_1 = |x_1| + \cdots + |x_n|$$
$$\|X\|_\infty = \max_k |x_k|.$$

Let us mention a more general class of norms on R^n. Let p be a number, $1 \leq p < \infty$, and define

(6.2)
$$\|X\|_p = (|x_1|^p + \cdots + |x_n|^p)^{1/p}.$$

We shall defer the proof that $\|\cdots\|_p$ is a norm, since to see that it satisfies the triangle inequality is somewhat more difficult than the verification in the special cases $p = 2$ (length) and $p = 1$. The notation for the maximum norm $|| \cdots ||_\infty$ is explained by the result that

$$\lim_{p \to \infty} || X ||_p = \max_k | x_k |$$

which we have left to the exercises.

EXAMPLE 3. *Complex Euclidean space* of dimension n is the complex linear space C^n, together with the norm

$$| X | = (| x_1 |^2 + \cdots + | x_n |^2)^{1/2}.$$

The functions $|| \cdots ||_p$ as defined above each provide a norm on C^n.

EXAMPLE 4. The *sup norm* on functions entered into our discussion of uniform convergence. In that discussion, usually we were dealing with continuous functions; however we might as well talk about sup norms in the following context. Let S be any non-empty set, and let $B(S)$ be the linear space of all bounded complex-valued functions on S. For f in $B(S)$ define

(6.3) $|| f ||_\infty = \sup \{| f(s) |; s \in S\}.$

It is a simple but important matter to verify that $B(S)$, with the sup norm, is a normed linear space.

(i) The norm $|| f ||_\infty$ is defined as the supremum (least upper bound) of the image of $| f |$, that is, as the sup of the following set of non-negative numbers

$$A_f = \{| f(s) |; s \in S\}.$$

The point of dealing with bounded functions f is that the set A_f is bounded and, accordingly, $|| f ||_\infty < \infty$. Note that A_f is non-empty because S is; hence $|| f ||_\infty \geq 0$.

(ii) If f and g are bounded functions on S, how do we show that

$$|| f + g ||_\infty \leq || f ||_\infty + || g ||_\infty?$$

There are various ways to describe why this is so. Perhaps the simplest is this. If $s \in S$, then

$$| (f + g)(s) | = | f(s) + g(s) |$$
$$\leq | f(s) | + | g(s) |$$
$$\leq || f ||_\infty + || g ||_\infty.$$

Thus $|| f ||_\infty + || g ||_\infty$ is an upper bound for the image of $| f + g |$ and is at least as large as $|| f + g ||_\infty$, the least upper bound.

(iii) The verification that $|| cf ||_\infty = | c | \, || f ||_\infty$ will be left as an exercise.

So $|| f ||_\infty$ is a semi-norm on $B(S)$. Why is it a norm? Well, suppose $|| f ||_\infty = 0$. Then $| f(s) | \leq 0$ at all points s, that is, $f(s) = 0$ for every s.

If K is a compact set in R^n, then one subspace of $B(K)$ is $C(K)$, the space of continuous complex-valued functions on K. (Remember it is a theorem that a continuous function on a compact set is bounded.) This provides us with another example of a normed linear space, namely, $C(K)$ with the sup norm.

EXAMPLE 5. Let I be a closed interval on the real line. Let $BV(I)$ be the space of complex-valued [real-valued] functions of bounded variation on I. There is a natural semi-norm on $BV(I)$, namely, total variation (p. 161):

$$(6.4) \qquad \|f\| = V(f).$$

The properties of a semi-norm are easily verified from the definition of the total variation of a function. In this case, some non-zero vectors do have norm zero. In fact $\|f\| = 0$ if and only if f is a constant function.

EXAMPLE 6. If I is a closed interval on the real line, let $C^1(I)$ be the space of functions of class C^1 on I. A norm on $C^1(I)$ which is frequently useful is

$$(6.5) \qquad \begin{aligned} \|f\| &= \|f\|_\infty + \|f'\|_\infty \\ &= \sup |f| + \sup |f'|. \end{aligned}$$

This verification is best carried out by noting that (1) the sup norm is a norm on $C^1(I)$, (2) $f \to \|f'\|_\infty$ is a semi-norm on $C^1(I)$, (3) the sum of a norm and a semi-norm is a norm. There is an analogous norm on $C^k(I)$, $k = 2, 3, \ldots$.

EXAMPLE 7. Consider $L(R^n, R^k)$, the space of linear transformations from R^n into R^k. In Example 21 of Chapter 3 we discussed a norm on $L(R^n, R^k)$:

$$(6.6) \qquad \|T\| = \sup \{|T(X)|; |X| = 1\}.$$

Note that $\|T\|$ is the sup norm of the restriction to the unit ball of the function $|T|$. This makes it easy to verify that (6.6) defines a semi-norm. The fact that it is a norm results from the observation that, if a *linear* transformation is 0 on the unit ball, it is 0.

EXAMPLE 8. Let I be a closed interval in R^1 and let $PC(I)$ be the space of piecewise-continuous functions on I. Then $PC(I)$ is a linear subspace of the space of all bounded functions on I. Therefore, the sup norm is a norm on $PC(I)$. There is another semi-norm on $PC(I)$ which has an important connection with integration:

$$\|f\|_1 = \int_I |f|.$$

The verification that this is a semi-norm is straightforward. We merely remark that it utilizes the fact that integration is linear. We can have $\|f\|_1 = 0$ without $f = 0$. It is not difficult to see that $\|f\|_1 = 0$ if and only if $f(x) = 0$ except at a finite number of points.

Note that the integral semi-norm $\|f\|_1$ is actually a norm on the space $C(I)$, because a continuous function for which

$$\int_I |f| = 0$$

must be 0 on I.

There is a class of semi-norms on $PC(I)$ which are analogous to norms on R^n:

(6.7) $$\|f\|_p = \left(\int |f|^p\right)^{1/p}, \qquad 1 \le p < \infty.$$

EXAMPLE 9. Let p be a number, $1 \le p < \infty$. Let ℓ^p be the space of all sequences $X = (x_1, x_2, \ldots)$ of complex [real] numbers for which

$$\sum_n |x_n|^p < \infty.$$

On ℓ^p we have a norm

(6.8) $$\|X\|_p = \left(\sum_n |x_n|^p\right)^{1/p}.$$

The space of bounded sequences is denoted ℓ^∞ and has the norm

(6.9) $$\|X\|_\infty = \sup\{|x_n|; n \in Z_+\}.$$

Our main point in mentioning these spaces ℓ^p is to strengthen the analogy (which we hope the reader has sensed) between Examples 2 and 8. In this Example (as in Example 8), concentrate mainly on the values $p = 1, 2, \infty$, where the verification of the triangle inequality is easier. Note that, until one has verified the triangle inequality, it is not apparent that ℓ^p is a subspace of the space of sequences.

EXAMPLE 10. Let K be a compact set in R^n. A complex-valued function f on K is said to satisfy a **Lipschitz condition** if there exists a constant $c > 0$ such that

$$|f(X) - f(Y)| \le c |X - Y|, \qquad X, Y \in K.$$

The smallest such c is called the **Lipschitz norm** of f:

$$\|f\| = \sup_{X \ne Y} \frac{|f(X) - f(Y)|}{|X - Y|}.$$

The space of functions which satisfy a Lipschitz condition on K is a subspace of $C(K)$ and the Lipschitz norm is a semi-norm on this subspace. Of course $\|f\| = 0$ if and only if f is constant.

EXAMPLE 11. This example will be a brief discussion of an important special class of normed linear spaces. If L is a real [complex] linear space, an **inner product** on L is a real- [complex-] valued function $\langle ., . \rangle$ on $L \times L$ such that

(i) $\langle X, X \rangle \ge 0$; if $\langle X, X \rangle = 0$, then $X = 0$;
(ii) $\langle Y, X \rangle = \langle X, Y \rangle^*$;

(iii) $\langle cX + Y, Z \rangle = c\langle X, Z \rangle + \langle Y, Z \rangle$.

An **inner product space** is a linear space L, together with an inner product on L. In any inner product space, we have the **Cauchy-Schwarz inequality**

$$|\langle X, Y \rangle|^2 \leq \langle X, X \rangle \langle Y, Y \rangle.$$

It is trivial when $\langle Y, Y \rangle = 0$, and when $\langle Y, Y \rangle \neq 0$, it results from applying the inequality $0 \leq \langle X + cY, X + cY \rangle$ to the case where c is the constant

$$c = -\frac{\langle X, Y \rangle}{\langle Y, Y \rangle}.$$

It follows from the Cauchy-Schwarz inequality that

(6.10) $$\|X\| = \langle X, X \rangle^{1/2}$$

satisfies the triangle inequality and is therefore a norm on L:

$$\begin{aligned}
\|X + Y\|^2 &= \langle X + Y, X + Y \rangle \\
&= \langle X, X \rangle + \langle X, Y \rangle + \langle Y, X \rangle + \langle Y, Y \rangle \\
&= \|X\|^2 + 2\mathrm{Re}\langle X, Y \rangle + \|Y\|^2 \\
&\leq \|X\|^2 + 2\|X\|\|Y\| + \|Y\|^2 \\
&= (\|X\| + \|Y\|)^2.
\end{aligned}$$

The norm on an inner product space is a very special type of norm, because it satisfies the **parallelogram law**:

$$\|X + Y\|^2 + \|X - Y\|^2 = 2[\|X\|^2 + \|Y\|^2].$$

If we start with a normed linear space, and if the norm satisfies the parallelogram law, then the norm comes from an inner product, i.e., there is one and only one inner product on the space which is related to the norm by (6.10). In the real case, the inner product is obtained from the norm by

(6.11) $$\langle X, Y \rangle = \tfrac{1}{4}[\|X + Y\|^2 - \|X - Y\|^2].$$

In the complex case, the inner product is

(6.12) $$\langle X, Y \rangle = \tfrac{1}{4} \sum_{n=1}^{4} i^n \|X + i^n Y\|^2.$$

In the examples we have discussed thus far, there were few inner product spaces. Of course, Euclidean space of dimension n is an inner product space. The space of continuous functions on a box B with the norm

$$\|f\|_2 = \left(\int_B |f|^2 \right)^{1/2}$$

is an inner product space. The inner product which gives that norm is

(6.13) $$\langle f, g \rangle = \int_B f g^*.$$

The space ℓ^2 (Example 9) is an inner product space:

(6.14) $\langle X, Y \rangle = \sum_n x_n y_n^*.$

Both it and the integration example are infinite-dimensional analogues of Euclidean space.

Exercises

1. Let k be a positive integer, and let S be the space of all sequences

$$X = (X_1, X_2, \ldots)$$

of vectors in R^k. Which of the following sets of sequences X are subspaces of S?

    (a) all X which are bounded;
    (b) all X such that $\{X_n\}$ converges;
    (c) all X such that $\sum_n X_n$ converges absolutely;
    (d) all X such that $|X_n|$ converges to 1.

2. True or false? The set of discontinuous functions is a subspace of the space of real-valued functions on $[0, 1]$. How about the set of convex functions?

3. Each semi-norm satisfies $\| X - Y \| \geq \| X \| - \| Y \|$.

4. On a linear space, the sum of a semi-norm and a norm is a norm.

5. If $\| \cdots \|$ is a semi-norm on L, then $M = \{X; \| X \| = 0\}$ is a subspace of L.

6. If $\| \cdots \|$ is a norm on the real linear space L and if it satisfies the parallelogram law

$$\| X + Y \|^2 + \| X - Y \|^2 = 2[\| X \|^2 + \| Y \|^2]$$

then

$$\langle X, Y \rangle = \tfrac{1}{4}[\| X + Y \|^2 - \| X - Y \|^2]$$

is an inner product on L.

7. By examining the proof of the Cauchy-Schwarz inequality, show that it is a strict inequality except in the case where the two vectors involved are linearly dependent (one is a scalar multiple of the other).

8. In a normed linear space $(L, \| \cdots \|)$, **distance** is defined by $d(X, Y) = \| X - Y \|$

    (a) Prove that, in an inner product space, the shortest distance between two points is (along) a straight line.

    (b) Give an example which shows that this property of straight lines fails in some normed linear space.

9. Show that, on a closed interval of the real line, every continuously-differentiable function satisfies a Lipschitz condition. Prove that the Lipschitz norm for such a function is the sup norm of its derivative.

10. Refer to the norms in Example 2. Prove that

$$\| X \|_\infty = \lim_{p \to \infty} \| X \|_p.$$

11. If f is continuous on $[0, 1]$, then

$$\lim_{p \to \infty} \left(\int_0^1 |f|^p \right)^{1/p} = \sup |f|.$$

12. Let $C^n(I)$ be the space of functions of class C^n on a closed interval I. Show that

$$\| f \| = \sum_{k=0}^{n} \frac{1}{k!} \| f^{(k)} \|_\infty$$

is a norm on $C^n(I)$ and that it satisfies $\| fg \| \leq \| f \| \| g \|$.

6.2. Norms on R^n

We shall give a geometrical description of all possible norms on R^n. This will make it easier to see that the functions $\| \cdots \|_p$, $1 \leq p < \infty$, are norms on R^n. Here is the basic fact.

Theorem 1. *Let* $\| \cdots \|$ *be a norm on* R^n. *There exist positive constants* k, K *such that*

(6.15) $$k|X| \leq \|X\| \leq K|X|, \qquad X \in R^n.$$

Proof. Let $E_1, \ldots, E_n$ be the standard basis vectors in R^n:

$$X = (x_1, \ldots, x_n)$$
$$= x_1 E_1 + \cdots + x_n E_n.$$

Then the triangle inequality tells us that

$$\|X\| \leq |x_1| \|E_1\| + \cdots + |x_n| \|E_n\|.$$

Let

$$M = \max_j \|E_j\|.$$

and we have

$$\|X\| \leq M \sum_j |x_j|.$$

Each $|x_j|$ is dominated by the length of X, so that

$$\sum_j |x_j| \leq n|X|.$$

Therefore, if $K = nM$,

(6.16) $$\|X\| \leq K|X|, \qquad X \in R^n.$$

Thus, we have half of what we want to know.

Now we want to prove that there exists $k > 0$ such that

$$k|X| \leq \|X\|, \qquad X \in R^n.$$

Since $|cX| = |c| |X|$ and $\|cX\| = |c| \|X\|$, we need only find $k > 0$ so that

$$k|X| \leq \|X\|, \qquad |X| = 1.$$

In other words, we wish to show that

(6.17) $$0 < \inf\{\|X\|; |X| = 1\}.$$

Now, by (6.16) $\|\cdots\|$ is a (uniformly) continuous function on R^n:

$$\left|\|X\| - \|Y\|\right| \le \|X - Y\|$$
$$\le K|X - Y|.$$

The unit sphere $\{X; |X| = 1\}$ is compact. Hence the infimum of $\|X\|$ over the unit sphere is attained at some point of the sphere, i.e., there exists Y with $|Y| = 1$ such that

$$\|Y\| = \inf\{\|X\|; |X| = 1\}.$$

Now $\|Y\| \ne 0$ because $Y \ne 0$. That establishes (6.17), and we are done.

Because of the fact that $\|cX\| = |c| \|X\|$, every norm on R^n is determined by its unit ball:

$$B = \{X; \|X\| < 1\}.$$

What does such a unit ball look like? First, B is an open set, because $\|\cdots\|$ is a continuous function on R^n. Second, B is symmetric about the origin: If $X \in B$ then $(-X) \in B$. Third, B is convex: If X, Y are in B, the line segment between X and Y is in B. Fourth, B is bounded: If $X \in B$ then $|X| \le 1/k$ (6.15).

Theorem 2. *Let* B *be a subset of* R^n *which is bounded, open, convex, and symmetric about the origin. Define*

(6.18) $$\|X\| = \inf\left\{t; t > 0 \quad and \quad \frac{1}{t}X \in B\right\}.$$

Then $\|\cdots\|$ *is a norm on* R^n *and* B *is the unit ball for this norm,* B = $\{X; \|X\| < 1\}$.

Proof. Let X be any vector in R^n. Consider the set of numbers

$$M_X = \left\{t; t > 0 \quad \text{and} \quad \frac{1}{t}X \in B\right\}$$

involved in defining the proposed norm of X. Then M_X is a very simple type of set, because it has the defining property of a half-line: If $t \in M_X$ and $s > t$, then $s \in M_X$. For

$$\frac{1}{s}X = cy, \quad \text{where } Y = \frac{1}{t}, \quad c = \frac{t}{s}.$$

Since B is a convex set which contains the origin, the facts that $Y \in B$ and $0 < c < 1$ imply that $cY \in B$.

It is important to show that M_X is non-empty. Since

$$\lim_{c \to 0} cX = 0$$

and B is a neighborhood of 0, we know that $cX \in B$ for all sufficiently small scalars c. Therefore $t \in M_X$ for all sufficiently large positive numbers t. So M_X is a non-empty open subset of the positive real line.

Now consider the proposed norm

$$\|X\| = \inf M_X.$$

Since M_X is non-empty,

$$0 \leq \|X\| < \infty.$$

Furthermore, from the first two paragraphs of the proof, we know that M_X is the open half-line

$$M_X = \{t; t > \|X\|\}.$$

Let c be a positive scalar. If $t > 0$, then

$$\frac{1}{t}X = \frac{1}{ct}(cX).$$

Hence, $t \in M_X$ if and only if $(ct) \in M_{cX}$. Thus $M_{cX} = cM_X$ and

$$\|cX\| = c\|X\|, \qquad c > 0.$$

Since B is symmetric about the origin, $M_X = M_{(-X)}$ and

$$\|-X\| = \|X\|.$$

We conclude that for all scalars c

$$\|cX\| = |c| \|X\|.$$

Now let us verify the triangle inequality. Given vectors X, Y and positive numbers s, t

(6.19) $$\frac{1}{s+t}(X+Y) = \frac{s}{s+t}\left(\frac{1}{s}X\right) + \frac{t}{s+t}\left(\frac{1}{t}Y\right).$$

Since

$$\frac{s}{s+t} + \frac{t}{s+t} = 1$$

we see from (6.19) that the vector $(s+t)^{-1}(X+Y)$ is on the line segment joining $(1/s)X$ to $(1/t)Y$. Since B is a convex set, this tells us that if $s \in M_X$ and $t \in M_Y$, then $(s+t) \in M_{X+Y}$. Therefore,

$$\inf M_{X+Y} \leq \inf M_X + \inf M_Y$$

that is,

$$\|X+Y\| \leq \|X\| + \|Y\|.$$

We have verified that $\|\cdots\|$ is a semi-norm. To see that it is a norm, we use the fact that B is bounded. Let X be any non-zero vector. Since B is bounded, there is a positive scalar c such that $cX \notin B$. Then $1/c$ is a lower bound for the set M_X so that $\|X\| \neq 0$.

This completes the proof, except for the fact that B is the unit ball of the norm. We have left the verification of this to the exercises.

Corollary. *Let* p *be a number,* $1 \le p < \infty$. *Then*

$$\|X\|_p = (\sum_k |x_k|^p)^{1/p}$$

is a norm on R^n.

Proof. Let

$$B_p = \{X \in R^n; |x_1|^p + \cdots + |x_n|^p < 1\}.$$

Since $f(x) = x^p$ is a continuous function on the positive real line, B_p is an open subset of R^n. Since

$$\lim_{x \to \infty} x^p = \infty$$

B_p is bounded. Clearly B_p is symmetric about the origin.

To see that B_p is a convex set, we verify that

$$f(x) = x^p, \qquad x \ge 0$$

is a convex function. This is an immediate consequence of the fact that (since $p \ge 1$) its derivative is an increasing function. See Example 3 and Exercise 11 of Section 4.1.)

All that remains to be done is to check that the norm on R^n associated with B_p in Theorem 2 is $\|\cdots\|_p$. If $X \in R^n$ and $t > 0$, then $(1/t)X \in B_p$ means

$$\sum_k \left|\frac{x_k}{t}\right|^p < 1$$

or

$$\frac{1}{t^p} \sum |x_k|^p < 1.$$

Therefore,

$$\inf\left\{t > 0; \frac{1}{t} \in B_p\right\} = \inf\{t; t^p > \sum_k |x_k|^p\}$$

$$= \inf\{s^{1/p}; s > \sum_k |x|^p\}$$

$$= (\sum_k |x_k|^p)^{1/p}.$$

It is sometimes useful to have in mind the relative shapes of the unit balls B_p, $1 \le p < \infty$. These are indicated in Figure 24, which also includes the limiting case B_∞, which is the unit ball for the sup norm.

Exercises

1. Can the unit ball for a norm on R^2 be a triangle? A pentagon? A hexagon?

2. Each ellipse centered at the origin defines a norm on R^2. Suppose the axes of the ellipse are the coordinate axes. From the equation of the ellipse, derive an algebraic formula for the norm on R^2 of which (the interior of) the ellipse is the unit ball.

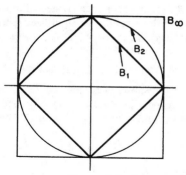

3. If $0 < p < 1$, show that the set

$$B_p = \{X \in R^n; \sum_k |x_k|^p < 1\}$$

is *not* convex.

4. Use the corollary in this section to show that the sequence spaces ℓ^p in Example 9 are normed linear spaces.

5. Use the corollary in this section to show that for each p, $1 \leq p < \infty$,

$$\|f\|_p = \left(\int |f|^p\right)^{1/p}$$

is a semi-norm on $PC(B)$. See Example 8.

6. Prove that, for $p > 1$, the function $f(x) = x^p$, $x > 0$, is **strictly convex**, i.e., if $x < y$ and $0 < t < 1$, then

$$f(tx + [1 - t]y) < tf(x) + (1 - t)f(y).$$

Use this to infer that the *closed* unit ball for the norm $\|\ \ \|_p$ on R^n is "strictly convex", in a sense which you define.

7. True or false? A norm on R^2 satisfies the parallelogram law (comes from an inner product) if and only if its unit ball is (the interior of) an ellipse?

8. Is there a relationship between "strict convexity" of the closed unit ball for a norm and the fact that the triangle inequality is strict when the two vectors involved are linearly independent?

6.3. Convergence and Continuity

In Chapters 2 and 3 we discussed convergence of sequences in R^n and the related concepts of open set, closed set, compact set, continuous function, etc. All of these ideas can be discussed in much greater generality; indeed, most of the results are valid in considerable generality. We are now going to describe that collection of ideas in the context of normed linear spaces. Most of the results are valid with precisely the same reasoning as was used in R^n. We shall conduct a review of those results. Of course, we shall also point out those results which utilized special properties of R^n.

Let L be a normed linear space. If $X \in L$, the **open ball** of radius ϵ about X is

$$B(X; \epsilon) = \{Y \in L; \|Y - X\| < \epsilon\}.$$

A **neighborhood** of X is a subset of L which contains some open ball about X. The sequence $\{X_n\}$ **converges to** X if every neighborhood of X contains X_n, except for a finite number of values of n. If $\{X_n\}$ converges to X, then we write

$$X = \lim_n X_n.$$

We have the elementary results that

(i) $X = \lim_n X_n$ if and only if $0 = \lim \|X - X_n\|$;

(ii) if $X = \lim_n X_n$ and $Y = \lim_n Y_n$ then $X + Y = \lim_n (X_n + Y_n)$ and $cX = \lim_n cX_n$.

The set U is **open** if it is a neighborhood of each of its points. The point X is a **cluster point** of the set K if every neighborhood of X contains infinitely many points of K. The set K is **closed** if it contains all of its cluster points. The set U is open if and only if $L - U$ is closed. The family of open sets is closed under the formation of arbitrary unions and finite intersections. There is a complemented statement about closed sets—arbitrary intersections and finite unions. The **closure** of a set is the intersection of all closed sets which contain it. The **interior** of a set is the union of all open sets which it contains. The set S is **dense** in the set T if S is a subset of T and the closure of S contains T. The set K is **compact** if each open cover of K has a finite subcover.

Relative neighborhoods, relatively open sets, relatively closed sets in S are defined by intersecting the corresponding families of sets with S. The set S is **connected** provided S and the empty set are the only subsets of S which are both open and closed relative to S.

Let L and M be normed linear spaces and suppose

$$D \xrightarrow{F} M, \qquad D \subset L.$$

If $X \in D$, then F is **continuous** at X provided, for each neighborhood V of $F(X)$, the inverse image $F^{-1}(V)$ is a neighborhood of X relative to D. We call F **continuous** if F is continuous at each point of D. The following are equivalent.

(i) F is continuous.

(ii) If $X \in D$, and $\epsilon > 0$, there exists $\delta > 0$ such that

$$\|F(T) - F(X)\| < \epsilon, \qquad T \in D, \quad \|T - X\| < \delta.$$

(iii) If V is any open set in M, then $F^{-1}(V)$ is open relative to D.

(iv) If $\{X_n\}$ is a sequence in D which converges to the point $X \in D$, then $F(X_n)$ converges to $F(X)$.

Sums and compositions of continuous functions are continuous. The image of a compact [connected] set under a continuous mapping is compact [connected].

The function F

$$D \xrightarrow{\ F\ } M, \qquad D \subset L$$

is **uniformly continuous** if, for each $\epsilon > 0$, there exists $\delta > 0$ such that

$$\| F(X_1) - F(X_2) \| < \epsilon, \qquad \| X_1 - X_2 \| < \delta.$$

A continuous function on a compact set is uniformly continuous.

Several of the results from Chapters 2 and 3 are conspicuously absent from the brief summary which we have just given. Primarily, the absentees are those results which depended upon properties of R^n which are not possessed by every normed linear space. We want now to discuss the special properties of R^n.

First, let us remark that (for the bulk of Chapters 2 and 3) it did not matter which norm we used on R^n. Theorem 1 shows us that *all norms on R^n define the same convergent sequences, hence, the same closed sets, the same open sets, the same continuous functions, etc.* Therefore, what we want to ask is this. Among normed linear spaces, what distinguishes the spaces which consist of R^n with some norm?

The linear space L is **finite-dimensional** if there exists in L an k-tuple of vectors $E_1, \ldots, E_n$ which spans L, i.e., which has the property that every $X \in L$ is a linear combination of those vectors:

(6.20) $$X = c_1 E_1 + \cdots + c_n E_n.$$

If L is finite-dimensional, we can always choose vectors $E_1, \ldots, E_n$ which span L and are linearly independent. (The number n is the **dimension** of L.) Via (6.20) we then have a 1:1 correspondence

$$X \longleftrightarrow (c_1, \ldots, c_n)$$

between vectors $X \in L$ and (all) vectors $(c_1, \ldots, c_n) \in R^n$. The norm on L corresponds to some norm on R^n. Thus, L behaves like R^n with some norm.

So, we see that, as we look back on Chapters 2 and 3, we should ask which of the results there made use of the fact that we were dealing with a finite-dimensional normed linear space. There were two principal results of that type.

1. In a finite-dimensional normed linear space, every Cauchy sequence converges.

2. In a finite-dimensional normed linear space, every bounded se-

quence has a convergent subsequence (hence, every closed and bounded set is compact).

The Bolzano-Weierstrass property (2) is more than a special property of finite-dimensional spaces, it is a characterization of such spaces. One can prove (see Section 6.5) that the normed linear space L is finite-dimensional if and only if each bounded sequence in L has a convergent subsequence.

On the other hand, the completeness property (1) is enjoyed by many spaces which are not finite-dimensional. Completeness is a weaker property than is sequential compactness (2); however, completeness itself has consequences which are sufficiently important that we introduce a special name for spaces with that property.

Definition. *The normed linear space* L *is* **complete** *if every Cauchy sequence in* L *converges. A complete normed linear space is (also) called a* **Banach space.**

We shall have more to say later about special properties of Banach spaces. We should state now the two results from Chapters 2 and 3 which exploited the completeness of R^n.

(i) If L is a Banach space and if $\sum_n X_n$ is an infinite series of points of X which converges absolutely,

$$\sum_n \|X_n\| < \infty$$

then the series converges.

(ii) Let L be a normed linear space, let M be a Banach space and let

$$D \overset{F}{\longrightarrow} M, \qquad D \subset L$$

be a uniformly continuous function. Then F can be extended to a continuous function on the closure of D:

$$\bar{D} \overset{\bar{F}}{\longrightarrow} M.$$

EXAMPLE 12. Perhaps the first normed linear space which we met after R^n was the space of continuous functions on a compact set in R^k. Now, we may start with K, a compact subset of the normed linear space L. Let $C(K)$ be the space of real- (or complex-) valued continuous functions on K. In many analysis problems, the "natural" norm on $C(K)$ is the sup norm

$$\|f\|_\infty = \sup_K |f|$$

because it measures when function values are uniformly close together: The norm $\|f - g\|_\infty$ is small if and only if $|f(X) - g(X)|$ is small, uniformly in X. As we saw earlier, convergence in the sup norm

$$\| f - f_n \|_\infty \longrightarrow 0$$

is precisely uniform convergence of $\{f_n\}$ to f.

The normed linear space which consists of $C(K)$ endowed with the sup norm is a Banach space. We repeat the skeleton of the argument, which should be familiar. Suppose $\{f_n\}$ is a Cauchy sequence:

$$\lim_{m,n} \| f_m - f_n \|_\infty = 0.$$

Then, for each $X \in K$, the sequence of numbers $\{f_n(X)\}$ is Cauchy, because

$$| f_m(X) - f_n(X) | \leq \| f_m - f_n \|_\infty.$$

By the completeness of the space of real [or complex] numbers, each of these numerical sequences converges. Let f be the function on K defined by the rule

$$f(X) = \lim_n f_n(X).$$

It is easy to see that $\{f_n\}$ converges uniformly to f: If

$$\| f_m - f_n \|_\infty < \epsilon, \qquad m, n \geq N$$

then (taking the limit on m)

$$\| f - f_n \|_\infty \leq \epsilon, \qquad n \geq N.$$

The uniform convergence of f_n to f tells us that f is continuous. Thus, we have shown that $\{f_n\}$ converges in the normed linear space $L = (C(K), \| \cdots \|_\infty)$.

EXAMPLE 13. Let us make two remarks about the special case of Example 12 in which $K = [a, b]$, a closed interval on the real line. First, the polynomial functions

$$p(x) = c_0 + c_1 x + \cdots + c_n x^n$$

form a linear subspace of $C([a, b])$ and this subspace is dense in the Banach space $(C, \| \cdots \|_\infty)$. This is merely a restatement of the Weierstrass approximation theorem: Each continuous function on $[a, b]$ can be uniformly approximated by polynomials.

Second, it is easy to show that the Bolzano-Weierstrass property fails for this Banach space. Let

$$(6.21) \qquad\qquad g(x) = \frac{x - a}{b - a}, \qquad a \leq x \leq b.$$

Then the sequence of powers $\{g^n\}$ is bounded:

$$\| g^n \|_\infty = 1, \qquad n = 1, 2, 3, \ldots$$

but, it has no uniformly convergent subsequence because

$$(6.22) \qquad\qquad \lim g^n(x) = \begin{cases} 1, & x = b \\ 0, & a \leq x < b. \end{cases}$$

This pointwise convergence tells us that the only function to which a subsequence could converge uniformly is the function defined by the right side of Equation (6.22). Since this function is discontinuous, such uniform convergence is impossible.

EXAMPLE 14. Let $C^1(I)$ be the space of functions of class C^1 on a closed interval, with the norm

$$\|f\| = \|f\|_\infty + \|f'\|_\infty.$$

This is a Banach space. Let us see why. Suppose $\{f_n\}$ is a Cauchy sequence. Since

$$\|f_n - f_k\| = \|f_n - f_k\|_\infty + \|f'_n - f'_k\|_\infty$$

the sequence $\{f_n\}$ converges uniformly *and* the sequence $\{f'_n\}$ converges uniformly. It follows from Theorem 5 of Chapter 5 that $f = \lim_n f_n$ is differentiable and

$$f' = \lim_n f'_n.$$

Since f'_n is continuous and converges uniformly to f' the function f' is continuous. Thus $f \in C^1(I)$ and $\|f - f_n\| \to 0$.

Now $C^1(I)$ is also a normed linear space using the sup norm. That normed linear space is *not* complete; it is (in fact) dense in $C(I)$.

EXAMPLE 15. Refer to the sequence spaces ℓ^p, from Example 9. These spaces are Banach spaces. Let's see how we would verify that for the case $p = 1$. The space ℓ^1 consists of the sequences

$$X = (x_1, x_2, \ldots)$$

of complex numbers such that

$$\|X\|_1 = \sum_n |x_n| < \infty.$$

Suppose we have in ℓ^1 a sequence $\{X_n\}$ which is Cauchy:

$$\lim \|X_m - X_n\|_1 = 0.$$

If $X_n = (x_{n1}, x_{n2}, x_{n3}, \ldots)$, then for each k the sequence of kth coordinates $\{x_{nk}\}$ is a Cauchy sequence of numbers:

$$|x_{mk} - x_{nk}| \leq \|X_m - X_n\|_1.$$

The completeness of the complex numbers tells us that for each k there exists $x_k \in C$ such that

$$\lim_n x_{nk} = x_k.$$

What we want to show is that the sequence $X = (x_1, x_2, \ldots)$ is in ℓ^1 and that

$$\|X - X_n\|_1 \longrightarrow 0.$$

To that end, consider the double sequence

$$s_{nk} = \sum_{j=1}^{k} |x_{nj}|.$$

For each n, the limit on k of the sequence $\{s_{nk}\}$ exists:

$$\lim_k s_{nk} = \|X_n\|_1.$$

For each k, the limit on n of the sequence $\{s_{nk}\}$ exists:

$$\tag{6.23} \lim_n s_{nk} = \sum_{j=1}^{k} |x_j|.$$

Furthermore, the convergence in (6.23) is uniform in k because $\{s_{nk}\}$ is Cauchy in n, uniformly in k:

$$|s_{mk} - s_{nk}| = \left| \sum_{j=1}^{k} |x_{nj}| - \sum_{j=1}^{k} |x_{nj}| \right|$$
$$\leq \sum_{j=1}^{k} |x_{mj} - x_{nj}|$$
$$\leq \|X_m - X_n\|_1.$$

By the Moore-Osgood double limit theorem (Theorem 3, Chapter 5),

$$\lim_n \sum_{j=1}^{k} |x_j|$$

exists and is equal to the limit on n of the sequence $\{\|X_n\|_1\}$. Thus $X \in \ell^1$ and

$$\|X\|_1 = \lim_n \|X_m\|_1.$$

To see that $\{X_n\}$ converges to X in the norm of ℓ^1, apply the same double sequence argument we just used to the sequence

$$t_{nk} = \sum_{j=1}^{k} |x_j - x_{nj}|.$$

For each n, the limit on k of this sequence is $\|X - X_n\|_1$. On the other hand,

$$\lim_n t_{nk} = 0$$

and this consequence is uniform in k. It follows that

$$\lim_n \|X - X_m\|_1 = \lim_k 0 = 0.$$

The sequence space ℓ^2 is called **Hilbert's space**. It has become common to call any complete inner product space a **Hilbert space**.

EXAMPLE 16. An important normed linear space is the one consisting of $C(B)$, the space of continuous functions on a closed box in R^n, together with the norm

$$\|f\|_1 = \int_B |f|.$$

In this space, two functions, f and g, are close together when the integral

$$\int_B |f - g|$$

is small. This is quite a different measure of approximation than that which is provided by the sup norm. Since

$$\int_B |f - g| \leq \|f - g\|_\infty \, m(B)$$

it is true that uniform convergence implies convergence in the norm $\|\cdots\|_1$; however, we may have

$$\|f - f_n\|_1 \longrightarrow 0$$

while $\{f_n\}$ fails to converge pointwise, much less uniformly. Indeed, it can happen that

(6.24) $\|f_n\|_1 \longrightarrow 0$ and $\|f_n\|_\infty \longrightarrow \infty$.

The phenomenon (6.24) is illustrated in the case $B = [0, 1]$ by the functions $f_n(x) = \sqrt{n} \sin nx$ or by the tent functions in Figure 25.

The space $C(B)$ is *not* complete relative to the norm $\|\cdots\|_1$. For the case $B = [0, 1]$, this is treated in the exercises.

We close this section with a remark. The little review of convergence, continuity, etc., which we carried out at the beginning of the section can be carried out just as easily in the context of metric spaces. A **metric space**

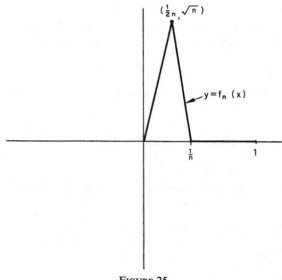

FIGURE 25

is a set S together with a real-valued function d on $S \times S$ which satisfies

    (i) $d(X, Y) \geq 0$; $d(X, Y) = 0$ if and only if $X = Y$;

    (ii) $d(X, Y) = d(X, Y)$;

    (iii) $d(X, Z) \leq d(X, Y) + d(Y, Z)$.

Such a function d is called a **metric** or distance function on S. Each normed linear space $(L, \|\cdots\|)$ has a natural metric on it: $d(X, Y) = \|X - Y\|$. By analogy with the normed linear space case, it should be clear how one defines ϵ-ball, open set, convergence and so on in a metric space. The interested reader can check to see that the basic results of Chapters 2 and 3 are valid in that setting. About the only thing to be gained by using metric spaces rather than subsets of normed linear spaces is that to a certain extent one avoids relatively open, relatively closed, etc.

Exercises

1. Let K be a compact set and let $C(K)$ be the space of *real* continuous functions on K. Equip $C(K)$ with the sup norm. Show that the set of positive continuous functions is an open subset of $C(K)$.

2. Equip (real) $C(B)$ with the norm

$$\|f\|_1 = \int_B |f|$$

(B is a box). Is the set of positive continuous functions open?

3. Equip $C([0, 1])$ with the sup norm. *Fix* a positive integer n, and let P_n be the subspace of polynomial functions of degree at most n:

$$p(x) = c_0 + c_1 x + \cdots + c_n x^n.$$

Show that P_n is a closed subspace of $C([0, 1])$.

4. If X is a Banach space, every absolutely convergent series in X converges.

5. In a normed linear space, every convex set is connected.

6. If a normed linear space has the Bolzano-Weierstrass property, it is complete.

7. Equip $C([0, 1])$ with the norm

$$\|f\|_1 = \int_0^1 |f(x)| \, dx.$$

Let L be the function on $C([0, 1])$

$$L(f) = f(0).$$

Is L continuous?

8. Let S be the function from $C([0, 1])$ into $C([0, 1])$ defined by squaring: $S(f) = f^2$. If $C([0, 1])$ is endowed with the sup norm, is S continuous? What if we use the integral norm $\|\cdots\|_1$?

9. If we pin down the constant of integration, then anti-differentiation defines a mapping (function)

$$C([a, b]) \xrightarrow{\ I\ } C([a, b])$$

$$(If)(x) = \int_a^x f(t)\, dt.$$

Is I continuous relative to the sup norm? The norm $\| \cdots \|_1$?

10. Differentiation maps functions of class C^1 into continuous functions. Think about whether differentiation is continuous for various norms on $C^1([a, b])$ and $C([a, b])$.

11. Let f be a piecewise-continuous function on $[0, 1]$ which is not continuous and cannot be made continuous by altering its values at a finite number of points, e.g., $f(x) = 0$ on $[0, \frac{1}{2}]$ and $f(x) = 1$ on $[\frac{1}{2}, 1]$.

(a) Show that there exist continuous functions f_n such that $\| f - f_n \|_1$ goes to 0. ($\| \cdots \|_1$ is the integral norm.)

(b) Show that no sequence $\{f_n\}$ as in (a) can converge in the normed linear space $(C, \| \cdots \|_1)$. *Hint:* If f_n converged in the norm $\| \ldots \|_1$ to a continuous function g, what could you conclude about $\| f - g \|_1$?

12. Define functions f_n on $[0, 1]$ by

$$f_n(x) = \begin{cases} n, & 0 \le x \le \dfrac{1}{n} \\[2mm] \dfrac{1}{x}, & \dfrac{1}{n} < x \le 1. \end{cases}$$

Show that $\{f_n\}$ is a Cauchy sequence in $(C, \| \cdots \|_1)$ but does not converge in that space.

13. True or false? If $C^1([a, b])$ is given the norm

$$\| f \| = \sup | f | + \sup | f' |$$

then the subspace of polynomial functions is dense in $C^1([a, b])$.

14. If X is a normed linear space, every finite-dimensional subspace of X is closed.

15. True or false? The space Lip (K) in Example 10 is complete.

16. Is ℓ^1 dense in ℓ^∞?

17. The space of functions of bounded variation (Example 5) is a complete semi-normed space.

18. Refresh your memory by giving the proof of this theorem: A uniformly continuous function from a (subset of a) normed linear space into a Banach space can be extended to a continuous function on the closure of its domain.

19. Let L and M be normed linear spaces and let T be a transformation from L into M which is *linear*: $T(cX_1 + X_2) = cT(X_1) + T(X_2)$. Prove that T is continuous if and only if there exists a non-negative constant k such that

$$\| T(X) \| \le k \| X \|, \quad \text{all } X \in L.$$

(As a matter of convenience, the same symbol has been used for the norms in L and M, although these norms are presumably different.)

20. In the notation of Exercise 19, let $B(L, M)$ be the set of all continuous linear transformations from L into M. Show that $B(L, M)$ is a linear space in a natural way and that

$$\| T \| = \sup_{X \neq 0} \frac{\| T(X) \|}{\| X \|}$$

is a norm on that linear space.

6.4. Completeness

There are several fundamental properties of Banach spaces which are not possessed by normed linear spaces in general. We do not have time to go into all of these. We shall content ourselves with presenting two basic results, along with some of their consequences. These results will illustrate the fact that completeness (like compactness) is used to prove that certain objects exist.

We shall be interested in the completeness of subsets of a normed linear space L, even in situations where L itself might fail to be complete. Of course, the subset K is called **complete** if each Cauchy sequence in K converges to a vector in K. Every complete subset is closed. If L is a Banach space, a subset K is complete if and only if it is closed.

Theorem 3. *If* L *is a normed linear space, every finite-dimensional subspace of* L *is complete and is, therefore, a closed subset of* L.

Proof. Let M be a subspace of (finite) dimension k. Let $E_1, \ldots, E_k$ be a basis for M. The map

$$R^k \xrightarrow{T} M \qquad \text{(use } C^k \text{ in the complex case)}$$

defined by $T(c_1, \ldots, c_k) = c_1 E_1 + \cdots + c_k E_k$ establishes a 1:1 correspondence between vectors in R^k and vectors in M which preserves linear operations. Thus

$$\||(c_1, \ldots, c_k)\|| = \| T(c_1, \ldots, c_k) \|$$

defines a norm $\|| \cdots \||$ on R^k. Since all norms on R^k define the same convergent sequences, $(R^k \||\cdots\||)$ is a complete normed linear space. Thus M is a complete space.

In order to show that a subset of a normed linear space is complete, it is not necessary to verify that every Cauchy sequence converges. It is sufficient, and usually it is technically simpler, to work with sequences of the following special type.

Definition. Let L *be a normed linear space. A* **fast Cauchy sequence** *in* L *is a sequence* $\{X_n\}$ *such that*

(6.25) $$\sum_n \| X_n - X_{n+1} \| < \infty.$$

The terminology requires some justification. We need to show that a sequence which satisfies (6.25) is a Cauchy sequence. To verify this, first note that

$$\lim_n \sum_{k=n}^{\infty} \|X_k - X_{k+1}\| = 0.$$

Thus, given $\epsilon > 0$, there is a positive integer N_ϵ such that

$$(6.26) \qquad \sum_{k=N_\epsilon}^{\infty} \|X_k - X_{k+1}\| < \epsilon.$$

Suppose that $n > m \geq N_\epsilon$. Then

$$X_m - X_n = (X_m - X_{m+1}) + (X_{m+1} - X_{m+2}) + \cdots + (X_{n-1} - X_n),$$

and hence

$$\|X_m - X_n\| \leq \sum_{k=m}^{n-1} \|X_k - X_{k+1}\|.$$

By (6.26)

$$\|X_m - X_n\| < \epsilon, \qquad m, n \geq N_\epsilon.$$

Recall that, in order for $\{X_n\}$ to be a Cauchy sequence, it is *not* sufficient that

$$\lim_n \|X_n - X_{n+1}\| = 0.$$

What we have just verified is that, if $\|X_n - X_{n+1}\|$ goes to 0 fast enough so that

$$\sum_n \|X_n - X_{n+1}\| < \infty$$

then the sequence is Cauchy.

Lemma. *A subset* K *of the normed linear space* L *is complete if and only if each fast Cauchy sequence in* K *converges to a point of* K.

Proof. One half of the lemma is trivial. Suppose we know that each fast Cauchy sequence in K converges in K. Let $\{Y_n\}$ be an arbitrary Cauchy sequence. For each positive integer k, there is a positive integer N_k such that

$$\|Y_m - Y_n\| < 2^{-k}, \qquad m, n \geq N_k.$$

Evidently, we can choose such integers N_k so that $N_1 < N_2 < N_3 < \cdots$. Let

$$X_k = Y_{N_k}.$$

Then

$$\sum_k \|X_k - X_{k+1}\| < \sum_k 2^{-k}$$
$$= 1.$$

Therefore $\{X_k\}$ converges to a vector $X \in K$. Since $\{Y_n\}$ is a Cauchy sequence and the subsequence $\{Y_{N_k}\}$ converges, the entire sequence converges to the same limit X.

Our first basic theorem deals with the existence of "fixed points" for certain special mappings.

Definition. *If* S *is a subset of a normed linear space, a* **contraction** *of* S *is a function*

$$S \xrightarrow{T} S$$

such that there exists c, $0 < c < 1$, *with*

(6.27) $\|TX - TY\| \le c\|X - Y\|,$ $X, Y \in S.$

Theorem 4. *If* K *is a complete non-empty subset of the normed linear space* L, *and if* T *is a contraction of* K, *then* T *has one and only one fixed point; i.e., there exists one and only one point* $Y \in K$ *such that* $T(Y) = Y$.

Proof. Let X be any point of K. We iterate the contraction T:

$$X$$
$$TX$$
$$T^2X = T(TX)$$
$$\cdot$$
$$\cdot \qquad \cdot$$
$$\cdot \qquad \cdot$$
$$T^nX = T(T^{n-1}X)$$
$$\cdot$$
$$\cdot \qquad \cdot$$
$$\cdot \qquad \cdot \qquad \cdot$$

The sequence T^nX will converge to the fixed point Y. We have

$$\|TX - T^2X\| = \|TX - T(TX)\|$$
$$\le c\|X - TX\|$$

and so

$$\|T^2X - T^3X\| \le c\|TX - T^2X\|$$
$$\le c^2\|X - TX\|$$
$$\cdot$$
$$\cdot \qquad \cdot$$
$$\cdot \qquad \cdot$$
$$\|T^nX - T^{n+1}X\| \le c^n\|X - TX\|.$$

Since $0 < c < 1$, we have

$$\sum_{n=1}^{\infty} \|T^nX - T^{n+1}X\| < \infty.$$

Thus $\{T^nX\}$ is a fast Cauchy sequence, and since K is complete there exists $Y \in K$ with

$$Y = \lim_n T^nX.$$

Since T is a contraction, T is (uniformly) continuous. Thus

$$TY = \lim_n T(T^n X)$$
$$= \lim_n T^{n+1} X$$
$$= Y.$$

So Y is a fixed point.

Why is Y the unique fixed point? If $TZ = Z$ then

$$\|Y - Z\| = \|TY - TZ\|$$
$$\le c\|Y - Z\|.$$

Since $0 < c < 1$, it must be that $\|Y - Z\| = 0$.

The result just established not only proves that a fixed point exists, but the proof says that iteration (repeated application of T) will lead from any point of K to an approximation of the fixed point. Such iterative processes occur frequently in mathematics and its applications.

EXAMPLE 17. Suppose p is a non-negative continuous function on a closed interval $[a, b]$ and we are trying to approximate $\sqrt{p}$. We can try Newton's method, which is to make an educated guess, g, and try to improve it by taking $h = g + \frac{1}{2}(p - g^2)$ as a second approximation, $k = h + \frac{1}{2}(p - h^2)$ as a third, etc. This method is based, therefore, on the mapping

$$C([a, b]) \xrightarrow{\;T\;} C([a, b])$$

defined by

$$Tf = f + \tfrac{1}{2}(p - f^2).$$

A fixed point of T is a continuous square root of p, since $Tf = f$ if and only if $p - f^2 = 0$. Under what circumstances might the contraction theorem be applicable? In other words, what might reasonably insure that

$$\|Tf - Tg\|_\infty \le c\|f - g\|_\infty$$

where $0 < c < 1$? Now

$$Tf - Tg = \tfrac{1}{2}(g^2 - f^2)$$
$$= \tfrac{1}{2}(g + f)(g - f)$$

so

$$\|Tf - Tg\|_\infty \le \tfrac{1}{2}\|g + f\|_\infty \|g - f\|_\infty.$$

Thus, if we can restrict T to a set of functions $f, g, \cdots$ whose sup norms are bounded by $c < 1$, we will have a contraction. Suppose, for example, that

$$\|p\|_\infty = c^2 < 1$$

This is not much of a restriction, since we can always multiply p by a constant. Now let

$$K = \{f \in C([a, b]); 0 \leq f(x) \leq \sqrt{p(x)}\}.$$

Evidently K is closed under uniform convergence, and all the functions in K are bounded by $c < 1$. The set K is mapped by T inside itself:

$$K \overset{T}{\longrightarrow} K.$$

For, if $0 \leq f \leq \sqrt{p}$, then

$$f + \tfrac{1}{2}(p - f^2) \geq f \geq 0$$

and

$$
\begin{aligned}
f + \tfrac{1}{2}(p - f^2) &= f + \tfrac{1}{2}(\sqrt{p} + f)(\sqrt{p} - f) \\
&\leq f + (\sqrt{p} - f) \\
&= \sqrt{p}
\end{aligned}
$$

because

$$0 \leq \tfrac{1}{2}(\sqrt{p} + f) < 1.$$

EXAMPLE 18. With the contraction theorem, we can establish the existence (and uniqueness) of solutions to certain first order differential equations:

$$\frac{dy}{dx} = f(x, y).$$

We could just as easily deal with Banach space valued functions; however, let us deal only with vector (R^n) valued functions on an interval. We suppose that we are given an interval I on the real line and a function F which maps (part of) R^{n+1} into R^n. We seek a function Y

$$R \overset{Y}{\longrightarrow} R^n$$

which satisfies the differential equation

(6.28) $$Y'(x) = F(x, Y(x)).$$

We must nail down some constants of integration; hence, we specify an initial condition

(6.29) $$Y(x_0) = Y_0$$

where x_0 is a given point of I and Y_0 is a given vector in R^n. If we wish to think of it that way, (6.28) is a system of differential equations

$$y'_1(x) = f_1(x, y_1(x), \ldots, y_n(x))$$
$$\vdots \qquad\qquad \vdots$$
$$y'_n(x) = f_n(x, y_n(x), \ldots, y_n(x))$$

which we have subjected to the initial conditions

$$y_j(x_0) = y_{j0}, \qquad 1 \leq j \leq n.$$

Now we need to be clear about our assumptions concerning the given function F. For simplicity we shall assume that $F(x, Y)$ is defined for all $x \in I$ and all vectors Y near Y_0:

$$I \times \bar{B}(Y_0; r) \xrightarrow{F} R^n.$$

We assume that F is continuous and that F satisfies a (uniform) Lipschitz condition in the second variable:

$$(6.30) \qquad |F(x, Y_1) - F(x, Y_2)| \leq M|Y_1 - Y_2|$$

where M is some constant. These conditions are certainly satisfied if F is a smooth function.

Here is the thing to observe. If the function Y satisfies (6.28), then the fundamental theorem of calculus tells us that

$$(6.31) \qquad Y(x) = Y_0 + \int_{x_0}^x F(t, Y(t))\, dt.$$

The expression on the right "operates" on (certain) functions Y to produce other functions, and what we are looking for is a fixed point for that operation. Where does that integral on the right operate? It operates on certain continuous functions from I into R^n. So, we consider the Banach space $C(I; R^n)$, which is the space of continuous functions from I into R^n, endowed with the sup norm (assume I is compact):

$$\|G\|_\infty = \sup \{|G(x)|; x \in I\}.$$

The integral in (6.31) will make sense for functions Y whose values are such that $(g, Y(t))$ is in the domain of F. So, let us consider the set of functions which are uniformly within r of the constant function Y_0:

$$(6.32) \qquad K = \{Y \in C(I; R^n); \|Y - Y_0\|_\infty \leq r\}$$

(Remember that Y_0 is a function in (6.32).) Then, we have an operation (function) T defined on K by

$$K \xrightarrow{T} C(I; R^n)$$

$$(TY)(x) = Y_0 + \int_{x_0}^x F(t, Y(t))\, dt.$$

Notice that

$$|(TY)(x) - Y_0| \leq |x - x_0| \sup |F|.$$

Therefore, the function TY is uniformly near Y_0 in an interval about x_0. If δ is any number such that

$$\delta \sup |F| \leq r$$

if J is the interval

(6.33) $$J = \{x \in I; |x - x_0| \leq \delta\},$$

and if we take

$$K = \{Y \in C(J; R^n)); \|Y - Y_0\|_\infty \leq r\},$$

then

$$K \xrightarrow{T} K.$$

(Here, the norm $\|Y - Y_0\|_\infty$ refers to the supremum over the interval $(J.)$ If δ is sufficiently small, then T is a contraction:

$$|(TY_1)(x) - (TY_2)(x)| = \int_{x_0}^x [F(t, Y_1(t)) - F(t, Y_2(t))] \, dt\,|$$
$$\leq |x - x_0| M \|Y_1 - Y_2\|_\infty$$

where M is the constant of (6.30). Specifically, if we require that $\delta M < 1$, then T is a contraction of K. By the fixed point theorem there is, on the interval J, one and only one function Y which satisfies (6.28) and (6.29).

Clearly then, we have a unique solution to our problem on the entire interval I. Take the solution on J; use its values at the end points of J as initial data; and solve (6.28) in an interval of length δ about each end point. Keep doing that. The point is that the δ does not depend on x_0, and the local solutions patch together correctly by the uniqueness.

It will clarify the proof of our second basic theorem if we reformulate completeness geometrically.

Lemma. *The subset* K *(of the normed linear space* L*) is complete if and only if it is closed and has the following property: If* $\{K_n\}$ *is a sequence of subsets of* L *such that*

(i) *each* K_n *is closed and non-empty,*
(ii) $K_1 \supset K_2 \supset K_3 \supset \cdots,$
(iii) $\lim_n \ diam \ (K_n) = 0,$

then the intersection $\bigcap_n K_n$ *is non-empty.*

Proof. Suppose that K is complete. Let $\{K_n\}$ be any sequence of subsets with properties (i), (ii), and (iii). Let X_n be any point of K_n. There is such a point, by (i). Since by (ii) the sets K_n decrease, we have

$$X_k \in K_N, \qquad k \geq N.$$

Therefore,

$$\|X_m - X_n\| \leq \operatorname{diam}(K_N), \qquad m, n \geq N.$$

Since (iii) tells us that $\lim_N \operatorname{diam}(K_N) = 0$, we have

$$\lim_{m,n} \|X_m - X_n\| = 0,$$

that is, $\{X_n\}$ is a Cauchy sequence. Since K is complete, there exists $K \in L$ with

$$\lim_n \|X - X_n\| = 0.$$

It is easy to see that X is in each of the sets K_N, because K_N is closed and $X_n \in K_N$ for $n \geq N$. (Note that the condition on the diameters guarantees that the intersection of all the sets K_n contains precisely the one point X.)

Conversely, suppose K is closed and has the stated property for sequences of closed subsets. Let $\{X_n\}$ be a Cauchy sequence in K. For each positive integer n, there exists a positive integer N_n such that

$$\|X_k - X_m\| < \frac{1}{n}, \qquad k, m \geq N_n.$$

We can choose these integers so that $N_1 \leq N_2 \leq N_3 \leq \cdots$. Let K_n be the intersection with K of the closed ball $\bar{B}(X_{N_n}; 1/n)$:

$$K_n = \left\{ X \in K; \|X - X_{N_n}\| \leq \frac{1}{n} \right\}.$$

Obviously each K_n is a closed, non-empty set, and diam $(K_n) \leq 2/n$. Since K_n contains every X_k with $k \geq N_n$, and since $N_1 \leq N_2 \leq N_3 \leq \cdots$, we have $K_1 \supset K_2 \supset K_3 \supset \cdots$. By hypothesis, there is a point X which lies in every K_n. We have

$$\|X - X_k\| \leq \frac{2}{n}, \qquad k \geq N_n$$

and thus

$$\lim_k X_k = X.$$

Theorem 5 (*Baire Category Theorem*). *Let* K *be a complete subset of the normed linear space* X. *Let* $\{U_n\}$ *be a sequence of subsets of* K *such that*

  (i) *each* U_n *is open relative to* K;
  (ii) *each* U_n *is dense in* K.

Then the intersection

$$\bigcap_n U_n$$

is dense in K.

Proof. First, we prove that the intersection of all the U_n is non-empty, as follows. Choose any point X_1 in U_1. Since U_1 is open relative to K, there exists r_1 so that the intersection of the ball

$$B_1 = B(X_1; r_1)$$

with K is contained in U_1. Shrink r_1 a little, and we may arrange that $K \cap \bar{B}_1 \subset U_1$. Since U_2 is dense in K, there is a point of U_2 in $K \cap B_1$. Use the fact that U_2 is open to find a closed ball:

$$K \cap \bar{B}_2 \subset U_2$$
$$\bar{B}_2 \subset \bar{B}_1.$$

We may arrange that $r_2 < \frac{1}{2}r_1$. By induction, we obtain a decreasing sequence of closed balls:

$$\bar{B}_1 \supset \bar{B}_2 \supset \bar{B}_3 \supset \cdots$$
$$\lim_n \operatorname{diam} (B_n) = 0$$
$$\bar{B}_n \cap K \subset U_n$$
$$\bar{B}_n \cap K \neq \varnothing.$$

Since K is complete, there is a point of K which lies in every $\bar{B}_n$. That point is (accordingly) in the intersection of all the U_n.

Now, it is apparent that the intersection of the sets U_n is dense in K. Just look at the proof. Given $X \in K$ and $\epsilon > 0$, use the fact that U_1 is dense in K to choose the first ball B_1 so that $\bar{B}_1 \supset B(X; \epsilon)$.

Corollary. *A Banach space is not the union of a countable number of proper closed subspaces of itself.*

Proof. Let L be a Banach space. Let $S_1, S_2, \cdots$ be linear subspaces of L, each of which is closed and proper: $S_n = \bar{S}_n \neq L$. The assertion is that

$$\bigcup_n S_n \neq L.$$

We want to apply the Baire theorem to the open sets $U_n = L - S_n$. Certainly U_n is open, but is U_n dense in L? If it were not, then the interior of S_n would be non-empty, i.e., S_n would contain some open ball $B(X; \epsilon)$. Since S_n is a linear subspace, it would therefore contain the open ball of radius ϵ about the origin:

$$B(X; \epsilon) = X + B(0; \epsilon)$$

and, hence, would contain all of L because $B(0; \epsilon)$ contains a non-zero scalar multiple of every vector. We conclude that U_n is dense in L. By the Baire theorem, the intersection of all the U_n is still dense in L, which means that the union

$$\bigcup_n S_n$$

has empty interior.

Some people prefer to state the Baire theorem this way: If K is complete and $\{K_n\}$ is a sequence of closed subsets of K such that

$$K = \bigcup_n K_n$$

then one of the sets has a non-empty interior relative to K. This is essentially the form we used in the proof of the corollary.

EXAMPLE 19. The Baire theorem makes it vividly clear that the set of real numbers is uncountable. The complement of any one point is a dense open subset of R. Thus, if we delete any countable set from R, the remaining set is dense in R.

EXAMPLE 20. Let L be the linear space of polynomial functions on some interval $[a, b]$. There are various interesting norms on L, e.g.,

$$\| f \|_\infty$$
$$\| f \|_\infty + \| f' \|_\infty$$
$$\int_a^b |f|, \quad \text{etc.}$$

But, the corollary tells us that there is *no* norm relative to which L is a Banach space, because L is the union of an increasing sequence of finite-dimensional subspaces

$$L = \bigcup_n S_n$$
$$S_n = \{ f \in L; \text{ degree } (f) \leq n \}$$

and finite-dimensional subspaces are always closed. (See Theorem 3.)

EXAMPLE 21. Here is one way to show that "practically every" continuous function is nowhere differentiable. Consider a closed interval $I = [a, b]$ on the real line and the space $C(I)$ of continuous (real-or complex-valued) functions on I. We equip $C(I)$ with the sup norm as usual. If m, n are positive integers, define U_{mn} to be the set of all functions f in $C(I)$ with this property: For each $x \in I$, there exists a point $t \in I$ such that

(i) $0 < |t - x| \leq 1/m$
(ii) $|f(t) - f(x)| > n |t - x|$.

Now each U_{mn} is open in $C(I)$. We leave the proof as an exercise. Also, U_{mn} is dense in $C(I)$. To see that, just approximate each f uniformly by a piecewise linear function which consists of steep line segments joining points near to one another. By the Baire theorem

$$S = \bigcap_{m, n} U_{mn}$$

is dense in $C(I)$. What does a function in S look like? If $f \in S$, then given any x in I and given any positive integers m, n, there exists $t \in I$ with

$$0 < |t - x| \leq \frac{1}{m}$$
$$\left| \frac{f(t) - f(x)}{t - x} \right| > n.$$

Clearly such an f is not differentiable at any point of I. We conclude (at least) that the set of continuous yet nowhere differentiable functions is dense in $C(I)$.

Exercises

1. The normed linear space L is complete if and only if its closed unit ball $\bar{B}(0; 1)$ is complete.

2. Consider ℓ^∞, the space of all sequences $X = (x_1, x_2, \ldots)$ of real [complex] numbers, with the sup norm. Let

$$K = \{X \in \ell^\infty; x_1 = 0 \quad \text{and} \quad \lim_n x_n = 1\}.$$

(a) Show that K is a closed subset of ℓ^∞.
(b) Let T be the shift mapping

$$T(X) = (0, x_1, x_2, x_3, \ldots).$$

Show that $T(K) \subset K$.

(c) Show that T is a weak contraction:

$$\| T(X) - T(Y) \|_\infty \leq \| X - Y \|_\infty$$

but that T has no fixed point in K.

3. If $0 < c < 1$ and I is the closed interval $[0, c]$, then

$$(Tf)(x) = 1 + \int_0^x f(t)\, dt$$

defines a contraction on $C(I)$ whose fixed point will satisfy the differential equation $f' = f$ with the initial condition $f(0) = 1$. What approximations to the solution do you obtain from the sequence $T(0)$, $T^2(0)$, $T^3(0), \ldots$?

4. What contraction operator on $C([0, c])$ would you use to obtain a solution of $f'' + f = 0$, $f(0) = a$, $f'(0) = b$?

5. Let S be a square in the plane. Remove from S any sequence of line segments. The set which remains is dense in S.

6. Prove that the set of rational numbers is not the intersection of any countable family of open subsets of the real line.

7. Prove that each set U_{mn} in Example 21 is open and dense in $C(I)$.

8. True or false? In solving the differential equation $Y' = F(x, Y)$ (Example 18), if F is a function of class C^∞ in each variable, then the solution is of class C^∞. How about the case where F is real-analytic?

9. Let K be a compact set in a normed linear space, and let $\{X_n\}$ be a sequence of distinct points in K (a countable subset). Prove that there exists on K a real-valued continuous function which takes on different values at the various points X_k:

$$f(X_k) \neq f(X_n), \qquad k \neq n.$$

6.5. Compactness

In a finite-dimensional normed linear space, every closed and bounded set is compact. In the general normed linear space there is no such simple characterization of compact sets. Of course, we might not expect one in an incomplete space since, evidently, any Cauchy sequence which fails to converge provides an example of a bounded sequence with no convergent subsequence. But, even in a Banach space L, the unit ball

$$\{X \in L; \|X\| \leq 1\}$$

is closed and bounded, yet is never compact unless L is finite-dimensional. Before we prove this, let us consider some concrete examples, illustrating how the non-compactness of the unit ball comes about in a Banach space in which there are infinitely many independent directions in which to move.

EXAMPLE 22. In any of the sequence spaces ℓ^p, $1 \leq p \leq \infty$, the vectors

$$X_1 = (1, 0, 0, \dots)$$
$$X_2 = (0, 1, 0, \dots)$$
$$X_3 = (0, 0, 1, \dots)$$
$$\cdot \qquad \cdot$$
$$\cdot \qquad \cdot$$
$$\cdot \qquad \cdot$$

constitute a sequence in the ball: $\|X_n\|_p = 1$ for every n. It has no convergent subsequence because

$$\|X_m - X_n\|_p = 2, \qquad m \neq n.$$

In other words, the sequence wanders about, with no two of the vectors in the sequence ever coming near to one another.

This same phenomenon can be reproduced easily in a Banach space such as $C([0, 1])$, with the sup norm. For instance, if $g_n = 1/\sqrt{n}\, f_n$, where f_n is the tent function in Figure 25, we have $\|g_n\|_\infty = 1$ for each n and $\|g_m - g_n\| \geq 1$ whenever $m \neq n$.

We are now going to show that "separated" sequences of the type exhibited in Example 22 exist in the unit ball of any normed linear space L which is not finite-dimensional. We shall show that, given r, $0 < r < 1$, there exists a sequence of vectors $X \in L$ such that $\|X_n\| \leq 1$ and $\|X_m - X_n\| \geq r$ whenever $m \neq n$. We shall obtain the vectors inductively, roughly as follows. Having chosen $X_1, \dots, X_n$, choose X_{n+1} such that, among points in the unit ball of L, it is as far away as possible from the subspace spanned by $X_1, \dots, X_n$.

We need some elementary facts about distances. If M is a non-empty subset of a normed linear space L and X is a vector in L, the **distance from X to M** is (of course)

$$d(X; M) = \inf_{Z \in M} \| X - Z \|.$$

Lemma. *Let* M *be a closed subspace of* L *and let* X $\in$ L.

(i) d(X; M) = 0 *if and only if* X $\in$ M.

(ii) *If* Y $\in$ M, *then* d(X + Y; M) = d(X; M).

(iii) d(cX; M) = |c| d(X; M) *for all scalars* c.

Proof. (i) is a restatement of the fact that M is a closed set. (ii) Let $Y \in M$. Then

$$d(X + Y; M) = \inf_{Z \in M} \| X + Y - Z \|.$$

Since M is a subspace and $Y \in M$, as Z ranges over M the vectors $Y - Z$ do also:

$$\{ Y - Z; Z \in M \} = M.$$

This proves (ii). Conclusion (iii) is trivial if $c = 0$; it merely says $0 \in M$. If $c \neq 0$,

$$d(cX, M) = \inf_{Z \in M} \| cX - Z \|$$

$$= |c| \inf_{Z \in M} \| X - \frac{1}{c} Z \|.$$

As Z ranges over M, so does $(1/c)Z$, that is, $(1/c)M = M$. Therefore, $d(cX; M) = |c| d(X; M)$.

Here is the key fact for our present purposes.

Lemma. *If* M *is a proper closed subspace of a normed linear space* L, *then*

$$\sup_{\|X\| \leq 1} d(X; M) = 1.$$

Proof. If $\| X \| \leq 1$, then $d(X; M) \leq \| X \|$ because M contains the zero vector. Therefore,

$$0 \leq \sup_{\|X\| \leq 1} d(X; M) \leq 1$$

Let $0 < r < 1$. We shall show that there is a vector X in the unit ball of L for which $d(X; M) = r$. That will prove the lemma.

Since M is proper and closed, there is a vector X_0 in L such that

$$d(X_0; M) = \delta > 0.$$

By part (iii) of the lemma above

$$d\left(\frac{r}{\delta} X_0; M\right) = r.$$

Since $r < 1$, there is a vector $Y \in M$ such that

$$\left\| \frac{r}{\delta} X_0 - Y \right\| < 1.$$

Let

$$X = \frac{r}{\delta} X_0 - Y.$$

Then $\| X \| < 1$ and, by part (ii) of the lemma above, $d(X; M) = r$.

Theorem 6. *Let* L *be a normed linear space which is not finite-dimensional. If* $0 < r < 1$, *there is a sequence of vectors* $\{X_n\}$ *in* L *such that*

(i) $\| X_n \| \leq 1$ *for each* n;

(ii) $\| X_m - X_n \| \geq r$ *whenever* m $\neq$ n.

Proof. Let X_1 be any (non-zero) vector in the unit ball of L. Let M_1 be the subspace spanned by X_1. Since L is not finite-dimensional, M_1 is a proper subspace of L. Theorem 3 tells us that M_1 is closed. By the last lemma, there is a vector $X_2 \in L$ such that $\| X_2 \| \leq 1$ and $d(X_2; M_1) \geq r$. Let M_2 be the subspace spanned by X_1 and X_2. Then M_2 is a proper closed subspace of L, so there is a vector $X_3 \in L$ with $\| X_3 \| \leq 1$ and $d(X_3; M_2) \geq r$. We continue inductively: Having chosen $X_1, \ldots, X_n$, choose X_{n+1} in the unit ball such that the distance from X_{n+1} to the subspace spanned by $X_1, \ldots, X_n$ is at least r.

Corollary. *The normed linear space* L *is finite-dimensional if and only if its closed unit ball* $\{X \in L; \| X \| \leq 1\}$ *is compact.*

Proof. If L is finite-dimensional then L may be identified with R^k [or C^k] under some norm. The equivalence of all norms on R^k [C^k] tells us that the closed unit ball for that norm is closed and bounded. By the Heine-Borel theorem it is compact.

If the unit ball for L is compact, each sequence in it must have an accumulation point. Thus no separated sequence can exist in the unit ball, as in Theorem 6. *Conclusion:* L must be finite-dimensional.

The non-compactness of the unit ball tells us that, in a normed linear space which is not finite-dimensional, the interior of every compact set is empty, i.e., every compact set is "thin". Nevertheless, compact sets in such normed linear spaces are sometimes important, and we wish to say something about them.

Theorem 7. *Let* K *be a subset of the normed linear space* L. *The following are equivalent:*

(i) K *is compact.*

(ii) K *is sequentially compact, i.e., every sequence in* K *has a subsequence which converges to a point of* K.

(iii) K *is complete and, for every* $\epsilon > 0$, K *can be covered by a finite number of balls of radius* ϵ.

Proof. Assume (i). Let $\{X_n\}$ be a sequence in K. If $\{X_n\}$ has no accumulation point in K, then each X in K has a neighborhood U_X which contains X_n for only a finite number of n's. Since K is compact, a finite number of the sets U_X covers K?

Assume (ii). Certainly every Cauchy sequence in K converges to a point of K, because the sequential compactness guarantees that a subsequence converges. (If a Cauchy sequence has any convergent subsequence, then the entire sequence converges to the limit of the subsequence.) Let $\epsilon > 0$, and let us show that K can be covered by a finite number of ϵ-balls. Assume the contrary. Select any point X_1 in K. The ball $B(X_1 ; \epsilon)$ does not cover K; hence, there is a point X_2 in K with $\|X_1 - X_2\| \geq \epsilon$. But $B(X_1, \epsilon)$ and $B(X_2 ; \epsilon)$ together do not cover K, so there is a point X_3 in K with $\|X_3 - X_2\| \geq \epsilon$ and $\|X_3 - X_1\| \geq \epsilon$. By induction, we obtain a sequence of points X_n in K such that (for each n)

$$\|X_n - X_k\| \geq \epsilon, \qquad k = 1, \ldots, n - 1.$$

Such a sequence (obviously) cannot have any convergent subsequence. Thus (ii) implies (iii).

We reverse our merry-go-round for a moment to show that (ii) follows from (iii). Assume (iii), and let $\{X_n\}$ be any sequence in K. Cover K by a finite number of (closed) balls of radius 1. One of those balls must contain X_n for infinitely many values of n. Choose one such closed ball K_1. Cover K_1 by a finite number of closed balls of radius $\frac{1}{2}$. Intersect each closed ball with K_1. One such intersection must contain X_n for infinitely many values of n. Select one such intersection K_2. Cover K_2 by balls of radius $\frac{1}{3}$, and repeat. By induction, we obtain a nested sequence of closed sets in K:

(i) $K_1 \supset K_2 \supset K_3 \supset \cdots$

(ii) $\lim_n \text{diam} (K_n) = 0$

(iii) K_n contains X_k for infinitely many values of k.

For each k, choose X_{n_k} in K_k. By (ii), $\{X_{n_k}\}$ is a Cauchy sequence. Since K is complete that (sub) sequence converges to a point of K.

Now we are ready to show that (i) follows from (iii). Assume (iii). Then (as we just proved) K is sequentially compact. That means that each countable open cover of K has a finite subcover. Therefore, in order to prove that K is compact, we need only show that every open cover of K has a countable subcover. We do that in (almost) the same way as we did in proving the Heine-Borel theorem, Theorem 12 of Chapter 2. We find a replacement for the balls with rational centers and rational radii which we used in that proof.

First, we must find a countable subset of K which is dense in K. We can do that as follows. For each n, choose a finite number of points in K such that the balls of radius $1/n$ about those points cover K. Put all those points together, and we have a countable subset of K. It is dense in K, because (for each n) every point of K is within distance $1/n$ of one of the points.

Second, enumerate the points in that countable set: $X_1, X_2, \ldots$. For each k, consider all open balls about X_k with rational radius. The collection of all those balls (as k varies) is countable. Enumerate them $B_1, B_2, B_3, \ldots$. The important property of this sequence of open balls is this: Every open set which intersects K contains one of the sets B_n.

Now, let $\{U_\alpha\}$ be any open cover of K. Let N be the set of those positive integers n such that B_n is contained in one of the sets U_α. For each $n \in N$ choose an index α_n so that B_n is contained in U_{α_n}. Then

$$\{U_{\alpha_n}; n \in N\}$$

is a countable (sub) cover of K.

We have established

Corollary. *Let* S *be a subset of a Banach space* X. *The following are equivalent.*

(i) *The closure of* S *is compact.*

(ii) *Each sequence in* S *has a subsequence which converges* (*in* X).

(iii) *For each* $\epsilon > 0$, S *can be covered by a finite number of balls of radius* ϵ.

We are now going to prove Ascoli's theorem, which characterizes the compact sets in one particular infinite-dimensional normed linear space—the space of continuous functions on a compact set. This is a very useful theorem, because it guarantees that certain sequences of continuous functions have uniformly convergent subsequences.

Let K be a compact set (in some normed linear space). Let $C(K)$ be the space of continuous complex-valued functions on K, endowed with the sup norm

$$\| f \|_\infty = \sup \{| f(X)|; X \in K\}.$$

Let S be a subset of $C(K)$, i.e., a family of continuous functions on K. Let's see what it means for S to be compact in the Banach space $C(K)$. Suppose that (the closure of) S is compact.

Let $\epsilon > 0$. We can cover S by a finite number of ϵ balls. In other words, there exist functions $f_1, \ldots, f_n$ in S such that every $f \in S$ is in one of the open balls:

(6.33) $\{f, \| f - f_k \|_\infty < \epsilon\}, \quad k = 1, \ldots, n.$

(There are finitely many functions in S such that every f in S is uniformly within ϵ of one of those functions.) As some wise man once observed, this means that the functions in S are "equicontinuous," that is, if X is near Y, then $f(X)$ is near $f(Y)$, uniformly for all f in S. To be specific, given ϵ, choose $f_1, \ldots, f_n$ as in (6.33). Each f_k is uniformly continuous, hence

$$|f_k(X) - f_k(Y)| < \epsilon, \quad X, Y \in K, \quad |X - Y| < \delta_k.$$

Let δ be the least of the numbers δ_k, so that

(6.34) $|f_k(X) - f_k(Y)| < \epsilon, \quad X, Y \in K, |X - Y| < \delta.$

Let f be any function in S. There exists k, $1 \leq k \leq n$, such that f is uniformly within ϵ of f_k. Therefore,

$$|f(X) - f(Y)| \leq |f(X) - f_k(X)| + |f_k(X) - f_k(Y)| + |f_k(Y) - f(Y)|$$
$$|f(X) - f(Y)| < 3\epsilon, \quad X, Y \in K, \quad |X - Y| < \delta.$$

Definition. *Let* S *be a family of functions on the set* K. *If* $X \in K$, *the family* S *is* **equicontinuous at** X *if, for each* $\epsilon > 0$, *there exists* $\delta > 0$ *such that*

$$|f(X) - f(Y)| < \epsilon, \quad Y \in K, |X - Y| < \delta, f \in S.$$

The family S *is* **(uniformly) equicontinuous** *if, for each* $\epsilon > 0$, *there exists* $\delta > 0$ *such that*

$$|f(X) - f(Y)| < \epsilon, \quad X, Y \in K, |X - Y| < \delta, f \in S.$$

We have just shown that each compact subset of $C(K)$ is (uniformly) equicontinuous. Ascoli's theorem is essentially the converse to that. Before we prove Ascoli's theorem, let us look at some examples, in order to be certain that we understand the meaning of equicontinuity. (The reader may prefer to go on to the proof of the theorem and then return to the examples.)

EXAMPLE 23. Let $K = [a, b]$, a closed interval on the real line. The simplest (interesting) equicontinuous family of functions on $[a, b]$ is obtained from estimates on derivatives. For smooth functions f

$$f(x) = c + \int_a^x f'(t)\, dt;$$

hence

$$|f(x) - f(y)| = \left| \int_y^x f'(t)\, dt \right|$$
$$\leq |x - y| \sup |f'|.$$

Therefore, consider a family such as the collection of all functions of class C^1 for which the derivatives are bounded by 3. For the functions in that family, we have a uniform estimate on the distance from $f(x)$ to $f(y)$:

$$|f(x) - f(y)| \le 3|x - y|.$$

So "given ϵ, we can choose one δ which will serve simultaneously for all f in that family." That is the meaning of equicontinuity. Obviously we can extend the example to the following case. Let $M > 0$ and let S be the family of all functions on $[a, b]$ which are differentiable and have derivatives bounded by M. Then S is an (uniformly) equicontinuous family of functions on $[a, b]$.

EXAMPLE 24. In a variety of problems (e.g., solving differential or integral equations) one meets integral operators (operations) of the following type. Suppose we are given (say) a continuous function K on the rectangle $[a, b] \times [c, d]$. Then we can define a transformation

$$C([c, d]) \overset{T}{\longrightarrow} C([a, b])$$

by

$$(Tf)(x) = \int_c^d f(t) K(x, t)\, dt.$$

Now

$$(Tf)(x) - (Tf)(y) = \int_c^d f(t)[K(x, t) - K(y, t)]\, dt.$$

Therefore,

$$|(Tf)(x) - (Tf)(y)| \le M(x, y) \int_c^d |f(t)|\, dt$$

where

$$M(x, y) = \sup \{|K(x, t) - K(y, t)|; c \le t \le d\}.$$

Since K is continuous,

$$\lim_{|x-y| \to 0} M(x, y) = 0.$$

Therefore, it should be clear that T transforms $\{f; \int |f| \le 1\}$ into an equicontinuous family, i.e.,

$$S = \{Tf; \int_c^d |f| \le 1\}$$

is equicontinuous.

EXAMPLE 25. Let D be the open unit disk in the complex plane. Let f be a complex-analytic function on D:

$$f(z) = \sum_{n=0}^{\infty} a_n z^n.$$

Suppose for the moment that f is analytic on the closed disk $\bar{D}$, i.e., that the power series for f converges in a disk of radius $1 + \epsilon$. As we have seen, we then have Cauchy's formula (5.38):

$$f(\alpha) = \frac{1}{2\pi i} \int_{|z|=1} \frac{f(z)}{z - \alpha} \, dz$$

$$= \frac{1}{2\pi} \int_{-\pi}^{\pi} \frac{f(e^{i\theta})}{e^{i\theta} - \alpha} e^{i\theta} \, d\theta.$$

Thus,

$$|f(\alpha) - f(\beta)| = \left| \frac{1}{2\pi} \int_{-\pi}^{\pi} f(e^{i\theta}) e^{i\theta} \left[\frac{1}{e^{i\theta} - \alpha} - \frac{1}{e^{i\theta} - \beta} \right] d\theta \right|$$

$$\leq \frac{1}{2\pi} \int_{-\pi}^{\pi} |f(e^{i\theta})| \left| \frac{\alpha - \beta}{(e^{i\theta} - \alpha)(e^{i\theta} - \beta)} \right| d\theta$$

$$\leq |\alpha - \beta| \, \|f\|_\infty \cdot \frac{1}{2\pi} \int_{-\pi}^{\pi} |e^{i\theta} - \alpha|^{-1} |e^{i\theta} - \beta|^{-1} \, d\theta.$$

Suppose we restrict α, β to a closed subdisk

$$D_r = \{\alpha; |\alpha| \leq r\}.$$

Then

$$|e^{i\theta} - \alpha| \geq 1 - |\alpha|$$
$$\geq 1 - r$$

so that

$$(6.35) \qquad |f(\alpha) - f(\beta)| \leq |\alpha - \beta| \frac{\|f\|_\infty}{(1 - r)^2}, \qquad \alpha, \beta \in D_r.$$

Obviously then, the same inequality holds if f is merely analytic on D.

From (6.35) we see immediately the following. If $M > 0$, the family of complex-analytic functions on D which are bounded by M is (uniformly) equicontinuous on each compact subdisk D_r. This equicontinuity is another of the special important properties of complex-analytic functions.

Theorem 8 (Ascoli). *Let* K *be a compact set (in a normed linear space). Let* S *be a subset of* C(K). *Then* S *is compact if and only if* S *is closed, bounded, and (uniformly) equicontinuous.*

Proof. We proved the "only if" part of the theorem in motivating the definition of equicontinuity. What remains to be proved is that a closed, bounded, equicontinuous family is compact in $C(K)$. It will suffice to establish sequential compactness. Let $\{f_n\}$ be a (uniformly) equicontinuous sequence of complex-valued functions on K which is bounded:

$$\|f_n\|_\infty \leq M, \qquad n = 1, 2, 3, \ldots.$$

We must extract from $\{f_n\}$ a subsequence which converges uniformly. Here is the *basic fact*: If $\{f_n\}$ is a bounded equicontinuous sequence and if $\epsilon > 0$, there exists a subsequence such that all the functions in the subsequence are uniformly within ϵ of one another (i.e., all the functions in the subsequence lie in one ϵ-ball in $C(K)$). Suppose we have established this basic fact. The proof of the existence of a uniformly convergent subsequence is then quite easy: Choose a ball of radius 1 in $C(K)$ which contains f_n for infinitely many values of n. Apply the "basic fact" to the f_n's which are inside that ball, and obtain a ball of radius $\frac{1}{2}$ which contains f_n for infinitely many n's, etc. By induction we obtain a nested sequence $B_1 \supset B_2 \supset \cdots$ in $C(K)$ such that (i) diam $(B_n) \rightarrow 0$, and (ii) each B_n contains f_k for infinitely many values of k. Choose f_{n_k} in B_k and we have the subsequence we seek.

So, all we need to do is prove the basic fact. Given ϵ, determine δ by the equicontinuity:

$$(6.36) \qquad |f_n(X) - f_n(Y)| < \frac{\epsilon}{3}, \qquad X, Y \in K, \quad \|X - Y\| < \delta.$$

Choose points $X_1, \ldots, X_N$ in K so that the open balls $B(X_k; \delta)$ cover K. In order to ensure that

$$\|f_m - f_n\|_\infty < \epsilon$$

it will suffice to have

$$(6.37) \qquad |f_m(X_k) - f_n(X_k)| < \frac{\epsilon}{3}, \qquad k = 1, \ldots, N.$$

That follows from (6.36) and the choice of $X_1, \ldots, X_N$:

$$|f_m(X) - f_n(X)| \leq |f_m(X) - f_m(X_k)| + |f_m(X_k) - f_n(X_k)| \\ + |f_n(X_k) - f_n(X)|.$$

But it is trivial to find a subsequence such that, for any f_m and f_n in the subsequence, the values are within $\epsilon/3$ at each of the finitely many points $X_1, \ldots, X_N$. That can be done for any bounded sequence of functions on any set, because $\{(f_n(X_1), \ldots, f_n(X_N))\}$ is a bounded subset of C^N.

Corollary. *If $\{f_n\}$ is a sequence of complex-valued continuous functions on a compact set* K *and if the sequence is bounded and (uniformly) equicontinuous, then there exists a subsequence $\{f_{n_k}\}$ which converges uniformly on* K.

Corollary. *Let* D *be an open set in the plane and let $\{f_n\}$ be a sequence of complex-analytic functions on* D. *If the sequence is bounded,*

$$\|f_n\|_\infty \leq M, \qquad n = 1, 2, 3, \ldots$$

then it has a subsequence which converges, uniformly on every compact subset of D.

Proof. As we showed in Example 25, a bounded sequence of complex-analytic functions is uniformly equicontinuous on each closed disk inside

D. It should then be clear from Ascoli's theorem that, if *K* is any compact subset of *D*, there is a subsequence which converges uniformly on *K*. We want to find one subsequence which does this, simultaneously for all compact sets *K* in *D*.

Write *D* as the union of an increasing sequence of compact sets. For instance, let K_n be the set of points *z* such that

(i) $z \in D$;

(ii) $|z| \leq n$;

(iii) the distance from *z* to the boundary of *D* is not less than $1/n$.

Then K_n is compact,

$$K_1 \subset K_2 \subset K_3 \subset \cdots$$
$$D = \bigcup_n K_n.$$

Suppose we are given a sequence of analytic functions f_n, and the sequence is bounded on each compact subset of *D*. Pick subsequences $S_1 \supset S_2 \supset S_3 \supset \cdots$ so that all the functions in the subsequence S_n are uniformly within 2^{-n} of one another on the set K_n. Then, any subsequence

$$f_{n_k} \in S_k$$

converges, uniformly on each compact subset of *D*.

Exercises

1. True or false? If $\{f_n\}$ is a sequence of functions of class C^1 on the closed interval [a, b], then $\{f_n\}$ is equicontinuous if and only if the sequence of derivatives $\{f'_n\}$ is bounded in sup norm.

2. True or false? If $\{F_n\}$ is a sequence of functions of bounded variation on [a, b] and if each F_n has total variation at most 1, then the sequence of functions

$$f_n(x) = \int_a^x dF_n(t)$$

is equicontinuous.

3. If you did not know Ascoli's theorem, would you regard it as obvious that any sequence of smooth functions on [0, 1] with derivatives bounded by 6 has a subsequence which converges uniformly on [0, 1]?

4. Let

$$C([0, 1]) \overset{T}{\longrightarrow} C([0, 1])$$

be the map defined by

$$(Tf)(x) = \int_0^1 f(t) K(x, t) \, dt$$

where *K* is a continuous function on the square [0, 1] × [0, 1]. If *B* is a bounded subset of $C([0, 1])$, then the set $T(B)$ has compact closure in $C([0, 1])$.

5. Let ℓ^1 be the space of (absolutely) summable sequences:

$$X = (x_1, x_2, x_3, \ldots)$$

$$\|X\|_1 = \sum_n |x_n| < \infty.$$

In the Banach space ℓ^1, consider the subset

$$S = \left\{ X; |x_n| \leq \frac{1}{n^2}, n = 1, 2, 3, \ldots \right\}.$$

Is S compact?

6. Let f be a continuous complex-valued function on the closed interval $[0, 1]$ such that $f(0) = 0$ and $\|f\|_\infty \leq 1$. Show that the sequence of powers f, f^2, f^3, ... is equicontinuous if and only if $\|f\|_\infty < 1$.

7. Give an example of a continuous function f on $[0, 1]$ such that $f(0) = 0$, $\|f\|_\infty = 1$ and the sequence of its powers $\{f^n\}$ has compact closure in the normed linear space consisting of $C([0, 1])$, together with the integral norm $\|\cdots\|_1$.

8. If S and T are compact subsets of the normed linear space L, then the algebraic sum $S + T$ is compact. (*Hint:* Addition is continuous.)

9. Let $C_B(R)$ be the space of bounded continuous complex-valued functions on the real line, equipped with the sup norm

$$\|f\|_\infty = \sup_R |f|.$$

If $f \in C_B(R)$ and $t \in R$, the t-**translate** of f is the function

$$f_t(x) = f(x + t).$$

The function f is **periodic** if $f_t = f$ for some $t \neq 0$. Show that, if f is periodic, then the set of all translates of f is a compact subset of $C_B(R)$. (*Hint:* Look at the map which sends each t into f_t.)

10. Refer to Exercise 9. The function $f \in C_B$ is called **almost periodic** if the set of all translates of f has compact closure in $C_B(R)$. Prove the following:

(a) The sum of two almost periodic functions is almost periodic; hence, every exponential polynomial

$$P(x) = \sum_{k=1}^n c_k e^{it_k x}, \qquad t_k \in R, \quad c_k \in C$$

is almost periodic.

(b) The product of two almost periodic functions is almost periodic.

(c) The space of almost periodic functions is a closed subspace of $C_B(R)$.

11. The function $f \in C_B(R)$ is almost periodic if and only if, for each $\epsilon > 0$, there exists $\delta > 0$ such that every interval of length δ contains a number t for which

$$\|f - f_t\|_\infty < \epsilon.$$

(*Hint:* Cover the set of translates by ϵ-balls.)

12. Let n be a positive integer and let $a_0, \ldots, a_n$ be continuous real-[complex-] valued functions on a closed interval I on the real line. Suppose that a_n has no zeros on I. Then any function f which satisfies the differential equation

$$a_0 f + a_1 f' + \cdots + a_n f^{(n)} = 0$$

on I will be of class C^n. Give $C^n(I)$ the norm

$$\|f\|_\infty + \|f'\|_\infty + \cdots + \|f^{(n)}\|_\infty.$$

(a) If M is the set of solutions of the differential equation, then M is a closed subspace of $C^n(I)$.

(b) Use Ascoli's theorem and the fact that a_n has no zeros to show that the closed unit ball of M is compact and, hence, that M is finite-dimensional.

6.6. Quotient Spaces

Suppose that L is a linear space and M is a (linear) subspace of L. We shall associate with the pair (L, M) a linear space L/M, known as the quotient (or difference) space of L modulo M. Then we shall use quotient spaces to discuss the completion of a normed linear space.

Loosely speaking, L/M will be the linear space obtained by doing linear operations in L under the agreement to regard two vectors as "equal" whenever they differ by a vector in the subspace M.

If X and Y are vectors in L, we say that X is **congruent to Y modulo** M and write

(6.38) $X \equiv Y, \qquad \mod M$

if the difference $(X - Y)$ is in M. Congruence modulo M has the three properties of equality which one uses repeatedly:

(a) For each X in L, $X \equiv X$.
(b) If $X \equiv Y$, then $Y \equiv X$.
(c) If $X \equiv Y$ and $Y \equiv Z$, then $X \equiv Z$.

These result from the definition, together with the facts that (a) the zero vector is in M, (b) the negative of any vector in M is again in M, (c) the sum of two vectors in M is again in M.

Congruences may be added or multiplied by scalars:

(d) If $X_1 \equiv Y_1$ and $X_2 \equiv Y_2$, then $X_1 + X_2 \equiv Y_1 + Y_2$.
(e) If $X \equiv Y$, then $cX \equiv cY$.

These follow immediately from the fact that M is closed under vector addition and under scalar multiplication. It follows from (d) and (e) that we may legitimately form linear combinations of congruences: If $X_i \equiv Y_i$, $i = 1, \ldots, n$ and if $c_1, \ldots, c_n$ are scalars, then

$$c_1 X_1 + \cdots + c_n X_n \equiv c_1 Y_1 + \cdots + c_n Y_n.$$

Properties (a)–(e) constitute the formal justification for carrying out linear operations in L, while disregarding vectors in M (vectors which are congruent to 0). Let us consider a simple case in which we might be interested in doing just this sort of thing.

EXAMPLE 26. Suppose we are interested in problems or questions which involve integration of piecewise-continuous functions over an interval $[a, b]$. From the point of view of integration, certain functions in the linear space $L = PC([a, b])$ may be disregarded, namely, the functions which have the value 0 except at a finite number of points. These are precisely the piecewise-continuous functions f such that

$$\int_a^b |f| = 0.$$

If we let M be the subspace consisting of these "null" functions for integration, then

$$f \equiv g, \quad \text{mod } M$$

means that $f(x) = g(x)$ except for a finite number of values of x and, in particular, implies that

$$\int_a^b f = \int_a^b g.$$

We might put it this way. If we chose to regard functions which are congruent modulo M as being the same, the integral would never know the difference.

There is another way of looking at linear operations modulo the subspace M, and this leads to the concept of the quotient space L/M. Suppose we ask whether calculations modulo M correspond to bona fide linear operations and relations in some vector space. If so, then a given vector X and all vectors congruent to X must be names for (must represent) the same vector in the new space. The vector which they represent will be the set

(6.39) $$Q(X) = \{Y \in L; X \equiv Y, \text{mod } M\}.$$

Note that $Q(X)$ is a very special type of subset of L. It is the X-translate of the subspace M:

$$\begin{aligned} Q(X) &= X + M \\ &= \{X + Z; Z \in M\}. \end{aligned}$$

In Figure 26 this is pictured for two choices of X in the case where L is R^2 and M is a straight line through the origin.

Note that the following conditions are equivalent:

(i) $Q(X) = Q(Y)$.
(ii) X and Y lie in the same translate of M.
(iii) $X + M = Y + M$.
(iv) Y is in the X-translate of M.
(v) $Y \in Q(X)$.

Now we let L/M be the collection of all (distinct) translates of M. The vector addition and scalar multiplication will be defined on L/M in such a way that

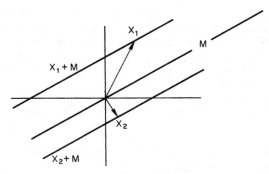

FIGURE 26

(6.40)
$$Q(X_1) + Q(X_2) = Q(X_1 + X_2)$$
$$cQ(X) = Q(cX).$$

In other words, the operations in L/M will be defined in such a way that the map

(6.41)
$$L \xrightarrow{\;Q\;} \frac{L}{M}$$

is a linear transformation.

Theorem 9. *The set* **L/M**, *consisting of all translates of the subspace* **M**, *can be given the structure of a linear space in such a way that the map*

$$Q(X) = X + M$$

is a linear transformation. This structure on **L/M** *is unique.*

Proof. Let T_1 and T_2 be translates of M. If Q is to be linear, the rule for adding T_1 and T_2 must be this: Choose any vectors $X_i \in T_i$, that is, any vectors X_i such that $Q(X_i) = T_i$, and let

(6.42)
$$T_1 + T_2 = Q(X_1 + X_2).$$

How do we know this is a well-defined rule for adding translates? We must verify that the translate $Q(X_1 + X_2)$ does not depend upon the choice of X_1 and X_2, i.e., that it is the same for all $X_1 \in T_1$, $X_2 \in T_2$. But, this is simply a restatement of property (d) of congruence modulo M—sums of congruent vectors are congruent.

If T is a translate of M and Q is to be linear, the rule which defines cT must be this: Choose any vector $X \in T$, that is, any X such that $Q(X) = T$, and let

$$cT = Q(cX).$$

Once again it must be verified that $Q(cX)$ is the same set for all vectors X in T. This is a restatement of property (e) of congruence modulo M—scalar multiplication preserves congruences.

We have shown that there is one and only one way to define a vector addition and scalar multiplication on L/M so that (6.40) is valid. In order to show that L/M with these operations is a linear space, we must verify properties (1)–(9) of Section 6.1, the defining properties of a linear space: Addition is associative; addition is commutative; scalar multiplication distributes over addition, etc. This is extremely easy to do, because we have the map Q satisfying (6.40). Each property follows from the corresponding property in the space L. Of course, the zero vector in L/M is $Q(0) = M$.

The linear space L/M is called the **quotient space** of L modulo M, and Q is called the associated **quotient map**.

If L is a normed linear space and M is a closed subspace of L, then there is a "natural" norm on the quotient space L/M. The norm of a translate of M is defined as its distance from the origin in L:

(6.43)
$$||| T ||| = \inf_{Y \in T} || Y ||.$$

Of course, this may be rewritten

$$||| X + M ||| = \inf_{Z \in M} || X + Z ||$$

which is equivalent to saying, "The norm of $Q(X)$ is the distance from X to M."

Lemma. *Let* L *be a linear space,* M *a subspace of* L, *and let* $|| \cdots ||$ *be a semi-norm on* L. *The function* $||| \cdots |||$ *defined by* (6.43) *is a semi-norm on* L/M. *It is a norm provided* M *is closed* (*under sequential convergence*).

Proof. We verify the three properties of a semi-norm.

(i) For each translate T we have $0 \leq ||| T ||| < \infty$, because $\{|| Y ||; \ Y \in T\}$ is a non-empty set of non-negative numbers.

(ii) To verify the triangle inequality, let T_1 and T_2 be translates of M. Let $X_1 \in T_1$ and $X_2 \in T_2$. Then $(X_1 + X_2) \in (T_1 + T_2)$. Therefore,

$$||| T_1 + T ||| \leq || X_1 + X_2 ||$$
$$\leq || X_1 || + || X_2 ||.$$

Since this is true for all $X_1 \in T_1$ and all $X_2 \in T_2$, we have

$$||| T_1 + T_2 ||| \leq ||| T_1 ||| + ||| T_2 |||.$$

(iii) If T is a translate of M and c is a scalar

$$||| cT ||| = \inf_{X \in T} || cX ||$$
$$= \inf_{X \in T} |c| \, ||| X ||$$
$$= |c| \inf_{X \in T} || X ||$$
$$= |c| \, ||| T |||.$$

To verify the assertion as to when $||| \cdots |||$ is a norm, we consider a translate T such that $||| T ||| = 0$. By definition, this means that there is a sequence $\{X_n\}$ in T such that

$$\lim_n || X_n || = 0.$$

Fix a vector X in T, so that $T = X + M$. Then each X_n has the form

$$X_n = X + Z_n, \qquad Z_n \in M.$$

Thus

$$\lim_n || X + Z_n || = 0$$

or

$$\lim_n (-Z_n) = X.$$

If M is closed, this implies that $X \in M$. But then $T = M$, the zero vector in L/M.

The [semi-] norm $||| \cdots |||$ is called the **quotient [semi-] norm** on L/M. If, in a particular context, L is understood to be endowed with a particular norm, then usually it is assumed that L/M is equipped with the quotient norm. It may be worth remarking that the quotient map

$$L \xrightarrow{\;Q\;} \frac{L}{M}$$

is then a continuous linear transformation from L onto L/M. The (uniform) continuity of Q is immediate from the fact that Q is linear and norm-decreasing, $||| Q(Z) ||| \leq || Z ||$:

$$||| Q(X) - Q(Y) ||| = ||| Q(X - Y) |||$$
$$\leq || X - Y ||.$$

The most important case of the last lemma is the one in which L is a normed linear space and M is a closed linear subspace. In a finite-dimensional normed linear space, every subspace is closed. In general this is not the case, e.g., M may be dense in L as the polynomials are dense in the space of continuous functions on an interval. In such cases, the quotient space L/M is of relatively little interest. We do not have time here to go into the why's and wherefore's of this. Suffice it to say that, in a normed linear space, it is the closed linear subspaces which are of prime importance.

There is one other (special) case of the last lemma in which we are interested.

EXAMPLE 27. Suppose that $|| \cdots ||$ is a semi-norm on L and M is the subset of null vectors:

$$M = \{X \in L; || X || = 0\}.$$

It is easy to see that M is a linear subspace of L and that M is closed under sequential convergence. Therefore, the quotient semi-norm on L/M is actually a norm. It has a very special property in this case:

$$||| Q(X) ||| = || X ||.$$

What the quotient structure on L/M does is to provide a natural way of "dividing out" the null vectors—forgetting about them, one might say.

We dealt with a special case of this situation in Example 26, where L was the space of piecewise-continuous functions, endowed with the integral semi-norm

$$|| f ||_1 = \int_a^b |f|.$$

Dividing out by the functions of semi-norm zero amounts to creating a new linear space by identifying functions which differ only at a finite number of points.

The vectors in the quotient space are not easy to describe, except to say they are certain classes of functions. Therefore, one usually handles this special type of quotient space as we did in Example 26: Carry out linear operations in L, using congruence modulo null vectors in place of equality.

One technical remark may be in order about the proof of the last lemma. We did not discuss convergence, open sets, closed sets, completeness, etc. in the context of a semi-normed linear space. One can do this easily and naturally, because things are formally the same as in a normed space. But there can be one confusing point: A sequence may have more than one limit. For

$$\lim_n X_n = X$$

$$\lim_n X_n = Y$$

only tells us that $|| X - Y || = 0$ and, consequently, that $Y = X + Z$, where Z is a null vector. Therefore, if a sequence converges, the set of its limit points is a translate of the subspace of null vectors. In particular, every closed linear subspace contains the space of null vectors. This is why, when we divide out by a closed subspace, the quotient semi-norm automatically becomes a norm.

Theorem 10. *If* L *is a complete semi-normed linear space and* M *is a closed linear subspace, then* L/M *is a Banach space.*

Proof. We want to show that L/M is complete. It suffices to show that every fast Cauchy sequence in L/M converges. Let $\{T_n\}$ be such a sequence:

$$(6.44) \qquad\qquad \sum_{n=1}^{\infty} ||| T_n - T_{n+1} ||| < \infty.$$

Let Q be the quotient map

$$L \xrightarrow{\;Q\;} \frac{L}{M}.$$

By the definition of the quotient norm, for each positive integer n there is a vector Z_n in L such that

(6.45)
$$Q(Z_n) = T_n - T_{n+1}$$
$$\|Z_n\| < \|\!|\!| T_n - T_{n+1} |\!|\!\| + 2^{-n}.$$

Select any vector X_1 such that $Q(X_1) = T_1$ and then define

$$X_n = X_1 - (Z_1 + \cdots + Z_{n-1}), \qquad n = 2, 3, 4, \ldots.$$

The definition of X_n is so constructed that

$$X_n - X_{n+1} = Z_n, \qquad n = 1, 2, 3, \ldots.$$

By (6.44) and (6.45) we have

$$\sum_{n=1}^{\infty} \|X_n - X_{n+1}\| < \sum_{n=1}^{\infty} \|\!|\!| T_n - T_{n+1} |\!|\!\| + \sum_{n=1}^{\infty} 2^{-n}$$
$$< \infty.$$

Since L is complete, the fast Cauchy sequence $\{X_n\}$ converges, say

$$\lim_n X_n = X.$$

Since Q is continuous,

$$\lim_n Q(X_n) = Q(X).$$

But, since $Q(X_1) = T_1$ and $Q(X_{n+1} - X_n) = Q(Z_n) = T_{n+1} - T_n$, we have $Q(X_n) = T_n$ for each n. Thus $\{T_n\}$ converges in L/M to the vector $T = Q(X)$.

Exercises

1. Endow R^3 with the standard norm (length). Let M be any linear subspace of R^3, e.g., a line or plane through the origin. Let $M^\perp$ be the set of all vectors which are orthogonal (perpendicular) to M

$$M^\perp = \{X; \langle X, Z \rangle = 0, \quad \text{all} \quad Z \in M\}.$$

(a) Show that $M^\perp$ is a subspace of R^3.

(b) Let $X \in R^3$. Show that the infimum

$$\inf_{Z \in M} \|X - Z\|$$

is attained for precisely one vector $Z \in M$ and (for this vector) $X - Z$ is in $M^\perp$.

(c) Each vector $X \in R^3$ is uniquely expressible in the form $X = Y + Z$, where $Y \in M^\perp$ and $Z \in M$.

(d) Identify the quotient space R^3/M with $M^\perp$, by showing that each translate of M contains precisely one vector Y in $M^\perp$ and that the (quotient) norm of the translate is the length of Y.

2. Let L be the space of continuous real- [complex-] valued functions on the closed interval $[-1, 1]$, endowed with the sup norm. Let M be the set of even functions in L:

$$f(-x) = f(x), \qquad -1 \leq x \leq 1.$$

(a) Show that M is a closed subspace of L.

(b) Show that each translate of M contains precisely one odd function:

$$g(-x) = -g(x), \qquad -1 \leq x \leq 1.$$

(c) Identify L/M with the space of odd continuous functions.

(d) Is there a simple relationship between the quotient norm on L/M and the sup norm on odd functions?

3. Let M be a subspace of the linear space L. Let T be a *linear transformation* from L into some linear space N:

$$L \xrightarrow{\;T\;} N$$
$$T(cX + Y) = cT(X) + T(Y).$$

Suppose that $T(Z) = 0$ for every $Z \in M$. Show that there is a linear transformation S, from L/M into N, such that T is the composition of S and the quotient map Q:

$$L \xrightarrow{\;Q\;} \frac{L}{M}$$
$$T \searrow \quad \downarrow S$$
$$N.$$

4. Let M be a closed subspace of the normed linear space L. Show that U is an open subset of L/M if and only if $Q^{-1}(U)$ is an open subset of L.

5. Let T be a linear transformation from a normed linear space L to a normed linear space N. Prove that the following are equivalent

(i) T is continuous.

(ii) There exists a constant $M > 0$ such that

$$\|T(X)\| \leq M\|X\|, \qquad \text{all } X \in L.$$

6. Let L be a normed linear space and let f be a linear functional on L, i.e., a linear transformation from L into the scalar field. Show that f is continuous if and only if the null space $\{X; f(X) = 0\}$ is a closed subspace of L.

7. Let M be a closed subspace of the normed linear space L. Prove that the image of the open unit ball $B(0, 1)$ under the quotient map

$$L \xrightarrow{\;Q\;} \frac{L}{M}$$

is precisely the open unit ball about the origin in L/M.

8. Let L be a linear space. There are natural ways to define sum and scalar multiple for arbitrary subsets of L:

$$A + B = \{X + Y; X \in A, Y \in B\}$$
$$cA = \{cX; X \in A\}.$$

(a) Look at properties (1)–(8) defining a linear space in Section 6.1. Show that the collection of all subsets of L, together with these operations, satisfies all these conditions except for the existence of a zero vector and the existence of negatives.

(b) Suppose M is a subspace of L. Show that L/M, the set of translates of M, is a vector space under the operations above (and, indeed, that these operations coincide with the linear operations we introduced for L/M).

(c) Now prove that (b) describes the only way that a family of subsets can be a linear space under the defined operations on sets. In other words, show that, if a family of subsets of L forms a linear space under these operations, it is the family of translates of some subspace of L.

6.7. The Completion of a Space

It is frequently useful to know that a normed linear space L can be completed to a Banach space, i.e., that we can embed L densely in a Banach space. If L is not complete, it means that L has certain "holes" where the limits of some Cauchy sequences are missing, and we want to know that it is possible to fill in all the holes.

If $S = \{X_n\}$ is a Cauchy sequence in L which does not converge, we want to adjoin to L a limit point for the sequence. We shall arrive at the completion of L by adjoining a limit point for each such sequence. But, we must be careful. If we have two Cauchy sequences $\{X_n\}, \{Y_n\}$ and if $\{X_n - Y_n\}$ converges to 0, then obviously we should not adjoin to X different limit points for the two sequences.

Given any normed linear space L, we form Cauchy(L), the space of all Cauchy sequences in L. If

$$S = \{X_n\}$$
$$T = \{Y_n\}$$

are two Cauchy sequences in L, their sum is

$$S + T = \{X_n + Y_n\}$$

and that is again a Cauchy sequence. Scalar multiplication in Cauchy(L) is defined by

$$cS = \{cX_n\}.$$

All that we are saying is this. Consider the space of all sequences of vectors in L, i.e., the space of all functions from Z_+ into L. Then Cauchy(L) is a linear subspace of that space of sequences.

We introduce a semi-norm on Cauchy(L):

(6.46) $$\|S\| = \lim_n \|X_n\|$$

That makes sense because, if $\{X_n\}$ is a Cauchy sequence, then $\{\|X_n\|\}$ is a Cauchy sequence in R. We said "semi-norm" because we can have

$\|S\| = 0$ without $S = 0$. In fact, $\|S\| = 0$ if and only if the sequence $S = \{X_n\}$ converges to 0.

Now we remark that Cauchy(L) is complete, that is, *every Cauchy sequence in* Cauchy(L) *converges in the space.* Let $\{S_n\}$ be a Cauchy sequence in Cauchy(L):

$$S_n = (X_{n1}, X_{n2}, X_{n3}, \ldots).$$

For each k, there is a positive integer N_k such that

(6.47) $$\|X_{km} - X_{kn}\| < \frac{1}{k}, \qquad m, n \geq N_k.$$

Choose any one of the points X_{kn} with $n \geq N_k$ and call that point Y_k. Then

(6.48) $$\|Y_k - X_{kn}\| < \frac{1}{k}, \qquad n \geq N_k.$$

Let T_k be the constant sequence

$$T_k = (Y_k, Y_k, Y_k, \ldots).$$

This is certainly a Cauchy sequence. Furthermore, according to (6.48),

(6.49) $$\|S_k - T_k\| \leq \frac{1}{k}.$$

Since $\{S_n\}$ is a Cauchy sequence, (6.49) makes it apparent that $\{T_n\}$ is a Cauchy sequence, i.e., $\{Y_n\}$ is a Cauchy sequence in L. But $\{T_n\}$ obviously converges in the space Cauchy(L); it converges to the Cauchy sequence

$$S = (Y_1, Y_2, Y_3, \ldots).$$

So S_n also converges to S.

Theorem 11. *Each normed linear space is a dense subspace of a Banach space.*

Proof. We have the complete space Cauchy(L), and it is quite clear how to embed L in Cauchy(L). We identify X with the constant sequence $(X, X, X, \ldots)$. The only problem is that we have only a semi-norm on Cauchy(L). We take care of that by dividing out the null sequences, as in Example 27.

Let Null(L) be the space of Cauchy sequences of semi-norm zero, that is, the space of null sequences in L (ones which converge to zero). By Theorem 10, we know that the quotient space associated with the subspace is a Banach space. Furthermore, L is embedded in that Banach space by

(6.50) $$L \xrightarrow{\ P\ } \text{Cauchy}(L) \xrightarrow{\ Q\ } \frac{\text{Cauchy}(L)}{\text{Null}(L)}$$

where $P(X)$ is the constant sequence $P(X) = (X, X, X, \ldots)$ and Q is the quotient map. The composition $Q \circ P$ is norm-preserving because both P and Q are. In particular $Q \circ P$ is $1 : 1$. It also preserves linear operations because both P and Q are linear transformations. We have left to the exercises the proof that the image of L is dense in the quotient Banach space.

Exercises

1. Verify that the natural injection of L into Cauchy(L) carries L onto a dense subspace.

2. Use the result of Exercise 5 of Section 6.6 to prove the following. If T is a continuous linear transformation from the normed linear space L into the Banach space M, then T can be extended uniquely to a continuous linear transformation from $\bar{L}$ to M. ($\bar{L}$ denotes the completion of L.)

3. If L is an inner product space, show that its completion to a Banach space is also an inner product space.

7. The Lebesgue Integral

7.1. Motivation

We are going to discuss a type of integral which is both a more general and a more powerful tool than the Riemann integral we dealt with in Chapter 4. This integral, developed by Henri Lebesgue shortly after the beginning of this century, is more general in the sense reflected in these two statements.

(i) Every function f which is Riemann-integrable is Lebesgue-integrable and the two processes assign the same integral to f.

(ii) The class of Lebesgue-integrable functions contains many functions, e.g., unbounded functions or functions on unbounded domains, which are not properly Riemann-integrable and which were treated traditionally by "improper integrals".

This increased range of application of the integration process would not by itself justify the effort we will invest in developing the Lebesgue integral, because it is not too difficult to embellish the Riemann process to be able to deal with such mathematical statements as:

$$\int_0^1 \frac{1}{\sqrt{x}}\, dx = 2, \qquad \int_{-\infty}^{\infty} e^{-x^2}\, dx = \sqrt{\pi}\,.$$

The significance and power of the more general method of integration derive from the fact that the class of Lebesgue-integrable functions is "complete", in several senses. This refers, in the first instance, to the concept of completeness discussed in Chapter 6.

(iii) The linear space of Lebesgue-integrable functions is complete relative to the (semi-) norm

$$\|f\|_1 = \int |f|$$

that is, it is a (semi-) Banach space.

The completeness can be expressed in other forms, but appreciation of their interest or significance requires adoption of a particular point of view. Instead of regarding integration as a process for calculating the integrals of specific functions, we regard integration as an operator (function) on the space (set) of all integrable functions,

$$I(f) = \int f.$$

This operation is **linear**

$$I(cf + g) = cI(f) + I(g)$$

and **positive**

$$I(f) \geq 0 \quad \text{if} \quad f \geq 0.$$

Lebesgue's brilliant idea was to focus on another critical property of integration, which we may call **weak continuity**: If $f_1 \geq f_2 \geq f_3 \geq \cdots$ is a monotone-decreasing sequence of (non-negative) integrable functions and if it converges pointwise to 0, then

$$\lim_n I(f_n) = 0.$$

The second form of "completeness" states, roughly speaking, that Lebesgue extended the integral to the largest class of functions to which it could reasonably be extended.

(iv) The space of Lebesgue-integrable functions is the largest subspace of the space of complex-valued functions to which the integral can be extended so that it remains linear, positive, and weakly continuous.

Strictly speaking, Lebesgue did not deal with the completeness in the form described in (iv). He phrased the completeness in terms of the concept of measure (length, area, volume, etc.). The language will be simpler if we deal only with area for the time being. Similar remarks are applicable to (k-dimensional) measure in R^k.

Area is related to the integration of functions on R^2 in this way. If E is a subset of R^2 and if k_E is the **characteristic function** of E,

(7.1) $$k_E(X) = \begin{cases} 1, & X \in E \\ 0, & X \notin E, \end{cases}$$

then the area of E should be the integral of its characteristic function

$$A(E) = \int k_E.$$

If k_E is integrable (in the sense of Lebesgue), this "definition" of area makes sense. Lebesgue focused on the properties of the area function which are analogous to the cited properties of the integral operator. Area is **additive**,

$$A(D \cup E) = A(D) \cup A(E) \quad \text{if} \quad D \cap E = \varnothing$$

it is **positive**,

$$A(E) \geq 0$$

and it is **weakly continuous**:

If $E_1 \supset E_2 \supset E_3 \supset \cdots$ and $\bigcap_n E_n = \varnothing$, then $\lim_n A(E_n) = 0$.

There are two ways to reformulate the weak continuity of area. The first reformulation was the heart of the ancient Greeks' "method of exhaustion": If $S_1 \subset S_2 \subset S_3 \subset \cdots$ is an increasing sequence of sets which "exhaust" the set S,

$$S = \bigcup_n S_n$$

then $A(S) = \lim_n A(S_n)$. The relationship to weak continuity is provided by the fact that $E_n = S - S_n$ is a decreasing sequence with empty intersection if and only if S_n is an increasing sequence with union S. Furthermore, if area is additive, we have $A(S_n) + A(S - S_n) = A(S)$, so that $A(E_n)$ tends to 0 if and only if $A(S_n)$ tends to $A(S)$. The second reformulation notes that if S is the union of the increasing sequence $\{S_n\}$ and we define $T_1 = S_1$, $T_n = S_n - S_{n-1}$, $n \geq 2$, we have

$$(7.2) \qquad S = \bigcup_n T_n, \qquad T_i \cap T_j = \varnothing, \quad i \neq j.$$

The additivity of area shows that $A(S_n) = A(T_1) + \cdots + A(T_n)$; hence $A(S_n)$ tends to $A(S)$ if and only if

$$(7.3) \qquad A(S) = \sum_n A(T_n).$$

This reformulation of weak continuity, which says that (7.3) follows from (7.2), is called the **countable additivity** of area. It is, perhaps, the most convenient form to work with because it includes the additivity property as a special case.

Of course, we have not stated to which class of sets the area function is being applied. We tend to take the concept of area more for granted than we do "integral", as if every subset of the plane could be assigned an (numerical measure of) area in a meaningful way. But there are some very wild subsets of the plane. Using the axiom of choice (see Appendix) sets can be identified to which no appropriate measure of area can be assigned; that is, there exist sets which are "non-measurable". These sets are so horrible that one might reasonably take the attitude that area should be discussed only for the simplest sets. But even the ancient Greeks seemed to be intrigued by the idea of discovering the answer to this question:

Starting from the simple sets, to what class of sets can the definition of area be extended? All of us can be impressed by what Lebesgue was able to prove, by developing his integral.

(v) The largest class of bounded subsets of the plane to which area can be extended so that it remains positive and countably additive is the class of bounded sets which have Lebesgue-integrable characteristic functions. Furthermore, for every such set, the area is the integral of the characteristic function.

We wish to stress the fact that the "completeness" statements (iv) and (v) are phrased rather loosely. They convey the correct general ideas, but a number of technicalities must be cleared up before they can be transformed into precise mathematical assertions. It is not necessary, nor would it be advisable, for us to stop to do this now.

If the completeness properties, (iii)–(v), describe a special property of Lebesgue integration, they do not convey adequately what its significance is. We shall try to give some indication of this.

The development of a complete theory of integration placed powerful new tools in the hands of mathematical analysts. Within a few years, significant advances were made in the study of complex-analytic functions. Complete theories became possible in the study of Fourier series and Fourier transforms. Lebesgue's methods lent themselves to further generalization, in which other "measures" were used in place of length, area, volume, etc. The formulation of modern probability theory would not have been possible without these developments. All of this has had its impact not only on the structure of mathematics as a discipline but also on the range of its applications. Let us close this introductory section with a few tangible examples of the tools and results which this more general integral will provide.

One of the recurring questions about integration is when "the integral of the limit (function) is the limit of the integrals." In other words, if f_1, $f_2, \ldots$ are integrable functions on, say, a box and if

$$f(X) = \lim_n f_n(X)$$

when can we conclude that f is integrable and

$$\int f = \lim_n \int f_n?$$

If each f_n is continuous and the convergence is uniform, there is no problem; but this is far from an adequate result for many purposes. For example, in any situation where each f_n is continuous but f is not, the convergence cannot be uniform. To contrast the results of this type which hold for Riemann vs. Lebesgue integration, let us list the following properties.

(a) Each f_n is integrable.
(b) The sequence $\{f_n\}$ is bounded.

(c) $\{f_n\}$ converges pointwise to f.

(d) f is integrable.

(e) $\int f = \lim_n \int f_n$.

The strongest theorem for Riemann integration on a box in R^k is:

$$(a) + (b) + (c) + (d) \Longrightarrow (e).$$

For Lebesgue integration, (d) can be moved out of the hypothesis and into the conclusion:

$$(a) + (b) + (c) \Longrightarrow (d) + (e).$$

Thus, for the Lebesgue integral on a box, bounded pointwise convergence allows us to interchange the order of the limits. This is a very powerful technical tool.

(vi) If $\{f_n\}$ is a bounded sequence of Lebesgue-integrable functions on a box in R^k which converges pointwise to the function f, then f is Lebesgue-integrable and

$$\int f = \lim_n \int f_n.$$

Parallel to the illustration just given is an equally powerful tool for dealing with unbounded functions:

(vii) If $f_1 \leq f_2 \leq f_3 \leq \cdots$ is an increasing sequence of (real-valued) Lebesgue integrable functions and the sequence of integrals $\int f_n$ is bounded above, then it follows that the limit

$$f(X) = \lim_n f_n(X)$$

is finite at "almost every" point X, that f is a Lebesgue-integrable function, and that

$$\int f = \lim_n \int f_n.$$

(We will define "almost every" in Section 7.3.)

Tools such as (vi) and (vii) are very closely interwoven with the completeness properties, (iii)–(v). They enable one to give complete theories of the following type.

In Chapter 5 we showed that if f is a complex-valued piecewise-continuous function on the interval $[-\pi, \pi]$ and if

$$\sum_{n=-\infty}^{\infty} c_n e^{inx}$$

is the Fourier series for f:

(7.4) $$c_n = \frac{1}{2\pi} \int_{-\pi}^{\pi} f(x) e^{-inx} \, dx$$

then

(7.5) $$\frac{1}{2\pi} \int_{-\pi}^{\pi} |f(x)|^2 \, dx = \sum_{n=-\infty}^{\infty} |c_n|^2$$

and (consequently) the partial sums

$$s_N(x) = \sum_{n=-N}^{N} c_n e^{inx}$$

converge to f in the "square-integral" norm, i.e.,

$$\|f - s_N\|_2 = \left(\int_{-\pi}^{\pi} |f(x) - s_N(x)|^2 \, dx \right)^{1/2}$$

tends to 0. Lebesgue's integral provides the following beautiful converse: If $\{c_n\}$ is any sequence of complex numbers such that

$$\sum_{n=-\infty}^{\infty} |c_n|^2 < \infty$$

then there is a complex-valued function f on $[-\pi, \pi]$ such that f and f^2 are Lebesgue integrable and $\{c_n\}$ is the sequence of Fourier coefficients of f. This gives a complete correspondence between L^2, the space of Lebesgue square-integrable functions on $[-\pi, \pi]$ and ℓ^2, the space of square-summable sequences of complex numbers. The correspondence is $1:1$ if we agree to disregard null functions, those which are zero at almost every point; and it preserves norms, (7.5).

The reader may like to keep these few examples in mind as reasons *why* we want to develop the Lebesgue integral. It is (iii) which we will use as the guiding principle for *how* we develop it.

7.2. The Setting

In this section we shall give a precise description of the setting in which we will develop the Lebesgue integral. We begin by reviewing the small amount of information we need about the (Riemann) integral of a continuous function. We shall change the context slightly from that of Chapter 4, in order to be able to work directly with integrals over (all of) R^k instead of over a fixed box.

Definition. *A complex-valued function* f *on* R^k *is said to have* **compact support** *if there exists a compact set* K *such that* f(X) = 0 *for each* X *not in* K.

Note that the sum of two functions of compact support is a function of compact support: If f_1 vanishes outside K_1 and f_2 vanishes outside K_2, then $f_1 + f_2$ vanishes outside $K_1 \cup K_2$. Therefore, the set of functions of compact support is a linear subspace of the space of all (complex) functions on R^k.

Notation. We denote by $C_c(R^k)$ the set of all continuous complex-valued functions on R^k which have compact support. Evidently this is also a linear subspace of the space of functions on R^k.

We want to define the Riemann integral over R^k of a continuous function of compact support. We need the following simple result.

Lemma. *Let* B *be a closed box in* R^k *and let* $\tilde{B}$ *be a closed box which contains* B. *Let* f *be a continuous function on* $\tilde{B}$ *such that* f $= 0$ *on* $\tilde{B} - B$. *Then*

$$\int_{\tilde{B}} f = \int_{B} f.$$

Proof. The integral of f over B is approximated by sums

$$S(f, P, T) = \sum_{j=1}^{n} f(T_j) m(B_j)$$

where $P = \{B_1, \ldots, B_n\}$ is a partition of B and $T = \{T_1, \ldots, T_n\}$ is a choice of points $T_j \in B_j$. Given any P, we can find a partition $\tilde{P}$ of $\tilde{B}$ such that

(a) $P \subset \tilde{P}$, i.e., each box B_j is one of the boxes of $\tilde{P}$;
(b) $\|\tilde{P}\| \leq \|P\|$.

Given a "choice" T, extend it to a choice $\tilde{T}$ of points in the boxes $\tilde{P}$, subject to the condition that, if a box in $\tilde{P}$ is not one of the boxes of P, then the point chosen in the box is outside B. Then, since $f = 0$ on $\tilde{B} - B$,

$$S(f, \tilde{P}, \tilde{T}) = S(f, P, T).$$

If the mesh $\|P\|$ (and therefore the mesh $\|\tilde{P}\|$) is small, then $S(f, P, T)$ will be near $\int_B f$ and $S(f, \tilde{P}, \tilde{T})$ will be near $\int_{\tilde{B}} f$. This establishes the lemma.

Definition. *If* f *is a continuous function on* R^k *which has compact support, the* **Riemann integral** *of* f *over* R^k *is*

$$\int f = \int_{B} f$$

where B *is any closed box such that* f $= 0$ *outside* B.

The point is, of course, that $\int_B f$ does not depend on B, as long as $f = 0$ outside B. In other words, if B_1 and B_2 are closed boxes outside of which f vanishes, then

$$\int_{B_1} f = \int_{B_2} f.$$

This follows immediately from the last lemma, which tells us that

$$\int_{B_1} f = \int_{B_1 \cap B_2} f$$

$$\int_{B_2} f = \int_{B_1 \cap B_2} f.$$

Theorem 1. *The Riemann integral on* $C_c(R^k)$ *has these properties.*

(i) *It is linear,*

$$\int (cf + g) = c \int f + \int g.$$

(ii) *It is positive; i.e., if* $f \geq 0$, *then*

$$\int f \geq 0.$$

(iii) *It is weakly continuous; i.e., if* $f_1 \geq f_2 \geq f_3 \geq \cdots$ *is a monotone-decreasing sequence of non-negative functions in* $C_c(R^k)$ *which tends pointwise to* 0, *then*

$$\lim_n \int f_n = 0.$$

Proof. It is only (iii) which is not obvious. To prove it, choose a closed box B such that $f_1 = 0$ outside B. Since $0 \leq f_n \leq f_1$, we see that $f_n = 0$ outside B. Accordingly,

$$\int f_n = \int_B f_n, \qquad n = 1, 2, 3, \ldots.$$

Since (a) each f_n is continuous, (b) $\{f_n\}$ converges monotonely downward to 0 on B, and (c) B is compact, Dini's theorem (Theorem 1 of Chapter 5) tells us that $\{f_n\}$ converges uniformly to 0 on B. Therefore,

$$\lim_n \int_B f_n = 0.$$

Now we can explain the idea behind what we intend to do. We consider the normed linear space which consists of $C_c(R^k)$ together with the norm

(7.6) $\|f\|_1 = \int |f|.$

This norm is called the **L¹-norm** on $C_c(R^k)$.

We are going to complete the normed linear space $(C_c, \|\cdots\|_1)$ to a Banach space. Since the integral is uniformly continuous on $C_c(R^k)$:

$$\left| \int f - \int g \right| \leq \int |f - g| = \|f - g\|_1$$

it will extend uniquely to a (linear) function on the completion. Essentially, this completion will be the space of Lebesgue-integrable functions and the extension of $\int$ will be the integral on this space.

In Chapter 6 we showed how to complete any given normed linear space to a Banach space. We shall *not* need to use this general result, but its method of proof is what we will keep in the backs of our minds to guide what we do. The basic idea of the completion process is very simple: With each Cauchy sequence in the normed linear space we associate an abstract vector to serve as the limit of the sequence, subject to the condition that, when the difference of two Cauchy sequences converges to 0, they should have the same associated "abstract" limit vector. Thus, in the case at hand, we want to associate a limit to each sequence $\{f_n\}$ in $C_c(R^k)$ which is Cauchy in the L^1-norm:

$$(7.7) \qquad\qquad \lim_{n,\,p} \|f_n - f_p\|_1 = 0.$$

But we want to assign a limit which is more than an abstract vector. We want the limit to be a function on R^k. Therefore, what we would hope is that the L^1-Cauchy condition (7.5) would imply that the sequence $\{f_n\}$ converges pointwise. Unfortunately, this is not always so. What is true, however, is that if $\{f_n\}$ is a fast L^1-Cauchy sequence:

$$\sum_{n=1}^{\infty} \|f_n - f_{n+1}\| < \infty$$

then $\{f_n\}$ converges pointwise at "almost every" point of R^k. Thus we obtain a function to employ as the limit of the Cauchy sequence. The functions which arise in this way will be the Lebesgue-integrable functions on R^k. That is, a Lebesgue-integrable function on R^k will be a (complex-valued) function defined at "almost every" point of R^k such that there exists a fast L^1-Cauchy sequence in $C_c(R^k)$ which converges to it pointwise at "almost every" point. The integral of such a function will (of course) be the limit of the integrals of the functions in the Cauchy sequence. It will require a bit of effort to formulate all of this precisely and to verify that it is valid.

Exercises

1. Let I be a closed (bounded) interval on the real line and let J be an open interval which contains I. Construct a continuous function f on R^1 such that $f = 1$ on I, $f = 0$ outside J, and $0 < f < 1$ on $J - I$.

2. Let K be a compact subset of R^k and let U be a bounded open set which contains K. Let $g(X) = d(X, R^k - U)$. Let I be the closed interval $[a, b]$, where

$$a = \inf_K g$$
$$b = \sup_K g.$$

By proper choice of a function f as in Exercise 1 show that the composition

$h = f \circ g$ satisfies

  (i) $h \in C_c(R^k)$;
  (ii) $h = 1$ on K;
  (iii) $h = 0$ outside U;
  (iv) $0 < h < 1$ on $U - K$.

3. Let f be a continuous function on R^k, e.g., a polynomial, and let K be any compact subset of R^k. Use the result of Exercise 2 to show that there is a continuous function on R^k which has compact support and agrees with f on K.

4. Let f be a function in $C_c(R^k)$. Prove that the real and imaginary parts of f are in $C_c(R^k)$, as is $|f|$.

5. Prove that the L^1-norm is a norm on $C_c(R^k)$.

6. Let K and U be as in Exercise 2 and let h be the function constructed there. Then $h \geq h^2 \geq h^3 \geq \cdots$ is a monotone-decreasing sequence of functions in $C_c(R^k)$. Does it converge uniformly? What would you call

$$\lim \int h^n ?$$

7. If f is a complex-valued function on R^k, the **support** of f is the closure of the set $\{X; f(X) \neq 0\}$.

  (a) Show that a function has compact support if and only if its support is compact.

  (b) Show that, if f has compact support, the support of f is the smallest compact set K such that $f = 0$ outside K.

8. Every function in $C_c(R^k)$ is bounded, hence the sup norm is a norm on $C_c(R^k)$.

9. Is the normed linear space $(C_c, \|\cdots\|_\infty)$ complete? If not, describe its completion (as a subspace of the bounded continuous functions on R^k).

*10. If K is a compact subset of R^k and f is a continuous complex-valued function on K, there exists $g \in C_c(R^k)$ such that $g(X) = f(X)$, $X \in K$.

7.3. Sets of Measure Zero

We have made several references to convergence at "almost every" point, and we have indicated that this phenomenon will of necessity arise in the discussion of Lebesgue-integrable functions. In this section, we will give a precise formulation of it, using "sets of measure zero". We will gain a certain technical advantage if we discuss the more general concept of "outer measure" and then relate both it and "measure zero" to Riemann integrals.

Definition. *If* S *is a subset of* R^k, *the* **outer measure** *of* S *is*

$$(7.8) \qquad\qquad m^*(S) = \inf_{\{B_n\}} \sum_{n=1}^{\infty} m(B_n)$$

where the infimum is taken over all countable coverings of S *by boxes.*

Some explanation is in order. Given the set S, there are various sequences of boxes B_1, B_2, B_3 which cover S:

$$S \subset \bigcup_{n=1}^{\infty} B_n.$$

The collection of all boxes which have rational vertices (corners) is one such covering. This tells us that $0 \le m^*(S) \le \infty$. This is not, however, the sort of countable covering which motivates the definition of outer measure. We have in mind, rather, a sequence of very tiny boxes such that

(i) B_1, B_2, B_3, . . . just barely cover S;
(ii) there is very little overlap of the various boxes B_n.

For such a sequence $\{B_n\}$ the sum $\sum_n m(B_n)$ ought to be just a little bit larger than the number which we would hope to call the measure (k-dimensional volume) of S. As we use smaller and smaller boxes, we can improve on (i) and (ii), getting closer and closer to the "measure" of S. The smallest number we can hope to attain as a limit is the infimum, which we have named $m^*(S)$. We call this the "outer measure" of S rather than just the "measure" of S, for two reasons. First, we obtained it by squeezing down on S from the outside. Second, we indicated in Section 7.1 that it is not possible to assign a measure to *all* subsets of R^k so that the properties we expect measure to have are preserved; i.e., there exist "non-measurable" sets.

To most people it would appear more natural to use finite coverings by boxes in defining outer measure. There is even a systematic way to go about this which has a "natural" appeal. We subdivide all of R^k into boxes using a gridwork of some given "mesh". Then to estimate the outer measure of a given set S, we add up the areas of those boxes which contain points of S. (See Figure 27.) By using grids of finer and finer mesh, one ought to obtain sums which converge downward to the measure (or

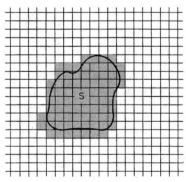

FIGURE 27

outer measure) of S. Obviously, for unbounded sets this would involve infinite coverings, but for bounded sets only a finite number of boxes formed by any given gridwork would touch S. For very "nice" sets this process is adequate, but, as Lebesgue realized, it is inadequate for many (bounded) sets which one must deal with in order to have an adequate theory of integration. (See Example 1.)

Lemma. *If* $S, S_1, S_2, \ldots$ *are subsets of* $\mathbf{R}^k$ *such that*

$$S \subset \bigcup_{n=1}^{\infty} S_n$$

then

(7.9) $$m^*(S) \leq \sum_{n=1}^{\infty} m^*(S_n).$$

Proof. If $m^*(S_n) = \infty$ for some n, the inequality is trivially true. If $m^*(S_n) < \infty$ for every n, we proceed as follows. Let $\epsilon > 0$. By the definition of $m^*(S_1)$, there is a sequence $\{B_n^1\}$ of boxes which covers S_1 and satisfies

$$\sum_{n=1}^{\infty} m(B_n^1) < m^*(S_1) + \frac{\epsilon}{2}.$$

Similarly, there is a sequence of boxes $\{B_n^2\}$ which covers S_2 such that

$$\sum_{n=1}^{\infty} m(B_n^2) < m^*(S_2) + \frac{\epsilon}{4}.$$

In general (for each $k = 1, 2, 3, \ldots$) we can find a sequence of boxes $\{B_n^k\}$ which covers S_k such that

$$\sum_{n=1}^{\infty} m(B_n^k) < m^*(S_k) + \frac{\epsilon}{2^k}.$$

Since the sequence of sets $B_1^k, B_2^k, B_3^k, \cdots$ covers S_k and the sequence $\{S_k\}$ covers S, the double sequence $\{B_n^k; k = 1, 2, 3, \ldots, n = 1, 2, 3, \ldots\}$ is a countable covering of S by boxes. Furthermore,

$$\sum_{k,n} m(B_n^k) = \sum_{k=1}^{\infty} \sum_{n=1}^{\infty} m(B_n^k)$$

$$< \sum_{k=1}^{\infty} \left[m^*(S_k) + \frac{\epsilon}{2^k} \right]$$

$$= \sum_{k=1}^{\infty} m^*(S_k) + \epsilon.$$

Therefore,

(7.10) $$m^*(S) < \sum_{k=1}^{\infty} m^*(S_k) + \epsilon.$$

Since (7.10) holds for every $\epsilon > 0$, the lemma is established.

Note that the lemma includes as a special case the (trivial) fact that $m^*(S) \leq m^*(T)$ whenever $S \subset T$, as one sees by taking $S_n = \varnothing$, $n \geq 2$.

It also includes the assertion that, for any sets S and T, we have

(7.11) $m^*(S \cup T) \leq m^*(S) + m^*(T)$.

It is not always true that (7.11) is an equality when $S \cap T = \varnothing$; that is, outer measure is *not* an additive function on the class of all subsets of R^k. We cannot prove this assertion now, because it would involve giving an example of a "non-measurable" set.

Let us hasten to point out that closed sets, open sets, and any sets we can generate from these will turn out to be "measurable" and m^* will provide their appropriate measures. This last assertion is not trivial to verify, even for the simplest of sets. For instance, suppose B is a box in R^k. Obviously it had better be the case that

$$m^*(B) = m(B).$$

But how do we know this is true? You will be guided through a proof of this in the exercises.

We need to use the following simple relationship between Riemann integrals and outer measure.

Lemma. *Let* f *be a non-negative continuous function on* R^k *which has compact support. For each number* c > 0

(7.12) $\int f \geq cm^*(\{X; f(X) \geq c\})$.

Proof. Let $c > 0$ and $S_c = \{X; m^*(X) \geq c\}$. Let B be any closed box such that $f = 0$ outside B and let $P = \{B_1, \ldots, B_N\}$ be a partition of B. We will show that there is a choice T of points $T_j \in B_j$ such that $S(f, P, T) \geq cm^*(S_c)$. We choose the points $T_1, \ldots, T_N$ as follows. If $B_j \cap S_c \neq \varnothing$, choose some point $T_j \in B_j \cap S_c$. If $B_j \cap S_c = \varnothing$, let T_j be any point in B_j. Since $f \geq 0$

(7.13)
$$S(f, P, T) \geq \sum_{S_c \cap B_j \neq \varnothing} f(T_j)m(B_j)$$
$$\geq c \sum_{S_c \cap B_j \neq \varnothing} m(B_j).$$

Since $c > 0$ and $f = 0$ outside B, the set S_c is contained in B. Thus the set of boxes B_j for which $S_c \cap B_j \neq \varnothing$ is a (finite) covering of S_c, and so

$$m^*(S_c) \leq \sum_{S_c \cap B_j \neq \varnothing} m(B_j).$$

If we combine this with (7.13) we have $S(f, P, T) \geq cm^*(S_c)$.

Definition. *A subset* S *of* R^k *is called a* **set of measure zero** *if* $m^*(S) = 0$.

Note that any subset of a set of measure zero is a set of measure zero. The following extends this observation a bit.

Lemma. *The following conditions on a set* S *are equivalent.*

(i) S *is a set of measure zero.*

(ii) S *is the union of a sequence of sets of measure zero.*

(iii) *For each $\epsilon > 0$ there is a sequence of sets $S_1, S_2, \ldots$ such that*

(7.14)
$$S \subset \bigcup_{n=1}^{\infty} S_n$$
$$\sum_{n=1}^{\infty} m^*(S_n) < \epsilon.$$

Proof. The implications (i) $\Rightarrow$ (ii) $\Rightarrow$ (iii) are trivial. If (iii) holds, then we know by the inequality (7.9):

$$m^*(S) \leq \sum_{n=1}^{\infty} m^*(S_n)$$

that $m^*(S) < \epsilon$, for every $\epsilon > 0$. Hence S is a set of measure zero.

EXAMPLE 1. The set consisting of a single point is a set of measure zero; indeed, any countable subset of R^k is a set of measure zero. Let's see why. Let S be a countable set and let $X_1, X_2, X_3, \ldots$ be an enumeration of the points of S. Let $\epsilon > 0$. For each n, choose an open box B_n such that

$$X_n \in B_n$$
$$m(B_n) < \frac{\epsilon}{2^n}.$$

Then the sequence $\{B_n\}$ covers S and

$$\sum_{n=1}^{\infty} m(B_n) < \epsilon.$$

Thus $m^*(S) < \epsilon$, for every $\epsilon > 0$.

As a special case of this example, we see that the set of points in R^k which have rational coordinates is a set of measure zero. In particular, the rational numbers comprise a set of measure zero on the real line. Every student of analysis should be acutely aware of the demonstration of this, a demonstration which can be rephrased as follows for the positive rational numbers. Enumerate these numbers (points) in some way, e.g., according to the scheme

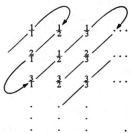

Suppose we have an interval of length $\epsilon > 0$. Divide the interval in half and use one half to cover the first rational point. Divide the remainder in half and use one half (one quarter of the original) to cover the second rational point. Divide in half again and cover the third rational point, etc. This describes a scheme, infinite though it may be, for covering the positive rational points by a sequence of intervals, the sums of whose lengths is ϵ. So we conclude that the "length" of the set of these rational points is 0. Note that if we cover the rational points in [0, 1] by any *finite* number of intervals, the sum of the lengths of those intervals is at least 1, because the rational points are dense in [0, 1]. This shows why, even for bounded sets, we use infinite coverings to define outer measure.

The fact that we must use countable coverings rather than just finite ones to see that the set of rational points is a set of measure zero is related to the fact that we must use Lebesgue's rather than Riemann's process to integrate the function

$$f(x) = \begin{cases} 1, & x \text{ irrational} \\ 0, & x \text{ rational.} \end{cases}$$

You will recall (Example 4 of Chapter 4) that this function f is *not* Riemann-integrable on [0, 1]. It is (or, will be) Lebesgue-integrable and we will have

$$\int_0^1 f(x)\, dx = 1$$

because $f(x) = 1$ except on a set of measure zero, which is a negligible set from the Lebesgue point of view.

EXAMPLE 2. The Cantor set (Example 15 of Chapter 2) is an uncountable set of measure zero in R^1. This set K is obtained by deleting from the closed unit interval a sequence of open intervals $I_1, I_2, I_3, \cdots$ such that $\sum_n m(I_n) = 1$. It seems plausible that $m^*(K) = 0$ because in forming K we removed from [0, 1] all of the length. In fact, this is easily made precise. After we have removed the open intervals $I_1, \cdots, I_n$, the part of [0, 1] which remains is the union of a finite number of non-overlapping intervals. (See, for example, Figure 5 on p. 59.) These remaining intervals cover K and the sum of their lengths is

$$1 - \sum_{j=1}^n m(I_j)$$

a sum which tends to 0 as n increases.

We come now to the first non-trivial result.

Theorem 2. *A subset* S *of* R^k *is a set of measure zero if and only if there exists a sequence* $\{f_n\}$ *such that*

(i) *each* f_n *is a real-valued continuous function of compact support on* R^k;

(ii) $f_1 \leq f_2 \leq f_3 \leq \cdots$;

(iii) *the sequence of integrals $\int f_n$ is bounded;*

(iv) $\lim\limits_n f_n(X) = \infty$ *for every point* $X \in S$.

Proof. Suppose we are given a sequence $\{f_n\}$ with properties (i)–(iv). By (ii) and (iii), the sequence of integrals $\left\{\int f_n\right\}$ is monotone-increasing and bounded above, hence it converges. We are now going to replace $\{f_n\}$ by a subsequence of itself. Let $g_j = f_{n_j}$, where $n_1 < n_2 < n_3 < \cdots$ and

$$\int f_{n_j} > \lim_n \int f_n - 4^{-j}$$

Then the sequence $\{g_n\}$ satisfies (i)–(iv) *and*

$$\int (g_{n+1} - g_n) < 4^{-n}, \qquad n = 1, 2, 3, \ldots.$$

Let $\epsilon > 0$. We will show that $m^*(S) < \epsilon$. Let

$$E_n = \left\{ X; \, g_{n+1}(X) - g_n(X) > \frac{1}{\epsilon 2^n} \right\}.$$

Then (iv) tells us that

$$S \subset \bigcup_{n=1}^{\infty} E_n.$$

Why? Because, if $X \notin E_n$ for every n, we have

$$g_{n+1}(X) - g_n(X) \leq \frac{1}{\epsilon 2^n}, \qquad n = 1, 2, 3, \ldots$$

and thus

$$
\begin{aligned}
g_n(X) &= g_1(X) + [g_2(X) - g_1(X)] + \cdots + [g_n(X) - g_{n-1}(X)] \\
&\leq g_1(X) + \sum_{j=1}^{\infty} [g_{j+1}(X) - g_j(X)] \\
&\leq g_1(X) + \sum_{j=1}^{\infty} \frac{1}{\epsilon 2^j} \\
&= g_1(X) + \frac{1}{\epsilon}
\end{aligned}
$$

for every n; hence $\lim\limits_n g_n(X) < \infty$.

Since $\int (g_{n+1} - g_n) < 4^{-n}$, the inequality (7.12) tells us that

$$\frac{1}{\epsilon 2^n} m^*(E_n) \leq 4^{-n}$$

$$m^*(E_n) \leq \epsilon 2^{-n}.$$

Therefore,

(7.15) $$\sum_{n=1}^{\infty} m^*(E_n) \leq \epsilon.$$

It follows from (7.9) and (7.15) that $m^*(S) \leq \epsilon$.

Now suppose that $m^*(S) = 0$. We shall construct functions f_n satisfying (i)–(iv). For each j there is a sequence of boxes $\{B_n^j\}$ such that

$$S \subset \bigcup_{n=1}^{\infty} B_n^j$$

$$\sum_{n=1}^{\infty} m(B_n^j) < 2^{-j}.$$

For each pair (j, n) choose a non-negative function $f_{jn} \in C_c(R^k)$ such that $f_{jn} \geq 1$ on B_n^j and $\int f_{jn} \leq 2m(B_n^j)$. (The verification that there is such a function f_{jn} is quite easy and has been left to the exercises.) Then the sequence of functions we seek is

$$f_p = \sum_{r=1}^{p} \sum_{j+n=r} f_{jn}.$$

Certainly $f_p \in C_c(R^k)$, and $f_1 \leq f_2 \leq f_3 \leq \cdots$ because each f_{jn} is non-negative. As for condition (iii), we have

$$\int f_p = \sum_{r=1}^{p} \sum_{j+n=r} \int f_{jn}$$
$$\leq \sum_{j,\,n} \int f_{jn}$$
$$\leq 2 \sum_{j,\,n} m(B_n^j)$$
$$\leq 2 \sum_{j} \sum_{j} 2^{-j}$$
$$= 2.$$

Moreover, since $f_{jn} \geq 1$ on B_n^j the value of $f_p(X)$ is at least as great as the number of pairs (j, n) such that $j + n \leq p$ and $X \in B_n^j$. If $X \in S$, then, for every j, there is an n such that $X \in B_n^j$. The point is that, if $X \in S$, then $X \in B_n^j$ for infinitely many pairs (j, n); hence,

$$\lim_p f_p(X) = \infty.$$

Definition. *Any phenomenon (associated with points in R^k) which happens except on a set of measure zero is said to happen* **almost everywhere** *(frequently abbreviated* **a.e.**) *or* **at almost every point**.

The last theorem states that if $f_1 \leq f_2 \leq f_3 \leq \cdots$ is a monotone-increasing sequence of (real-valued) functions in $C_c(R^k)$ such that the integrals $\int f_n$ are bounded, then the limit $\lim f_n(X)$ is finite at almost every point,

$$\lim_n f_n(X) < \infty, \qquad \text{a.e.}$$

As another illustration of the use of the term "almost everywhere", consider the function on R^1 defined by

$$f(x) = \begin{cases} 0, & \text{if } x \text{ is irrational or } x = 0 \\ \dfrac{1}{q}, & \text{if } x = \dfrac{p}{q}, \text{ a fraction in "lowest terms".} \end{cases}$$

You will recall that f is continuous at 0 and at each irrational point. Thus f is continuous almost everywhere. In fact, f is discontinuous at each non-zero rational point, but that has no bearing on the validity of the last sentence. Each continuous function on R^1 is also continuous almost everywhere. For a third illustration, consider the Cantor function (Example 24 of Chapter 3). This is a continuous function on $[0, 1]$ which is differentiable at each point of $[0, 1]$ not in the Cantor set K. Since K is a set of measure zero, the Cantor function is differentiable almost everywhere on $[0, 1]$. Incidentally, we will prove later that a complex-valued function on the interval $[a, b]$ is Riemann-integrable if and only if it is (a) bounded, and (b) continuous almost everywhere.

Exercises

1. Let B be a closed box in R^k. Show that there exists a non-negative f in $C_c(R^k)$ such that $f \geq 1$ on B and $\int f \leq 2m(B)$. (*Hint:* Refer to Exercise 1 of Section 7.2.)

2. If S is a subset of R^n, then

$$m^*(S) = \inf \sum_k m(B_k)$$

where the infimum is taken over all countable coverings of S by *open* boxes. (*Hint:* The inf obtained by using only open boxes could only be larger than $m^*(S)$. If $\{B_k\}$ is any sequence of boxes, we can fatten up B_k to an open box without increasing its measure by more than $\epsilon 2^{-k}$.)

3. If K is a compact subset of R^n, then

$$m^*(K) = \inf \sum_{k=1}^{N} m(B_k)$$

where the infimum is taken over all finite (open) covers of K by boxes.

4. If B is a closed box, then $m^*(B) = m(B)$.

5. If B is a box, the boundary of B is a set of outer measure zero.

6. If B is a box, then $m^*(B) = m(B)$.

7. If $B_1, \ldots, B_n$ are boxes contained in the open set U, and if the interiors $B_1^o, \ldots, B_n^o$ are pairwise disjoint, then

$$\sum_{j=1}^{n} m(B_j) \leq m^*(U).$$

(*Hint:* It suffices by shrinking, to prove it when $B_1, \ldots, B_n$ are pairwise disjoint closed boxes.)

8. Let U be an open subset of the real line R^1. Then (you are supposed to know that) U is the union of a countable collection of pairwise disjoint open intervals:

$$U = \bigcup_k I_k, \quad I_i \cap I_j = \varnothing, \quad i \neq j.$$

Show that

$$m^*(U) = \sum_k m(I_k).$$

(Assume that U is bounded, if you wish.)

9. Show that the circle $S^1 = \{(x, y); x^2 + y^2 = 1\}$ is a set of (2-dimensional) measure zero.

10. Find a compact set K in R^1 such that

(a) the interior of K is empty;

(b) $m^*(K) > 0$.

(*Hint:* Construct K as you would the Cantor set, but take out smaller open intervals.)

*11. Let ψ be a complex-valued function (operator) on $C_c(R^k)$ with these properties.

(i) ψ is linear, $\psi(cf + g) = c\psi(f) + \psi(g)$.

(ii) If B is a closed box and $f = 0$ outside B, then

$$|\psi(f)| \leq m(B)\|f\|_\infty.$$

(iii) If $f \in C_c(R^k)$ is *non-negative*, then

$$\psi(f) \geq m(B) \inf_B f$$

for *every* closed box B.

Prove that $\psi(f) = \int f$ for every $f \in C_c(R^k)$.

7.4. The Principal Propositions

We now proceed with the tasks (i) to assign a limit (function) to each sequence in $C_c(R^k)$ which is (fast) Cauchy relative to the L^1-norm, (ii) to assign an integral to each such limit function, (iii) to verify that the resulting integral is linear, positive, etc. on the space of "integrable" functions.

Theorem 3. *If $\{f_n\}$ is a sequence of continuous functions of compact support which is a fast Cauchy sequence in the* L^1-*norm,*

(7.16) $$\sum_{n=1}^\infty \int |f_{n+1} - f_n| < \infty$$

then $\{f_n\}$ converges pointwise almost everywhere on R^k.

Proof. Let

$$g_n = \sum_{j=1}^n |f_{j+1} - f_j|.$$

Then the functions g_n satisfy conditions (i)–(iv) of Theorem 2. Therefore

$$\lim_n g_n(X) < \infty, \qquad \text{a.e.}$$

that is,

(7.17) $$\sum_{n=1}^{\infty} |f_{n+1}(X) - f_n(X)| < \infty, \qquad \text{a.e.}$$

At every point X where (7.17) holds, $\{f_n(X)\}$ is a fast Cauchy sequence of complex numbers and, therefore, converges.

Definition. *A* **Lebesgue-integrable function** *on* $\mathbf{R}^k$ *is a function* f *such that*

   (i) f *is a complex-valued function defined almost everywhere on* $\mathbf{R}^k$;
   (ii) *there is a fast* L^1*-Cauchy sequence* $\{f_n\}$ *in* $C_c(\mathbf{R}^k)$ *such that*

$$f(X) = \lim_n f_n(X), \qquad \text{a.e.}$$

We have just seen that each fast L^1-Cauchy sequence $\{f_n\}$ in $C_c(R^k)$ converges pointwise almost everywhere to a Lebesgue-integrable function f. We would like to define

$$\int f = \lim_n \int f_n.$$

Certainly the sequence $\left\{ \int f_n \right\}$ converges, because it is a fast Cauchy sequence of numbers:

$$\sum_{n=1}^{\infty} \left| \int f_{n+1} - \int f_n \right| \le \sum_{n=1}^{\infty} \int |f_{n+1} - f_n| < \infty.$$

What we must show now is that the limit

$$\lim_n \int f_n$$

depends only on the limit function f and not on the particular sequence $\{f_n\}$ which we use to approximate f. In other words, we must show that if $\{f_n\}$ and $\{g_n\}$ are fast L^1-Cauchy sequences in $C_c(R^k)$ such that

$$\lim_n f_n(X) = \lim_n g_n(X), \qquad \text{a.e.}$$

then

$$\lim_n \int f_n = \lim_n \int g_n.$$

Here is the key result.

Theorem 4. *Let* $\{f_n\}$ *and* $\{g_n\}$ *be two monotone-increasing sequences of non-negative continuous functions of compact support. If*

$$\lim_n f_n(X) \ge \lim_n g_n(X), \qquad \text{a.e.}$$

then

$$\lim_n \int f_n \ge \lim_n \int g_n.$$

Proof. Let

$$f = \lim_n f_n$$

$$g = \lim_n g_n.$$

Since $f_1 \leq f_2 \leq f_3 \leq \cdots$ and $g_1 \leq g_2 \leq g_3 \leq \cdots$, these limits exist at every point, provided we allow $+\infty$ as a value. Since $f(X) \geq g(X)$ almost everywhere, Theorem 2 tells us that there is a sequence of non-negative functions in $C_c(R^k)$ such that

(i) $h_1 \leq h_2 \leq h_3 \leq \cdots$;

(ii) $\int h_n \leq 1$, for every n;

(iii) $\lim_n h_n(X) = \infty$, if $f(X) < g(X)$.

Let $\epsilon > 0$. Condition (iii) tells us that

(7.18) $$\lim_n [f_n(X) + \epsilon h_n(X)] \geq g(X), \qquad \text{all } X \in R^k.$$

Now fix a positive integer p. Since $g \geq g_p$, we have

(7.19) $$\lim_n [f_n(X) + \epsilon h_n(X)] \geq g_p(X), \qquad \text{all } X \in R^k.$$

Let B_p be a closed box such that $g_p = 0$ outside B_p. The sequence of functions

$$\varphi_n = \max(0, g_p - f_n - \epsilon h_n)$$

converges monotonely downward to 0. Since each φ_n is continuous and B_p is compact, this convergence to 0 must be *uniform on* B_p (Dini's theorem, Theorem 1 of Chapter 5). Thus for some N we have

$$\varphi_N(X) \leq \frac{\epsilon}{m(B_p)}, \qquad X \in B_p,$$

that is

$$f_N(X) + \epsilon h_N(X) \geq g_p(X) - \frac{\epsilon}{m(B_p)}, \qquad X \in B_p.$$

From this it follows that

$$\int f_N + \epsilon \geq \int f_N + \epsilon \int h_N$$

$$\geq \int_{B_p} (f_N + \epsilon h_N)$$

$$\geq \int_{B_p} g_p - \epsilon$$

$$= \int g_p - \epsilon.$$

Accordingly,

(7.20) $$\lim_n \int f_n \geq \int g_p - 2\epsilon.$$

Since (7.20) holds for every p and ϵ, we have

$$\lim_n \int f_n \geq \lim_p \int g_p.$$

Theorem 5. *Let $\{f_n\}$ and $\{g_n\}$ be fast L^1-Cauchy sequences of real-valued continuous functions of compact support. If*

$$\lim_n f_n(X) \geq \lim_n g_n(X), \qquad a.e.,$$

then

$$\lim \int f_n \geq \lim \int g_n.$$

Proof. Define $f_0 = g_0 = 0$. Then let

$$\alpha_n = \max (0, f_n - f_{n-1} - g_n + g_{n-1})$$
$$\beta_n = \max (0, g_n - g_{n-1} - g_n + f_{n-1})$$

so that

$$\alpha_n - \beta_n = f_n - f_{n-1} - g_n + g_{n-1}$$
$$\sum_{j=1}^n (\alpha_j - \beta_j) = f_n - g_n.$$

The sequences

$$F_n = \sum_{j=1}^n \alpha_n$$
$$G_n = \sum_{j=1}^n \beta_n$$

then satisfy the hypotheses of Theorem 4. Thus

$$0 \leq \lim_n \int (F_n - G_n)$$
$$= \lim_n \int \sum_{j=1}^n (\alpha_j - \beta_j)$$
$$= \lim_n \int (f_n - g_n).$$

Corollary. *If $\{f_n\}$ and $\{g_n\}$ are fast L^1-Cauchy sequences in $C_c(R^k)$ such that*

$$\lim_n f_n(X) = \lim_n g_n(X), \qquad a.e.,$$

then

$$\lim \int f_n = \lim \int g_n.$$

Proof. Apply Theorem 5 separately to the sequences of real parts of f_n and g_n and to the sequences of imaginary parts.

Definition. *If* f *is a Lebesgue-integrable function, the (Lebesgue)* **integral** *of* f *is*

$$\int f = \lim_n \int f_n$$

where $\{f_n\}$ *is any fast* L^1*-Cauchy sequence in* $C_c(R^k)$ *which converges to* f *pointwise almost everywhere.*

It is (of course) the last corollary which ensures that the integral is well-defined. Let us summarize the elementary properties of integration and integrable functions.

Theorem 6. *Let* f *be a complex-valued function defined almost everywhere on* R^k.

(i) *If* f *is Lebesgue-integrable, then* (cf + g) *is Lebesgue-integrable for every complex number* c *and every Lebesgue-integrable function* g, *and*

$$\int (cf + g) = c \int f + \int g.$$

(ii) *If* $u = \mathrm{Re}\,(f)$ *and* $v = \mathrm{Im}\,(f)$, *that is, if* $f = u + iv$ *where* u *and* v *are real-valued, then* f *is Lebesgue-integrable if and only if both* u *and* v *are Lebesgue-integrable.*

(iii) *If* f *is Lebesgue-integrable, then* $|f|$ *is Lebesgue-integrable and*

$$\left| \int f \right| \leq \int |f|.$$

(iv) *If* f *is Lebesgue-integrable and* $f(X) \geq 0$ *almost everywhere, then*

$$\int f \geq 0.$$

Proof. We have left the proofs of the various statements to the exercises.

Let us close this section by asking ourselves what concrete examples we know of Lebesgue-integrable functions, other than continuous functions of compact support.

EXAMPLE 3. The most important device we know for "constructing" Lebesgue-integrable functions from functions in $C_c(R^k)$ is provided by Theorem 2: If $f_1 \leq f_2 \leq f_3 \leq \cdots$ *is a monotone-increasing sequence of continuous functions of compact support and the sequence of integrals* $\left\{ \int f_n \right\}$ *is bounded, then*

$$f(X) = \lim_n f_n(X) < \infty, \quad \text{a.e.,}$$

f *is a Lebesgue-integrable function, and*

$$\int f = \lim_n \int f_n.$$

Of course, the same is true for decreasing sequences. It is a fact (which we shall not stop to prove) that *every* real-valued Lebesgue-integrable function f is a sum $f = f_1 + f_2 + f_3$, where f_1 is the limit of a monotone-increasing sequence in $C_c(R^k)$, f_2 is the limit of a decreasing sequence in $C_c(R^k)$, and $f_3 = 0$ almost everywhere. At this point, let us note some things we can do easily with the monotone convergence device.

Let K be any compact subset of R^k. There exists a function $h \in C_c(R^k)$ such that

(7.21)
$$h = 1, \qquad \text{on } K$$
$$0 \leq h < 1, \qquad \text{off } K.$$

(See Exercise 2, Section 7.2) The sequence of powers $h \geq h^2 \geq h^3 \geq \cdots$ is decreasing and converges pointwise (everywhere) to the characteristic function of K:

$$\lim h(X)^n = k_K(X) = \begin{cases} 1, & X \in K \\ 0, & X \notin K. \end{cases}$$

By the monotone convergence property, we see that *the characteristic function of any compact set is Lebesgue-integrable*. (The boundedness of the sequence of integrals is trivially satisfied in this case because $0 \leq h^n \leq h$.)

By essentially the same argument one can show that, *if f is any function in $C_c(R^k)$ and K is any compact set, then the restricted function*

$$k_K f = \begin{cases} f, & \text{on } K \\ 0, & \text{off } K \end{cases}$$

is Lebesgue-integrable. By taking real and imaginary parts and writing real-valued functions in $C_c(R^k)$ as differences of non-negative functions in $C_c(R^k)$, it is enough to verify this result when $f \geq 0$. In this case

$$k_K f = \lim_n h^n f$$

where h is a function satisfying (7.21); and the convergence is monotone. Now, it is a fact that, if g is any continuous complex-valued function on a compact set K, there is an f in $C_c(R^k)$ such that $f(X) = g(X)$, $X \in K$. (See Exercise 10, Section 7.2.) Therefore, *if g is any continuous function on the compact set K, the function*

(7.22)
$$f = \begin{cases} g, & \text{on } K \\ 0, & \text{off } K \end{cases}$$

is Lebesgue-integrable. Of course, we also know how to calculate the integral of f, using the function h (7.21) and any extension of g to a function in $C_c(R^k)$.

Since the integrals of continuous functions of compact support were defined using Riemann integrals on boxes, it should be clear that *in case K is a closed box, the Lebesgue integral of the function f* (7.22) *over R^k is equal to the Riemann integral of g over the box K.* By taking linear combinations (sums),

$$g = g_1 k_{B_1} + \cdots + g_n k_{B_n}$$

we see that the same result is valid when g is a piecewise-continuous function on a box. We have given (in the text, as distinguished from the exercises) a complete proof when g is a step function, i.e., a linear combination of characteristic functions of boxes:

$$g = c_1 k_{B_1} + \cdots + c_n k_{B_n}.$$

We can also analyze the integrability of continuous functions which are not of compact support. Let f be a continuous complex-valued function on R^k. When is f Lebesgue-integrable? Suppose first that $f \geq 0$. We do the obvious thing. We choose an increasing sequence of closed boxes B_n which exhaust R^k:

$$\bigcup_n B_n = R^k$$

and examine the increasing sequence of integrals

$$\int_{B_1} f \leq \int_{B_2} f \leq \cdots.$$

In order to be careful, let us assume that $\bar{B}_n$ is contained in the interior of B_{n+1}, $n = 1, 2, 3, \ldots$. As in Exercise 2, Section 7.2, choose $h_n \in C_c(R^k)$ so that

$$h_n(X) = 1, \qquad X \in B_n$$
$$0 < h_n(X) < 1, \qquad X \in B^o_{n+1} - B_n$$
$$h_n(X) = 0, \qquad X \notin B^o_{n+1}.$$

Then

$$h_1 f \leq h_2 f \leq h_3 f \leq \cdots$$
$$\lim_n h_n(X) f(X) = f(X), \qquad \text{for all } X$$
$$\int_{B_n} f \leq \int h_n f \leq \int_{B_{n+1}} f.$$

Thus, the sequence of integrals $\left\{ \int h_n f \right\}$ is bounded if and only if the sequence $\left\{ \int_{B_n} f \right\}$ is bounded. When these sequences are bounded, the non-negative function f will be integrable. What does this tell us when f is complex-valued? We know that, if f is integrable, then $|f|$ is integrable and (by what we just did) the integrals of $|f|$ over the various boxes B_n must be bounded. Write $f = u + iv$ where u and v are real-valued. Since $|u| \leq |f|$ and $|v| \leq |f|$, we know that $|u|$ and $|v|$ will be integrable if $|f|$

is. Furthermore $u = u^+ - u^-$, where $u^+ = \max(u, 0)$, $u^- = -\min(u, 0)$. These are continuous non-negative functions and $|u| = u^+ + u^-$. Therefore, by the test we just gave for the integrability of non-negative continuous functions, if $|u|$ is integrable, then u^+ and u^- are integrable and hence u is integrable. Similar comments apply to v. We conclude that (since f is continuous) f is integrable if and only if $|f|$ is. So, *if f is a continuous complex-valued function on R^k, then f is Lebesgue-integrable if and only if*

$$\lim_n \int_{B_n} |f| < \infty$$

where $\{B_n\}$ is any increasing sequence of closed boxes which exhaust R^k.

Exercises

1. If f and g are Lebesgue-integrable functions and c is a complex number, then $(cf + g)$ is a Lebesque-integrable function and $\int (cf + g) = c \int f + \int g$.

2. If f is a Lebesque-integrable function, then the real part, the imaginary part, and the absolute value of f are Lebesque-integrable functions. Furthermore $\left| \int f \right| \leq \int |f|$.

3. The "L^1-norm"

$$\|f\|_1 = \int |f|$$

is a semi-norm on the space of Lebesque-integrable functions.

4. If f is a Lebesque-integrable function and $\int |f| = 0$, then $f(X) = 0$ almost everywhere.

5. If f is any function on R^k which is [has the value] 0 almost everywhere, then f is Lebesque-integrable and $\int f = 0$.

6. If f is a continuous function of compact support, then f is Lebesgue-integrable and the Lebesque integral of f is equal to the Riemann integral of f.

7. If f is a Lebesque-integrable function and $f(X) \geq 0$ almost everywhere, then $\int f \geq 0$.

8. Let f and g be real-valued Lebesgue-integrable functions. Let

$$(f \vee g)(X) = \max(f(X), g(X))$$
$$(f \wedge g)(X) = \min(f(X), g(X)).$$

Then $f \vee g$ and $f \wedge g$ are Lebesgue-integrable functions. (*Hint*: The max and min of two continuous functions are continuous.)

9. Prove that the function

$$f(x) = e^{-|x|}$$

is Lebesgue-integrable on the real line and find its integral.

10. Prove that the function

$$f(x) = \frac{1}{x^2}, \qquad x \neq 0$$

is not Lebesgue-integrable on the real line but that the function

$$g(x) = \begin{cases} x^{-2}, & x \geq 1 \\ 0, & |x| < 1 \end{cases}$$

is Lebesgue-integrable. Replace x^{-2} by $|x|^{-1/2}$ and what happens?

11. The function

$$f(x) = \begin{cases} \dfrac{1}{x}, & 0 < |x| \leq 1 \\ 0, & \text{otherwise} \end{cases}$$

is not Lebesgue-integrable on R^1. The function defined on the complex numbers by

$$f(z) = \begin{cases} \dfrac{1}{z}, & 0 < z \leq 1 \\ 0, & \text{otherwise} \end{cases}$$

is Lebesgue-integrable on R^2.

12. True or false? Every integrable function agrees almost everywhere with some function which is nowhere continuous.

7.5. Completeness and Continuity

We have extended the integral from $C_c(R^k)$ to a much larger space. In this short section we shall first verify that we have accomplished our immediate aim of completing $C_c(R^k)$ relative to the L^1-norm, and then we shall discuss the relationship between integrability and continuity.

Notation. We denote by $L^1 = L^1(R^k)$ the space of (complex-valued) Lebesgue-integrable functions on R^k.

Theorem 7. *Under the semi-norm*

$$\|f\|_1 = \int |f|$$

$L^1(R^k)$ *is a complete semi-normed linear space, which contains* $C_c(R^k)$ *as a dense subspace.*

Proof. If $f \in L^1$, we have a fast L^1-Cauchy sequence $\{f_n\}$ in $C_c(R^k)$ such that

$$\lim_n f_n(X) = f(X), \qquad \text{a.e.}$$

How do we know that $\lim_n \|f - f_n\|_1 = 0$? Fix a positive integer p. Then $\{|f_n - f_p|\}$ is a fast L^{-1}Cauchy sequence in $C_c(R^k)$ which converges pointwise almost everywhere to $|f - f_p|$. Thus

$$\|f - f_p\|_1 = \int |f - f_p|$$

$$= \lim_n \int |f_n - f_p|.$$

But $\{f_n\}$ is L^1-Cauchy; in particular

$$\lim_{n, p} \|f_n - f_p\|_1 = 0;$$

hence,

$$\lim_p \|f - f_p\|_1 = \lim_p \lim_n \|f_n - f_p\|_1$$

$$= 0.$$

We want to show that each sequence in L^1 which is Cauchy relative to the L^1-norm (the L^1-semi-norm, to be precise) converges in L^1-norm to an integrable function. It is enough to show that each fast L^1-Cauchy sequence in L^1 converges. If $\{f_n\}$ is such a sequence:

$$\sum_{n=1}^{\infty} \|f_{n+1} - f_n\|_1 < \infty$$

we can choose (for each n) a function $g_n \in C_c(R^k)$ such that

$$\|f_n - g_n\|_1 < 2^{-n}.$$

Then $\{g_n\}$ is a fast L^1-Cauchy sequence in $C_c(R^k)$. By the reasoning in (i), there exists $f \in L^1$ such that $\lim_n \|f - g_n\|_1 = 0$. Plainly then $\lim_n \|f - f_n\|_1 = 0$.

Definition. *A* **null function** *is a complex-valued function which*

(a) *is defined almost everywhere on* R^k;
(b) *has the value* 0 *almost everywhere on* R^k.

In the exercises of the previous section it was pointed out that the null functions are precisely the Lebesgue-integrable functions f for which $\int |f| = 0$. Thus the set of null functions is the subspace of L^1 consisting of the functions of L^1-semi-norm zero. As we explained in Chapter 6, we can (therefore) form the quotient space L/N where N is the space of null functions, and this will be a Banach space which contains $C_c(R^k)$ as a dense linear subspace. Rather than carry out such a formal construction it is customary to say that L^1 *is* a Banach space (complete, normed linear space) provided we agree to identify any two functions which differ by a null function. This amounts to adopting the following.

Convention. *If* f, g *are functions in* L^1, *then* $f = g$ *will mean that* $f(X) = g(X)$, *a.e., that is, that* $f - g$ *is a null function.*

Note that with this convention it does follow that $f = 0$ when $\|f\|_1 = 0$. We should also remark that $f \geq g$ will mean $f(X) \geq g(X)$ almost everywhere, etc. Where there is any chance for confusion, we will reinsert the term "almost everywhere".

Now we turn our attention to the question: In passing from $C_c(R^k)$ to $L^1(R^k)$, how far away from continuity have we moved? How discontinuous can an integrable function be? One answer is, of course, that an integrable function can be nowhere continuous, because we can alter the values on a countable dense set in any way whatsoever, without affecting integrability (or the integral) of the function. On the other hand, we know that any integrable function can be approximated by continuous functions via pointwise convergence. This approximation can be strengthened to reveal that each integrable function displays a surprising amount of continuity, if we know where to look for it.

Theorem 8. *Let* f *be a Lebesgue-integrable function on* R^k. *If* ϵ *and* δ *are any positive numbers, there exist a set* S *and a continuous function* g *of compact support such that*

$$m^*(S) < \delta$$

(7.23)

$$|f(X) - g(X)| < \epsilon, \qquad X \notin S.$$

Proof. The proof is quite similar to one we have given before. We begin by choosing a sequence $\{f_n\}$ in $C_c(R^k)$ such that

$$\|f_{n+1} - f_n\|_1 < \frac{1}{4^n}$$

$$\lim_n f_n(X) = f(X), \qquad \text{a.e.}$$

For each n, let

$$S_n = \{X; f_{n+1}(X) - f_n(X)| > 2^{-n}\}.$$

By (7.12)

$$\int |f_{n+1} - f_n| \geq 2^{-n}\, m^*(S_n)$$

and since $\|f_{n+1} - f_n\|_1 < 4^{-n}$, we have

$$m^*(S_n) < 2^{-n}.$$

For each positive integer N define

$$E_N = \bigcup_{n=N}^{\infty} S_n.$$

Note that

$$m^*(E_N) \leq \sum_{n=N}^{\infty} m^*(S_n)$$

$$< \sum_{n=N}^{\infty} 2^{-n}$$

so that

$$\lim_N m^*(E_N) = 0.$$

Now, given $\epsilon > 0$ and $\delta > 0$, choose a positive integer N such that $m^*(E_N) < \delta$. If $X \notin E_N$, then

$$|f_{n+1}(X) - f_n(X)| \leq 2^{-n}, \qquad n \geq N.$$

Therefore the sequence $\{f_n\}$ converges uniformly on the complement of E_N. Let T be the set of points X such that $\{f_n(X)\}$ does not converge to $f(X)$. Then $m^*(T) = 0$, and if

$$S = T \cup E_N$$

we have $m^*(S) < \delta$ and $\lim_n f_n(X) = f(X)$, *uniformly* on the complement of S. The function g in the theorem can be taken to be any f_n such that $|f - f_n| < \epsilon$ on the complement of S. We should remark that S can always be replaced by a slightly larger open set having the same properties. (See next corollary.)

Corollary. *If f is a Lebesgue-integrable function on* R^k, *there exists a decreasing sequence of open sets* $\{U_n\}$ *such that*

(i) $\lim m^*(U_n) = 0$;

(ii) *for each* n, *the restriction of* f *to the complementary closed set* $K_n = R^k - U_n$ *is a continuous function on* K_n.

Proof. This result is a corollary of the proof of Theorem 8 rather than of its statement. In the proof we have the decreasing sequence of sets $\{T \cup E_N\}$, the outer measures of which tend to 0:

(7.24) $$m^*(T \cup E_N) < 2^{-(N-1)}.$$

We can cover $T \cup E_N$ by a sequence of *open* boxes, the sum of whose measures is less than $2^{-(N-1)}$. (See Exercise 2, Section 7.3.) Let V_N be the union of those open boxes, so that V_N is an open set and $m^*(V_N) < 2^{-(N-1)}$. Let $U_N = V_1 \cap V_2 \cap \cdots \cap V_N$. Then

$$U_N \text{ is open}$$
$$T \cup E_N \subset U_N$$
$$m^*(U_N) < 2^{-(N-1)}$$
$$U_1 \supset U_2 \supset U_3 \supset \cdots.$$

Furthermore, for each N, $\{f_n\}$ converges to f uniformly on the complement $K_N = R^k - U_N$. Thus the restriction of f to K_N is a continuous function on K_N (Theorem 2, Chapter 5).

The theorem states that if f is integrable, then, given any ϵ and δ, there is a continuous function of compact support which is uniformly within ϵ of f, except on a set of outer measure less than δ. Note what the first corollary does not say. It *does not* say that, given $\delta > 0$, there is a closed set K with $m^*(R^k - K) < \delta$ such that f is continuous at each point of K. We know that f need not be continuous at any point. It *does say* that if we are willing to totally disregard the values of f on an open set of small measure, then we obtain a continuous function on the complement of that set.

As we indicated earlier a Riemann-integrable function comes much closer to being continuous. Every such function is continuous almost everywhere. We could prove this now. In order to avoid repetition, we will defer the proof until the next section.

Exercises

1. (Translation-invariance of the integral) If f is a function on R^k and Y is a vector in R^k, the Y-translate of f is the function $T_Y f$ defined by

$$(T_Y f)(X) = f(X + Y).$$

Prove that if $f \in L^1(R^k)$, then $(T_Y f) \in L^1(R^k)$ and

$$\int T_Y f = \int f.$$

2. Prove that if $f \in L^1(R^k)$

$$\lim_{Y \to 0} \| f - T_Y f \|_1 = 0.$$

Hence, for a fixed f, the map

$$R^k \xrightarrow{\ T\ } L^1(R^k)$$

$$T(Y) = T_Y f$$

is continuous. (*Hint*: Prove $\lim_{Y} T_Y f = f$ first for functions in $C_c(R^k)$, and approximate.)

3. Let f be an integrable function on the real line. Let

$$F(x) = \int k_{(-\infty, x]} f, \qquad x \in R.$$

Prove that F is a continuous function on R^1. Prove that, at any point x where f is continuous, the function F is differentiable and $F'(x) = f(x)$.

4. Use Theorem 8 to show that every integrable function f can be expanded in a series

$$f = \sum_{n=1}^{\infty} f_n + h$$

where $f_1, f_2, f_3, \ldots$ are continuous functions of compact support and h is a null function.

5. If f is a non-negative integrable function, show that the series expansion in Exercise 4 *cannot* necessarily be done with non-negative functions f_n.

7.6. *The Convergence Theorems*

Now we present the several basic convergence theorems which make the Lebesgue integral such a powerful tool. First we summarize for sequences of integrable functions the results which we exploited for sequences of continuous functions of compact support.

Theorem 9. *If* $\{f_n\}$ *is a sequence of* (*Lebesgue*) *integrable functions such that*

(i) $\{f_n\}$ *is* L^1-*Cauchy;*
(ii) *the limit*

$$f(X) = \lim_n f_n(X)$$

exists almost everywhere;

then the function f *is integrable and* $\{f_n\}$ *converges to* f *in* L^1-*norm.*

Proof. The first statement need only be proved when $\{f_n\}$ is a fast L^1-Cauchy sequence. Thus we begin with such a sequence:

$$\sum_{n=1}^{\infty} \|f_{n+1} - f_n\|_1 < \infty.$$

We will show that it converges almost everywhere to an integrable function f and that $\|f - f_n\|_1 \to 0$. We use Theorem 8 to find functions $g_n \in C_c(R^k)$ and sets S_n with these properties:

(a) $\|f_n - g_n\|_1 < n^{-2}$;
(b) $m^*(S_n) < n^{-2}$;
(c) $|f_n(X) - g_n(X)| < n^{-1}, \qquad X \notin S_n.$

From (a) and the fact that $\{f_n\}$ is a fast L^1-Cauchy sequence, we conclude that $\{g_n\}$ is a fast L^1-Cauchy sequence in $C_c(R^k)$. By Theorem 3, the limit

$$f(X) = \lim_n g_n(X)$$

exists almost everywhere, f is an integrable function, and $\{g_n\}$ converges to f in L^1-norm. By condition (a), we see that $\{f_n\}$ also converges to f in L^1-norm. So, all we have to show is that $f_n(X)$ converges to $f(X)$ almost everywhere.

For each N, let

$$E_N = \bigcup_{n=N}^{\infty} S_n.$$

Let S and T be, respectively, the sets on which $\{f_n(X)\}$ and $\{g_n(X)\}$ do *not* converge to $f(X)$. Condition (c) guarantees that

$$S \subset T \cup E_N, \qquad N = 1, 2, 3, \ldots$$

because,   if $X \notin (T \cup E_N)$,   then $\{g_n(X)\}$ converges   to $f(X)$ and $\{f_n(X) - g_n(X)\}$ converges to 0. Since

$$m^*(S) \leq m^*(T) + m^*(E_N)$$

$$m^*(T) = 0$$

$$m^*(E_N) \leq \sum_{n=N}^{\infty} m^*(S_n) \leq \sum_{n=N}^{\infty} n^{-2}$$

we have

$$m^*(S) \leq \sum_{n=N}^{\infty} n^{-2}, \qquad \text{for all } N.$$

Thus $m^*(S) = 0$.

Theorem 10 (*Monotone Convergence Theorem*). *Let* $\{f_n\}$ *be a sequence of real-valued integrable functions on* $\mathbf{R}^k$ *which is monotone-increasing:*

$$f_1 \leq f_2 \leq f_3 \leq \cdots.$$

If the sequence of integrals $\left\{ \int f_n \right\}$ *is bounded, then the limit*

$$f(X) = \lim_n f_n(X)$$

is finite almost everywhere, f *is an integrable function, and*

$$\int f = \lim_n \int f_n.$$

Proof. Choose $n_1 < n_2 < n_3 < \cdots$ so that

$$\int f_{n_j} > \lim_n \int f_n - 2^{-j}.$$

Then $f_{n_1} \leq f_{n_2} \leq \cdots$ is a fast L^1-Cauchy sequence. So the result follows immediately from application of Theorem 9 to this subsequence.

Corollary (*Beppo Levi's Theorem*). *Let* $\{h_n\}$ *be a sequence of non-negative integrable functions and suppose that*

$$\sum_{n=1}^{\infty} \int h_n < \infty.$$

Then

$$h(X) = \sum_{n=1}^{\infty} h_n(X) < \infty, \qquad \text{a.e.}$$

h *is an integrable function, and*

$$\int h = \sum_{n=1}^{\infty} \int h_n.$$

Corollary (*Weak Continuity of Integration*). *If* $f_1 \geq f_2 \geq f_3 \geq \cdots$ *is a monotone-decreasing sequence of non-negative integrable functions and*

$$\lim_n f_n(X) = 0, \qquad \text{a.e.}$$

then

$$\lim_n \int f_n = 0.$$

The following result appeared in the 1906 doctoral dissertation of Pierre de Fatou. In the framework we have adopted, it is a most significant piece of information.

Theorem 11 (Fatou's Lemma). *Let* $\{f_n\}$ *be a sequence of non-negative integrable functions and suppose that the limit*

$$f(X) = \lim_n f_n(X)$$

exists almost everywhere and that the sequence of integrals

$$\int f_n$$

is bounded. Then f *is an integrable function and*

(7.25) $$\int f \le \liminf_n \int f_n.$$

Proof. If $\{x_n\}$ is a sequence of real numbers, let us recall the definition of

$$\liminf_n x_n.$$

This is the smallest limit point of the sequence. In other words, it is the smallest number which is the limit of some convergent subsequence of $\{x_n\}$. Another way to define the lim inf is this: For each n let

$$a_N = \inf_{n \ge N} x_n.$$

Then

$$a_1 \le a_2 \le a_3 \le \cdots$$

and so $\{a_n\}$ converges in the extended real number system and that defines the lim inf:

$$\liminf_n x_n = \lim_N a_N.$$

Since $\{f_n\}$ converges almost everywhere to f, we have

$$f(X) = \liminf_n f_n(X), \qquad \text{a.e.}$$

Why? Because, if a sequence converges, it converges to its lim inf ($=$ its lim sup). Define

$$g_N(X) = \inf_{n \ge N} f_n(X)$$

so that

$$g_1 \le g_2 \le g_3 \le \cdots$$

and

$$f(X) = \lim_N g_N(X), \qquad \text{a.e.}$$

Now, we'll show that each g_N is an integrable function. Assume, for the moment, that we have shown that. Since $g_N \leq f_N$, we have

$$\int g_N \leq \int f_N.$$

One of the hypotheses of this theorem is that the integrals of the f_N's are bounded. So, the integrals of the g_N's are bounded. Apply the monotone convergence theorem to the sequence $\{g_N\}$. *Conclusion:* The function f is integrable and

$$\int f = \lim_N \int g_N.$$

We shall be finished if we prove that

(i) each g_N is integrable;
(ii) $\lim_N \int g_N \leq \lim \inf_n \int f_n$.

Fix an N. Let

$$h_n^{(N)} = f_N \wedge \cdots \wedge f_{N+n}$$
$$= \min(f_N, \ldots, f_{N+n}).$$

Then

$$h_0^{(N)} \geq h_1^{(N)} \geq h_2^{(N)} \geq \cdots$$

and

$$\lim h_n^{(N)} = \inf_{n \geq N} f_n$$
$$= g_N.$$

Each $h_n^{(N)}$ is integrable (Exercise 8, Section 7.4). By the monotone convergence theorem, g_N is integrable and

$$\int g_N = \lim_n \int h_n^{(N)}.$$

Now

$$h_n^{(N)} \leq f_j, \qquad j = N, \ldots, N+n$$

so that

$$\int h_n^{(N)} \leq \int f_j, \qquad j = N, \ldots, N+n.$$

In other words,

$$\int h_n^{(N)} \leq \min \left\{ \int f_N, \ldots, \int f_{N+n} \right\}.$$

Consequently,

$$\int g_N = \lim_n \int h_n^{(N)} \leq \inf_{j \geq N} \int f_j.$$

Now let N get large:

$$\int f = \lim_N \int g_N \leq \lim \inf_n \int f_n.$$

Theorem 12 (Dominated Convergence Theorem). *Let* $\{f_n\}$ *be a sequence of integrable functions which converges pointwise almost everywhere to a function* f. *Suppose that there is an integrable function* g *such that* $|f_n| \leq |g|$ *for every* n. *Then* f *is an integrable function and*

$$\int f = \lim_n \int f_n.$$

Proof. We may as well assume $g \geq 0$, since we can always replace g by $|g|$. Furthermore, we may as well assume the functions f_n are real-valued, for if $f_n = u_n + iv_n$, then the sequences $\{u_n\}$ and $\{v_n\}$ satisfy the same conditions as does $\{f_n\}$.

Now we have

$$-g \leq f_n \leq g, \qquad k = 1, 2, 3, \dots .$$

Since g and f_n are integrable, $g - f_n$ is integrable. Furthermore,

$$g - f_n \geq 0$$

$$\int (g - f_n) \leq 2 \int f.$$

Also

$$\lim_n (g(X) - f_n(X)) = g(X) - f(X)$$

almost everywhere. By Fatou's lemma, $g - f$ is an integrable function and

$$\int (g - f) \leq \lim_n \inf \int (g - f_n).$$

Thus f is an integrable function and

$$\int f \geq \lim \sup \int f_n.$$

We have used the fact that

$$\lim_n \inf (-x_n) = - \lim_n \sup x_n.$$

Now apply Fatou's lemma to the sequence $g + f_n$ and you obtain

$$\int f \leq \lim_n \inf \int f_n.$$

Conclusion:

$$\lim_n \sup \int f_n \leq \int f \leq \lim_n \inf \int f_n$$

so that the sequence $\left\{ \int f_k \right\}$ converges and

$$\int f = \lim_n \int f_n.$$

Corollary (Bounded Convergence Theorem). *Let* $\{f_n\}$ *be a bounded sequence of integrable functions which converges pointwise almost everywhere to the function* f. *If there is a set of finite (outer) measure outside of which* *every* f_n *vanishes, then* f *is integrable and*

$$\int f = \lim_n \int f_n.$$

Proof. Suppose $m^*(S) < \infty$ and, for every n, $f_n = 0$ outside S. Then there is an open set U such that $S \subset U$ and $m^*(U) < \infty$. (Take U to be the union of a sequence of open boxes which cover S and have measures totalling $m^*(S) + \epsilon$.) Then k_U, the characteristic function of U, is an integrable function. (Why? Because $U = \bigcup_n K_n$ where K_n is an increasing sequence of compact sets. Apply the monotone convergence theorem.) If $|f_n| \leq M$ for all n, the function $g = Mk_U$ may be used in the dominated convergence theorem.

The dominated convergence theorem gives just about the best possible result enabling us to conclude that

$$\int (\lim f_n) = \lim \int f_n.$$

One must not be deluded into thinking that the interchange of the integral and the limit of the sequence is always possible using the Lebesgue integral. Some control over the convergence, e.g., as provided by the dominating function g, is necessary. Consider the sequence of functions of R^1 defined by

$$f_n(x) = \begin{cases} 1, & k \leq x \leq k + 1 \\ 0, & \text{otherwise.} \end{cases}$$

Then $\{f_n\}$ converges pointwise (boundedly) to 0, but

$$\int f_n = 1, \qquad \text{for all } n.$$

As an illustration of the utility of the monotone convergence theorem, let us prove the following. Although it is called a theorem, it should be taken in the spirit of an example.

Theorem 13. *A complex-valued function* f *on the interval* [a, b] *is Riemann-integrable if and only if*

(a) *it is bounded;*
(b) *it is continuous at almost every point of* [a, b].

If f *is Riemann-integrable, then the function*

$$g = \begin{cases} f & on \text{ [a, b]} \\ 0 & elsewhere \end{cases}$$

is Lebesgue-integrable on R^1 *and* $\int g = \int_a^b f(x)\, dx$.

Proof. Obviously we may assume that f is real-valued. We know that every Riemann-integrable function is bounded. Hence, our task is to show that a bounded, real-valued function f on $[a, b]$ is Riemann-integrable if and only if it is continuous at almost every point of $[a, b]$.

First, suppose f is Riemann-integrable. This means that the Riemann sums

$$(7.26) \qquad S(f, P, T) = \sum_{j=1}^{n} f(t_j)(x_j - x_{j-1}), \qquad t_j \in [x_{j-1}, x_j]$$

converge to $\int_a^b f(x)\, dx$ as the mesh

$$\|P\| = \max_j (x_j - x_{j-1})$$

goes to 0. Or, as we reformulated it, it means that

$$\lim_{\delta \to 0} \operatorname{diam} \sum_\delta (f) = 0$$

where $\sum_\delta (f)$ is the set of all sums (7.26) which arise from partitions P with $\|P\| \le \delta$. For any given partition P we have

$$(7.27) \qquad \sum_{j=1}^{n} m_j(x_j - x_{j-1}) \le S(f, P, T) \le \sum_{j=1}^{n} M_j(x_j - x_{j-1})$$

where

$$M_j = \sup_{t \in [x_{j-1}, x_j]} f(t)$$

$$m_j = \inf_{t \in [x_{j-1}, x_j]} f(t).$$

Furthermore by appropriate choice of $T = (t_1, \ldots, t_n)$ we can make $S(f, P, T)$ as close as we wish to the sum on the left or the sum on the right. These sums depend only on the partition P. Let us denote the lower sum (the one using m_j's) by $\underline{S}(f, P)$ and the upper sum (the one using M_j's) by $\bar{S}(f, P)$. Obviously, then,

$$\operatorname{diam} \sum_\delta(f) = \sup_{\|P\| \le \delta} [\bar{S}(f, P) - \underline{S}(f, P)].$$

Accordingly, since f is Riemann-integrable, we have

$$\lim_{\|P\| \to 0} [\bar{S}(f, P) - \underline{S}(f, P)] = 0$$

Choose a sequence of partitions $\{P_n\}$ such that

    (a) for each n, P_{n+1} is a refinement of P_n;
    (b) $\lim_n \|P_n\| = 0$;
    (c) $\lim_n [\bar{S}(f, P_n) - \underline{S}(f, P_n)] = 0.$

Now

$$\bar{S}(f, P_n) = \int_a^b u_n(x)\, dx$$

$$\underline{S}(f, P_n) = \int_a^b v_n(x)\, dx$$

where u_n is the step function which on each subinterval I defined by P_n takes on the constant value $\sup_I f$, and v_n is the corresponding step function using $\inf_I f$ on each subinterval. Evidently $u_n \geq v_n$ for each n and, because of the refinement condition (a), we have $u_n \geq u_{n+1}$, $v_n \leq v_{n+1}$. Thus,

(7.28) $\qquad v_1 \leq v_2 \leq v_3 \leq \cdots \leq f \leq \cdots \leq u_3 \leq u_2 \leq u_1.$

Since u_n is a step function, we know that if we extend it to be 0 outside $[a, b]$ we obtain a Lebesgue-integrable function on R^1. Let us call the extended function u_n as well. Do the same for each v_n. Then we also know (Example 3) that

$$\int u_n = \int_a^b u_n(x)\,dx = \bar{S}(f, P_n)$$

$$\int v_n = \int_a^b v_n(x)\,dx = \underline{S}(f, P_n).$$

Therefore, the Riemann-integrability, condition (b), tells us that

$$\lim_n \int (u_n - v_n) = 0.$$

By (7.28), $\{(u_n - v_n)\}$ is a monotone-decreasing sequence of non-negative integrable functions. Since their integrals tend to 0, the monotone convergence theorem (a corollary of it) tells us that

$$\lim_n [u_n(x) - v_n(x)] = 0, \qquad \text{a.e.}$$

Now we assert that f is continuous at each point x such that $u_n(x) - v_n(x)$ converges to 0. First, consider a point x which is not an endpoint of any of the subintervals defined by any of the partitions P_n. If $u_n(x) - v_n(x) < \epsilon$, then there is a (small) open interval about x on which $u_n(x) \leq f(t) \leq v_n(x)$, i.e., there is a $\delta > 0$ such that

$$|x - t| < \delta \quad \text{implies} \quad |f(x) - f(t)| \leq v_n(x) - u_n(x) < \epsilon.$$

So we see that f is continuous at x. A small variation on this argument works even if x is one of the countable number of end points which occur. If you don't want to fuss with that case, just throw this countable set (which has measure zero) in with the set of x's such that $u_n(x) - v_n(x)$ does not converge to 0, and you will see that we have already proved that f is continuous almost everywhere.

The argument we have just given is essentially reversible. Suppose f is known to be continuous almost everywhere. Choose a sequence of partitions which satisfy conditions (a) and (b) as before. If x is any point where f is continuous (and x is not an endpoint of any of the subintervals), then

$$\lim_n [u_n(x) - v_n(x)] = 0.$$

Therefore the sequence $\{(u_n - v_n)\}$ converges monotonely to a function which is 0 almost everywhere. Consequently,

$$\lim_n \int (u_n - v_n) = 0.$$

(Which theorem do you need to use to conclude this?) So we have

$$\lim [\bar{S}(f, P_n) - \underline{S}(f, P_n)] = 0$$

and this holds for any sequence of partitions $\{P_n\}$ which satisfies conditions
(a) and (b). It should be apparent that f is Riemann-integrable.

Exercises

1. If S is a set of finite outer measure, then

$$m^*(S) = \inf \left\{ \int f; f \in L^1, f \geq 0, f \geq 1 \text{ on } S \right\}.$$

2. If g is a bounded continuous function and f is an integrable function, then
fg is integrable.

3. (Continuity of the integral with respect to a parameter) Let $f(X, Y)$ be a
function which is continuous as a function of Y and integrable as a function of
X (for each fixed Y). Suppose that $|f(X, Y)| \leq g(X)$ for some integrable function g (independent of Y). Prove that

$$h(Y) = \int f(X, Y) \, dX$$

is continuous.

4. If $f \in L^1(R^1)$ the **Fourier transform** of f is the function

$$\hat{f}(t) = \int f(x) e^{-ixt} \, dx.$$

Show that $\hat{f}$ is uniformly continuous, for each f in L^1.

5. Let f be a continuous function on the real line which is everywhere differentiable and has a *bounded* derivative. Prove that, if $-\infty < a < b < \infty$, the
function $k_{[a,b]}f'$ is Lebesgue-integrable and

$$\int k_{[a,b]}f' = f(b) - f(a).$$

(*Hint*: Show that the difference quotient $n[f(x + (1/n)) - f(x)]$ converges pointwise boundedly to $f'(x)$ on $[a, b]$.)

6. The function $f(x) = x^2 \cos x^{-2}$ is everywhere differentiable on the real line.
Is its derivative integrable?

7. Let f be a non-negative integrable function of compact support. Prove that
$\sqrt{f}$ is an integrable function.

8. Let f be a bounded integrable function. Prove that f^2 is an integrable function.

9. Let $\{f_n\}$ be a sequence of integrable functions such that

(i) $0 \leq f_n \leq 1$;

(ii) $\lim\limits_n \int f_n = 1$;

(iii) there is some compact set K outside of which every f_n vanishes.

Prove that

$$\lim_n \int (1 - f_n)^2 = 0.$$

7.7. Measurable Functions and Measurable Sets

In this section we shall define the concept of measurable function, then use it to introduce measurable sets and study the behavior of measure (length, area, volume, etc.) as a function on the class of measurable sets. This will bring into focus a great deal of what we have done in this chapter.

Definition. A **measurable function** on R^k is a complex-valued function f such that

(i) f is defined almost everywhere on R^k;

(ii) there is a sequence of continuous functions (of compact support) which converges pointwise almost everywhere to f.

We should explain right away why we put the phrase "of compact support" in parentheses. This is because it is relatively easy to see that we get the same class of "measurable" functions whether we use continuous functions or continuous functions of compact support. Suppose $f_1, f_2, f_3,$ ... are continuous functions on R^k and

$$\lim_n f_n(X) = f(X), \qquad \text{a.e.}$$

Let $K_1 \subset K_2 \subset K_3 \subset \cdots$ be an increasing sequence of compact sets such that

$$\bigcup_n K_n = R^k.$$

For each n choose a function $h_n \in C_c(R^k)$ such that $h_n = 1$ on K_n. Let $g_n = f_n h_n$. Then $g_n \in C_c(R^k)$ and

$$\lim_n g_n(X) = \lim_n f_n(X)$$

at any point where $\{f_n(X)\}$ converges. This is because, given X, there exists an N such that $X \in K_N$ and hence

$$g_n(X) = f_n(X), \qquad n \geq N.$$

The measurable functions are our candidates for "reasonable" or "non-pathological" functions. They comprise quite a large class of functions. Let us note several basic things about this class.

(i) If f and g are measurable functions and c is a complex number, then $(cf + g)$ is a measurable function. In other words, the collection of measurable functions forms a linear space in the usual way.

(ii) The function f is measurable if and only if its real and imaginary parts are measurable functions.

(iii) If f is measurable, then $|f|$ is measurable.

(iv) If f and g are real-valued measurable functions, then

$$(f \vee g)(X) = \max(f(X), g(X))$$

$$(f \wedge g)(X) = \min(f(X), g(X))$$

are measurable functions.

(v) If f and g are measurable functions, then the product fg is a measurable function.

(vi) Every null function is measurable.

We have left the proofs of these six assertions to the exercises. The first five follow immediately from the corresponding facts about continuous functions. Statement (vi) requires a little bit more thought (but not much more).

The reader should compare very carefully the definition of measurable function with the definition of integrable function. Integrability appears to require a great deal more. With the machinery we have built up, we can show that integrability just amounts to "measurability plus some control on size".

Theorem 14. *If* f *is a measurable function, then* f *is integrable if and only if there exists an integrable function* g *such that* $|f| \leq |g|$. *In particular, every bounded measurable function of compact support is integrable.*

Proof. As is frequently the case, the only non-trivial part is the "if" half of the theorem. For the proof, once again we work only with the case of real-valued functions. We are told that (real-valued) f is measurable, so we have real continuous functions of compact support f_n such that

$$f(X) = \lim_n f_n(X), \qquad \text{a.e.}$$

We also have $|f| \leq g$ where g is a non-negative integrable function. In other words,

$$-g \leq f \leq g.$$

Let

$$h_n = (f_n \wedge g) \vee (-g)$$

that is

$$h_n(X) = \begin{cases} f_n(X), & \text{if } -g(X) \leq f_n(X) \leq g(X) \\ g(X), & \text{if } f_n(X) > g(X) \\ -g(X), & \text{if } f_n(X) < -g(X). \end{cases}$$

Since g is integrable (and f_n is), each h_n is integrable. Furthermore.

$$-g \leq h_n \leq g$$
$$\lim_n h_n(X) = f(X), \qquad \text{a.e.}$$

By the dominated convergence theorem, f is integrable.

The assertion about measurable functions which are bounded and have compact support follows from the observation that such a function is dominated by (a constant multiple of) the characteristic function of a compact set.

Corollary. *If* f *is an integrable function and* g *is a bounded measurable function, then* fg *is integrable.*

Proof. This follows immediately from Theorem 14 as soon as one notes that, since the product of two continuous functions is continuous, the product of two measurable functions is measurable.

Corollary. *If* $\{f_n\}$ *is a sequence of measurable functions which converges pointwise almost everywhere to the function* f, *then* f *is a measurable function.*

Proof. We may assume that the f_n's are real-valued. We now truncate the function f by chopping off both its head and its feet. Let $K_1 \subset K_2 \subset K_3 \subset \cdots$ be an increasing sequence of compact sets which exhaust R^k:

$$\bigcup_n K_n = R^k.$$

For each n let

$$h_n = (-n) \vee (k_{K_n} f) \wedge n$$

that is,

$$h_n(X) = \begin{cases} 0, & X \notin K_n \\ f(X), & \text{if } X \in K_n \text{ and } -n \leq f(X) \leq n \\ n, & \text{if } X \in K_n \text{ and } f(X) > n \\ -n, & \text{if } X \in K_n \text{ and } f(X) < -n. \end{cases}$$

Then h_n is a measurable function. In fact, h_n is an integrable function, because the sequence of functions

$$-n \vee (k_{K_n} f_j) \wedge n, \qquad j = 1, 2, 3, \ldots$$

converges pointwise boundedly to h_n, and all these functions vanish outside K_n (bounded convergence theorem). Since $h_n \in L^1$ we can find a continuous function of compact support g_n and a set S_n such that

$$m^*(S_n) < n^{-2}$$
$$|g_n(X) - h_n(X)| < \frac{1}{n}, \qquad X \notin S_n.$$

We assert that the sequence $\{g_n\}$ converges pointwise to f almost everywhere. Of course we let

$$E_N = \bigcup_{n=N}^{\infty} S_n.$$

Let T be the set of points where f is not defined, and let

$$E = \bigcap_N E_N.$$

Then T is a set of measure zero, and so is E because

$$E \subset E_N, \qquad N = 1, 2, 3, \ldots$$

$$m^*(E_N) \le \sum_{n=N}^{\infty} n^{-2}$$

We can see rather easily that $\{g_n\}$ converges pointwise to f outside the set of measure zero $(T \cup E)$. If $X \notin (T \cup E)$, then $f(X)$ is some real number (and X is in some K_n), so there is a positive integer M such that

$$X \in K_n, \qquad n \ge M$$

$$|f(X)| \le M.$$

Then we have

(7.29) $$h_n(X) = f(X), \qquad n \ge M.$$

Since $X \notin E$, there exists an N such that $X \notin E_N$. Accordingly,

(7.30) $$|g_n(X) - h_n(X)| < \frac{1}{n}, \qquad n \ge M.$$

It is apparent from (7.29) and (7.30) that $\lim_n g_n(X) = f(X)$.

Corollary. *If* $f_1 \le f_2 \le f_3 \le \cdots$ *is a monotone-increasing sequence of real-valued measurable functions and the limit*

$$f(X) = \lim_n f_n(X)$$

is finite almost everywhere, then f *is a measurable function.*

It is customary (and quite useful) to extend the integral to all non-negative measurable functions. The only way such a function can fail to be integrable is if its values are "too big, too often". In that case its integral will be $+\infty$.

Definition. *If* f *is a non-negative integrable function, the* **integral of** f *is the supremum of the integrals of all integrable functions* g *such that* $g \le f$:

$$\int f = \sup_{\substack{g \in L^1 \\ g \le f}} \int g.$$

Note that we have *not* defined the integral for all measurable functions. We have done so only for the non-negative ones (and the integrable ones, of course.) With the definitions we have used, the following is a very important summary result.

Theorem 15. *If* f *is a complex-valued function defined almost everywhere on* R^k, *the following are equivalent.*

(i) f *is integrable.*

(ii) f *is measurable and* $\int |f| < \infty$.

(iii) *There exists a sequence of continuous functions of compact support* $\{f_n\}$ *such that*

$$\lim_n f_n(X) = f(X), \qquad \text{a.e.}$$

$$\sup_n \int |f_n| < \infty.$$

Proof. Use Fatou's lemma and the monotone convergence theorem.

Definition. *A subset of* R^k *is called a* **measurable set** *if its characteristic function is a measurable function.*

The following properties of measurable sets are virtually immediate from what we know about measurable functions.

(i) If S and T are measurable sets, the union $S \cup T$ and the intersection $S \cap T$ are measurable sets.

(ii) The set S is measurable if and only if its complement $R^k - S$ is a measurable set.

(iii) If $\{S_n\}$ is a sequence of measurable sets, then the union $\bigcup_n S_n$ and the intersection $\bigcap_n S_n$ are measurable sets.

(iv) Every compact set is measurable.

(v) Every set of measure zero is a measurable set.

What measurable sets can we identify? Since every compact set is measurable, every closed set is measurable, because such a set is the union of an increasing sequence of compact sets. Every open set is such a union also; hence every open set is measurable. Of course we can also see this because the complement of an open set is closed. So the class of measurable sets contains all closed sets (and all open sets) and any set that we can generate from such sets using a countable number of operations involving union, intersection, and complementation. In addition it contains all sets of measure zero and sets we can generate by union, intersection, and complement, using either closed sets or sets of measure zero. That is about all we will try to say for now, except for the following illustrative example.

EXAMPLE 4. Let U be an open set. Let us describe how to calculate its measure $m(U)$. Start with a gridwork of mesh 1 as in Figure 27. To be precise, subdivide each coordinate axis into the non-overlapping semi-closed intervals $[n, n + 1)$ for $n = 0, \pm 1, \pm 2, \pm 3, \ldots$. This will partition

R^k into an infinite number of non-overlapping boxes. As a first approximation to $m(U)$ we take

$$\sum_n m(B_n)$$

where $B_1, B_2, \ldots$ are those boxes (formed by the gridwork) whose *closures* lie in U. There may be none of these, there may be a finite positive number, or there may be infinitely many. Let K_1 be the closure of the union of these boxes B_n. It should be clear that K_1 is also the union of the closures of the boxes B_n, because they are all of fixed size so that no sequence of points from *distinct* boxes can converge. Since

$$\bigcup_n B_n \subset K_1 = \bigcup_n \bar{B}_n$$

we have

$$\sum_n m(B_n) \le m(K_1) \le \sum_n m(\bar{B}_n) = \sum_n m(B_n)$$

so that

$$m(K_1) = \sum_n m(B_n).$$

Now apply the same process to the open set $U - K_1$, using a gridwork of mesh $\frac{1}{2}$. One obtains a closed set $K_2 \subset U - K_1$ which is the union of a (finite or infinite) sequence of closed boxes of size $\frac{1}{2}$, the sum of whose measures is $m(K_2)$. Apply the process to $U - K_1 - K_2$, using a gridwork of size $\frac{1}{4}$, and continue. The result is that

$$U = \bigcup_n K_n$$

where $K_1, K_2, K_3, \ldots$ are pairwise disjoint closed sets, and each K_n is the union of a sequence of closed boxes whose measures total $m(K_n)$. Put all the boxes together and U is expressed as the union of a sequence of boxes whose measures total $m(U)$. (See Theorem 16.) With a bit more care, one can express U as the union of a sequence of non-overlapping boxes (not closed ones, of course); but, from the point of view of measure, this is not a significant refinement.

Definition. *If S is a measurable set, the **measure** of S is the integral of the characteristic function of* S:

$$m(S) = \int k_S.$$

Lemma. *Let S and T be measurable sets.*

  (i) $0 \le m(S) \le \infty$.
  (ii) *If* $S \subset T$, *then* $m(S) \le m(T)$.
  (iii) $m(S \cup T) \le m(S) + m(T)$ *and equality holds if* $S \cap T = \varnothing$.

Proof. (i) and (ii) are immediate. For (iii) simply note that

$$k_{S \cup T} \le k_S + k_T$$

and equality holds if $S \cap T = \varnothing$.

Theorem 16. *Let* $\{S_n\}$ *be a sequence of measurable sets and let*

$$S = \bigcup_n S_n.$$

Then S *is measurable and*

(i) *(continuity of measure) if* $S_1 \subset S_2 \subset S_3 \subset \cdots$, *then*

$$m(S) = \lim_n m(S_n)$$

(ii) *(countable additivity of measure) if* $S_i \cap S_j \neq \varnothing$, $i \neq j$, *then*

$$m(S) = \sum_{n=1}^{\infty} m(S_n).$$

Proof. (i) and (ii) are equivalent assertions, because if S is the union of an increasing sequence $\{S_n\}$, we also have

$$S = S_1 \cup (S_2 - S_1) \cup (S_3 - S_2) \cup \cdots$$

to which (iii) is applicable; and we know that

$$m(S_n) = m(S_{n-1}) + m(S_n - S_{n-1}).$$

To verify (i), let k_n be the characteristic function of S_n, so that

$$k_1 \leq k_2 \leq k_3 \leq \cdots$$
$$\lim_n k_n(X) = k(X)$$

where $k = k_S$. Consider the limit

$$\lim_n m(S_n) = \lim_n \int k_n.$$

If this limit is finite, the monotone convergence theorems tells us that k is integrable and its integral is the limit, i.e.,

(7.31) $$m(S) = \lim_n m(S_n).$$

If the limit is $+\infty$, then $m(S) = \infty$ because $m(S) \geq m(S_n)$ for every n. Thus (7.31) is trivially satisfied.

We will have a very complete picture of "measure" as soon as we accomplish one more task. We must show that $m(S)$ is what it ought to be for all simple sets S. First let us note that our notation is consistent with what we did earlier.

(i) S is a set of measure zero if and only if S is measurable and $m(S) = 0$. This is immediate because

$$\int k_S = 0$$

if and only if k_S is a null function.

(ii) If B is a box, then $m(B)$ is the k-dimensional volume which we previously denoted by $m(B)$. See Exercises 4, 5, and 6 of Section 7.3.

Theorem 17. *If* S *is a measurable set, then*

$$m(S) = \inf\{m(U);\ U\ \text{open},\ S \subset U\}.$$

Consequently, $m(S) = m^*(S)$.

Proof. First note that $m(T) \leq m^*(T)$ for all measurable sets T. For, if $\{B_n\}$ is a sequence of boxes which cover T,

$$m(T) \leq \sum_n m(B_n).$$

Now, if $m(S) = \infty$, then $m(U) = \infty$ for all open U such that $S \subset U$. Similarly, $m^*(S) = \infty$. So the theorem is trivial in this case.

Suppose (then) that $m(S) < \infty$. This means that k_S is an integrable function. Let $\epsilon > 0$. By the corollary to Theorem 8, we can find an open set W and a real-valued function $f \in C_c(R^k)$ such that

(a) $m^*(W) < \epsilon$ and hence $m(W) < \epsilon$;
(b) $|f(X) - k_S(X)| < \epsilon,\ X \notin W$.

Let $V = \{X;\ f(X) > 1 - \epsilon\}$. Condition (b) tells us two things:

$$X \notin W,\ X \in S \Longrightarrow X \in V$$
$$X \notin W,\ X \notin S \Longrightarrow X \notin V.$$

In other words,

$$S \subset (V \cup W)$$
$$V \subset (S \cup W).$$

From the second relation we have

$$m(V) \leq m(S) + m(W) < m(S) + \epsilon.$$

If we let $U = V \cup W$, then U is an open set which contains S and

$$m(U) \leq m(V) + m(W)$$
$$< m(S) + 2\epsilon.$$

This establishes the first assertion of the theorem.

To see that $m(S) = m^*(S)$ just note (Example 4) that each open set U is the union of a sequence of boxes $\{B_n\}$ so that

$$m(U) = \sum_{n=1}^{\infty} m(B_n).$$

If $S \subset U$, then

$$m^*(S) \leq \sum_{n=1}^{\infty} m(B_n) = m(U)$$

and, if we take the infimum over open sets U, we see that $m^*(S) \leq m(S)$.

Corollary. *If* S *is a measurable set, then*

$$m(S) = \sup\{m(K);\ K\ \text{compact},\ K \subset S\}.$$

Proof. Apply the theorem to $R^k - S$.

In many situations we are interested in the integration of functions which are defined on a subset of R^k rather than on the whole space. For example, in the study of Fourier series, we are interested in functions on some interval of the real line, e.g., the interval $[-\pi, \pi]$. We want to have complete theories of integration on such subsets. For instance we want to assert that completion of the space of continuous functions on an interval $[a, b]$ relative to the norm

$$\int_a^b |f(x)| \, dx$$

yields the space of Lebesgue-integrable functions on $[a, b]$. Fortunately, with the theory of integration on R^k in hand, none of this presents us with any difficulties. All that we do is restrict the functions in $L^1(R^k)$ to the subset and forget everything that happens outside it.

Definition. *If* S *is a measurable subset of* R^k, *and* f *is an integrable function (or a non-negative measurable function) on* R^k, *the* **integral of** f **over** S *is*

$$\int_S f = \int k_S f.$$

The following properties are immediate from the definition and simple results which we know.

(i) If S and T are disjoint measurable sets,

$$\int_{S \cup T} f = \int_S f + \int_T f.$$

(ii) If $m(S) = 0$,

$$\int_S f = 0.$$

(iii) If f is real-valued,

$$m(S) \inf_S f \leq \int_S f \leq m(S) \sup_S f.$$

Definition. *If* S *is a measurable subset of* R^k, *a* **measurable function on** S *is a complex-valued function* f *such that*

(a) f *is defined at almost every point of* S;

(b) *there is a sequence of continuous functions* $\{f_n\}$ *of compact support on* R^k *such that* $\lim_n f_n(X) = f(X)$ *for almost every* X *in* S.

An **integrable function on** S *is one which satisfies* (a), (b), *and*

(c) *the sequence of integrals* $\int_S |f_n|$ *is bounded.*

Lemma. *Let* f *be a complex-valued function defined almost everywhere on* S. *The following are equivalent.*

(i) f *is a measurable [integrable] function on* S.

(ii) *There is a measurable [integrable] function* g *on* R^k *such that the restriction of* g *to* S *is* f, *i.e.,* $g(X) = f(X)$ *for all* $X \in S$.

(iii) *The function*

$$g = \begin{cases} f, & on\ S \\ 0, & off\ S \end{cases}$$

is a measurable [integrable] function on R^k.

Proof. Exercise.

Armed with this simple lemma we have available on S all of the convergence theorems we developed for measurable and integrable functions on R^k. Of course we denote by $L^1(S)$ the space of integrable-functions on S and endow it with the norm

$$\|f\|_1 = \int_S |f|.$$

It contains the restriction to S of every continuous function of compact support, and is complete relative to the L^1-norm. All of this is apparent because $L^1(S)$ is (can be identified with) the subspace of $L^1(R^k)$ consisting of those functions which vanish outside of S. Of course we will not use the notation $\|\ldots\|_1$ for the L^1-norm on S unless it is clear from the context that we have restricted our attention to functions on S.

Let us establish one non-trivial result about integrals over subsets.

Theorem 18. *Let* f *be an integrable function on* R^k. *For each* $\epsilon > 0$ *there exists* $\delta > 0$ *such that*

$$\int_S |f| < \epsilon$$

for every measurable set S *such that* $m(S) < \delta$.

Proof. Suppose the result fails for some ϵ. Then there is a sequence of measurable sets $\{S_n\}$ such that

$$m(S_n) < 2^{-n}$$

$$\int_{S_n} |f| \geq \epsilon.$$

Let

$$E_N = \bigcup_{n=N}^{\infty} S_n$$

$$E = \bigcap_{N=1}^{\infty} E_N.$$

Since $E_1 \supset E_2 \supset E_3 \supset \cdots$, the dominated convergence theorem states that

$$\int_E |f| = \lim_N \int_{E_N} |f| \geq \epsilon.$$

On the other hand, $m(E) = \lim_N m(E_N) = 0$, a contradiction.

Exercises

In view of the large number of exercises in this section, we have divided them into three general categories.

A. Measurable Functions

1. Be sure that you see how statements (i)–(v) on p. 337 follow directly from the corresponding statements about continuous functions.

2. Prove that every null function is a measurable function.

3. If f is a non-negative measurable function and $\{g_n\}$ is any sequence of integrable functions such that

$$g_1 \leq g_2 \leq g_3 \leq \cdots$$
$$\lim_n g_n(X) = f(X), \qquad \text{a.e.}$$

prove that $\lim_n \int g_n = f$.

4. If f and g are non-negative measurable functions and $c > 0$, then

$$\int (cf + g) = c \int f + \int g.$$

5. If f is a non-negative integrable function, discuss

$$\int \log f.$$

Show that it is always sensibly defined if we allow $-\infty$ as a value. (*Hint:* $\log x \leq x$ if $x \geq 1$.)

6. Let ω be a fixed non-negative integrable function. Let $L(\omega)$ be the set of measurable functions f such that $f\omega$ is integrable. For $f \in L(\omega)$ define

$$\|f\| = \int |f| \, \omega.$$

Prove the following:

(a) $(L(\omega), \|\ldots\|)$ is a semi-normed linear space.
(b) That space is complete.

7. Let

$$f(x) = \sin \frac{1}{x}, \qquad 0 < |x| \leq 1.$$

Is f an integrable function on $[-1, 1]$?

8. Let f be a fixed real-valued integrable function on R^k. For each measurable set S define

$$F(S) = \int_S f.$$

Let

$$b = \sup_S F(S).$$

True or false? There exists a measurable set S such that

$$F(S) = b.$$

True or false? The image of F is connected.

9. Let K be the set of measurable functions f such that $0 \leq f \leq 1$.

(a) K is a convex set.

(b) f is an extreme point of K if and only if f is the characteristic function of a measurable set.

10. If f is a real measurable function on R^k and g is a continuous function on the real line, then the composition $g \circ f$ is a measurable function.

11. If f is a continuous function on R^1, if g is a measurable function on R^1, *and if* $f^{-1}(E)$ *is measurable for every set of measure zero* $E \subset R^1$, then the composition $g \circ f$ is measurable.

*12. Define a function f on the interval $[0, 1]$ as follows: If

$$x = \sum_{n=1}^{\infty} a_n 2^{-n}, \qquad a_n = 0 \text{ or } 1$$

is the binary expansion of x, define

$$f(x) = \sum_{n=1}^{\infty} a_n 3^{-n}.$$

(a) Show that f is well-defined almost everywhere and that f is a measurable function on $[0, 1]$.

(b) Show that f maps $[0, 1]$ into the Cantor set.

(c) True or false? If g is a continuous function (on the Cantor set), then $g \circ f$ is a measurable function.

B. *Measurable Sets*

13. Prove that the set $S \subset R^k$ is measurable if and only if $R^k - S$ is measurable.

14. If $\{S_n\}$ is a sequence of measurable sets, show that $\bigcup_n S_n$ and $\bigcap_n S_n$ are measurable sets.

15. If S is a measurable set and $m(S) > 0$, there exists a point $X \in S$ such that $m(N \cap S) > 0$ for every neighborhood N of X.

16. (Translation-invariance of measure) If S is a measurable set, each translate of S

$$Y + S = \{Y + X; \ X \in S\}$$

is measurable and $m(Y + S) = m(S)$.

17. If B is a ball in R^k, then (B is measurable) and $m(B)$ depends only on the radius of B.

18. Let θ be an angle, $0 \leq \theta \leq 2\pi$. Let S be a rectangle in R^2 and let S_θ be the rectangle obtained by rotating S through the angle θ about the origin. Show that S can be decomposed into a finite number of triangles which can be translated to new positions in R^2 and reassembled to form S_θ. Conclude that Lebesgue measure on R^2 is rotation-invariant.

19. True or false? If S and T are measurable sets, the algebraic sum

$$S + T = \{X + Y;\ X \in S,\ Y \in T\}$$

is a measurable set.

20. Construct a set S in the plane such that $m(S) = 0$ but uncountably many vertical lines meet S in a set of positive 1-dimensional measure.

21. Describe how to construct a measurable subset S of the interval $[0, 1]$ such that, in each subinterval of $[0, 1]$, both S and its complement have positive measure, i.e., if $0 \le a < b \le 1$, then

$$0 < m(S \cap (a, b)) < b - a.$$

22. There does not exist a measurable subset of $[0, 1]$ such that

$$m(S \cap (a, b)) = \tfrac{1}{2}(b - a)$$

for each subinterval (a, b).

23. Let f be a real-valued function on R^k. Show that f is measurable if and only if the graph of f is a measurable set in R^{k+1}.

*24. If S is a linear subspace of R^k and if $m(S) > 0$, then $S = R^k$.

25. If S is a measurable set and f is a real-valued integrable function, then

$$m(S) \inf_S f \le \int f \le m(S) \sup_S f.$$

26. Define the **inner measure** of the (arbitrary) set S to be

$$m_*(S) = \sup \{m(K);\ K \text{ compact},\ K \subset S\}.$$

Prove that S is measurable if and only if $m^*(S) = m_*(S)$.

27. A family $\mathcal{S}$ of subsets of R^k is called a **sigma-algebra** of sets if it satisfies

    (i) the empty set and R^k are in $\mathcal{S}$;
    (ii) if $S \in \mathcal{S}$, then the complement $R^k - S$ is in $\mathcal{S}$;
    (iii) if $\{S_n\}$ is any sequence of sets in $\mathcal{S}$, the union $\bigcup_n S_n$ is in $\mathcal{S}$.

Prove the following.

    (a) The intersection of any collection of sigma-algebras is a sigma-algebra.

    (b) Given any family of sets in R^k there is a smallest sigma-algebra of subsets of R^k which contains the family. (This is called the sigma-algebra generated by the family.)

    (c) The sigma-algebra generated by the family of compact subsets of R^k is the same as the sigma-algebra generated by the family of closed boxes in R^k. (This sigma-algebra is called the family of **Borel sets** in R^k.)

    (d) A set S in R^k is measurable if and only if it differs from some Borel set by a set of measure zero.

28. (Invariance of measure under rigid motions) If S is a measurable subset of R^k, show that

$$m(S) = \inf_{\{B_n\}} \sum_{n=1}^{\infty} m(B_n)$$

where the infimum is taken over all countable coverings of S by open balls. Use the result of Exercise 17 to show that Lebesgue measure on R^k is invariant under (unchanged by) rigid motions of R^k.

29. (Carathéodory) The subset S of R^k is measurable if and only if

$$m^*(E) = m^*(E \cap S) + m^*(E \cap (R^k - S))$$

for every set $E \subset R^k$.

C. Simple Functions and the Lebesgue Process

30. Let f be a measurable function. If K is a compact set in the complex plane, then $f^{-1}(K)$ is a measurable set. Do the same for $f^{-1}(U)$ where U is open. (*Hint:* It's true for continuous functions.)

31. A **simple function** is a function s of the form

$$s = \sum_{n=1}^{N} a_n k_{E_n}$$

where $a_1, \ldots, a_n$ are constants and $E_1, \ldots, E_n$ are measurable sets. In other words, a simple function is one which is a linear combination of characteristic functions of measurable sets. Prove the following.

(a) The simple functions constitute a linear subspace of the (bounded) measurable functions.

(b) If f is a bounded measurable function, then f can be *uniformly* approximated by simple functions. (*Hint for* (b): Take a square in the plane which contains the image of f. Cut it up into small squares, and let the E_n's be the inverse images (under f) of those squares.)

32. If f is integrable and $\epsilon > 0$, there is a simple function s such that

$$\int |f - s| < \epsilon.$$

33. Let f be a real-valued function defined almost everywhere on R^k. Suppose that the set

$$E_t = \{X; f(X) > t\}$$

is measurable, for every real number t. Prove that f is a measurable function. (*Hint:* Use complements, intersections, etc. of the sets E_t for various t's to see that $\{X; f(X) < t\}$, $\{X; a < f(X) \leq b\}$, etc. are measurable. Approximate f by simple functions.)

34. (Lebesgue's ladder) Let f be a non-negative integrable function. Let

$$E_{11} = \{X; 0 \leq f(X) < \tfrac{1}{2}\}$$
$$E_{12} = \{X; \tfrac{1}{2} \leq f(X) < 1\}.$$

Let

$$S_1 = 0 \cdot \mu(E_{11}) + \tfrac{1}{2}\mu(E_{12}).$$

Now divide the interval $[0, 2)$ into 8 subintervals, $[0, \tfrac{1}{4})$, $[\tfrac{1}{4}, \tfrac{1}{2})$, $\ldots$, $[\tfrac{7}{4}, 2)$ and let $E_{21}, \ldots, E_{28}$ be the inverse images under f. Let

$$S_2 = 0 \cdot \mu(E_{21}) + \tfrac{1}{4}\mu(E_{22}) + \cdots + \tfrac{7}{4}\mu(E_{28}).$$

Now, divide $[0, 3)$ twice as finely, i.e., into 24 subintervals, form S_3 and continue. Prove that

$$\lim_k S_k = \int_B f.$$

(Be sure to draw a picture of what's going on. Show that there is nothing sacred about the particular partitions of the real line we used.)

D. A Non-Measurable Set

35. Lebesgue measure on the real line induces an analogous measure (arc length) on the unit circle C. The measure of any arc

$$\{e^{i\theta};\ \theta_1 \leq \theta \leq \theta_2\}$$

is $(\theta_2 - \theta_1)$, the outer measure of a subset of the circle is the infimum of the sum of the lengths of the arcs in the various countable coverings by arcs, etc. Just as Lebesgue measure on the line is translation-invariant, this measure on the circle is rotation-invariant: If S is a measurable subset of the unit circle, then

$$m(e^{i\theta}S) = m(S).$$

Let G be the set of all points e^{in}, $n = 0, \pm1, \pm2, \ldots$. These points are distinct because $e^{in} = e^{ik}$ is not possible with n and k distinct integers. Consider the various rotations of G:

$$e^{i\theta}G = \{e^{i(n+\theta)};\ n = 0, \pm1, \pm2, \ldots\}.$$

(a) Prove that $e^{i\theta}G = e^{i\psi}G$ if and only if $e^{i(\theta-\psi)} \in G$, i.e., if and only if $\theta - \psi = n + 2k\pi$ where n, k are integers.

(b) Prove that G has uncountably many distinct rotations.

(c) According to the axiom of choice (Appendix) there is a set S comprised of exactly one point from each of the distinct rotations of G. For such an S, show that the rotations $e^{in}S$ are pairwise disjoint and exhaust the circle:

$$\bigcup_n e^{in}S = C.$$

(d) From (c) show that S cannot be a measurable set. (*Hint*: By the rotation-invariance, the various sets $e^{in}S$ must all have the same measure.)

7.8. Fubini's Theorem

We come now to one of the most important theorems in mathematical analysis. It is the extension to the present context of the result that an integral over a rectangle can be reduced to iterated integrals over intervals, i.e., that we can integrate one variable at a time:

$$\int_a^b \int_c^d f(x, y)\, dx\, dy = \int_a^b \left\{ \int_c^d f(x, y)\, dy \right\} dx.$$

We treated this for continuous functions on boxes in Chapter 4. The result for Lebesgue-integrable functions on R^k–known as Fubini's theorem –is considerably more difficult to prove, but the effort is worth it, because what emerges is an extremely powerful technical tool.

Let us begin by restating the basic result for continuous functions of compact support.

Theorem 19. *If* f *is a continuous function of compact support on* $R^{k+\ell}$ *then*

(i) *for every* X *in* R^k *the function* $f(X, \cdot)$ *is a continuous function of compact support on* R^ℓ;

(ii) *the function* g *defined by*

$$g(X) = \int_{R^\ell} f(X, \cdot)$$

is a continuous function of compact support on R^k;

(iii) $$\int_{R^k} g = \int_{R^{k+\ell}} f.$$

Proof. Let B be some closed box outside of which $f = 0$, and apply Theorem 10 of Chapter 4.

In presenting Fubini's theorem, we shall depart from our usual format. We shall first state the result and then present the heart of the proof in the form of three lemmas.

Theorem 20 (Fubini's Theorem). *If* f *is an integrable function on* $R^{k+\ell}$, *then*

(i) *for almost every* X *in* R^k *the function* $f(X, \cdot)$ *is an integrable function on* R^ℓ;

(ii) *the function* g *defined (almost everywhere) by*

$$g(X) = \int_{R^\ell} f(X, \cdot)$$

is an integrable function on R^k;

(iii) $$\int_{R^k} g = \int_{R^{k+\ell}} f.$$

Notice the difficulty we are in right away. If f is continuous, then for each X the function $f(X, \cdot)$, the value of which at Y is $f(X, Y)$, is a perfectly well-defined function on R^ℓ. But if f is (merely) integrable there are a great many points at which the value of f is not defined, and we must *prove* that for almost every X in R^k value $f(X, Y)$ is defined for almost every Y in R^ℓ. Thus we must know something about the relationships between sets of measure zero in R^k, R^ℓ, and $R^{k+\ell}$.

Lemma 1. *If* S *is a set of measure zero in* $R^{k+\ell}$, *then, for almost every* X *in* R^k, *the section*

$$S_X = \{Y \in R^\ell; (X, Y) \in S\}$$

is a set of measure zero in R^ℓ.

Proof. Since $m(S) = 0$, Theorem 2 tells us that there is a sequence of real-valued functions $\{f_n\}$ of compact support on $R^{k+\ell}$ such that

(i) $f_1 \leq f_2 \leq f_3 \leq \cdots$;

(ii) the sequence of integrals $\{\int f_n\}$ is bounded;

(iii) $\lim_n f_n(X, Y) = \infty$ for every point $(X, Y) \in S$.

For each X in R^k the sequence $\{f(X, \cdot)\}$ is increasing in $C_c(R^\ell)$. Further-more the (values of the) sequence diverge at every point Y which is in the section S_X. Therefore (Theorem 2) if we can show that

$$\lim_n \int_{R^\ell} f_n(X, \cdot) < \infty$$

for almost every X, we will have proved the lemma. To this end, let

$$g_n(X) = \int_{R^\ell} f_n(X, \cdot), \qquad n = 1, 2, 3, \ldots .$$

According to Theorem 19,

$$\int_{R^k} g_n = \int_{R^{k+\ell}} f_n,$$

so that

$$\lim \int_{R^k} g_n = \lim \int_{R^{k+\ell}} f_n < \infty.$$

Since $\{g_n\}$ is an increasing sequence, this implies that

$$\lim_n g_n(X) < \infty, \qquad \text{a.e.}$$

and the lemma is established.

Lemma 2. *If* $f_1 \leq f_2 \leq f_3 \leq \cdots$ *is a monotone-increasing sequence of functions in* L^1 *for which Fubini's theorem holds, and if the limit function*

$$f = \lim_n f_n$$

is integrable, then Fubini's theorem is valid for f.

Proof. We begin by applying Lemma 1 to the set of points (X, Y) such that $f_n(X, Y)$ does *not* converge to $f(X, Y)$. We conclude that there is a set of measure zero $E \subset R^k$ such that if $X \notin E$, then $f_n(X, Y)$ converges to $f(X, Y)$, for almost every Y in R^ℓ. Associated with each f_n is a set of measure zero $E_n \subset R^k$ via statement (i) of Fubini's theorem: If $X \notin E_n$, then $f_n(X, \cdot)$ is an integrable function on R^ℓ. The union

$$E \cup E_1 \cup E_2 \cup \cdots$$

is a set of measure zero. Therefore for almost every X in R^k we have

(a) $\lim_n f_n(X, Y) = f(X, Y)$, for almost every Y in R^ℓ;

(b) for every n, $f_n(X, \cdot)$ is an integrable function on R^ℓ.

For such an X, the sequence

$$g_n(X) = \int_{R^\ell} f_n(X, \cdot)$$

is (defined and) an increasing sequence. Since Fubini's theorem holds for f_n, each g_n is in $L^1(R^k)$ and

$$\int_{R^k} g_n = \int_{R^{k+\ell}} f_n$$

so that

$$\lim_n \int_{R^k} g_n = \lim_n \int_{R^{k+\ell}} f_n < \infty.$$

Since $\{g_n\}$ is an increasing sequence, the monotone convergence theorem implies that for almost every X

(c) $\lim_n g_n(X) < \infty.$

If $X \in R^k$ satisfies conditions (a), (b), and (c), then the monotone convergence theorem guarantees that $f(X, \cdot)$ is an integrable function and

$$\int_{R^\ell} (f(X, \cdot) = \lim_n \int f_n(X, \cdot)$$
$$= \lim_n g_n(X)$$

which we know to be an integrable function of X. We also know that if

$$g(X) = \lim_n g_n(X)$$

then

$$\int_{R^k} g = \lim_n \int_{R^k} g_n$$
$$= \lim_n \int_{R^{k+\ell}} f_n$$
$$= \int_{R^{k+\ell}} f.$$

Lemma 3. *Fubini's theorem is valid for every bounded measurable function of compact support.*

Proof. Let f be a measurable function, bounded by 1, and suppose f has compact support. There is a sequence $\{f_n\}$ in $C_c(R^{k+l})$ which converges pointwise to f, almost everywhere. We can easily arrange that each f_n is bounded by 1 and that there is a fixed compact set outside of which every f_n vanishes. We know that Fubini's Theorem is valid for each f_n. The proof that it is valid for f is now virtually identical with the proof of Lemma 2, except that one uses Lemma 1 and the bounded convergence theorem rather than Lemma 1 and the monotone convergence theorem.

Proof of Fubini's Theorem. The family of integrable functions for which Fubini's theorem holds is obviously a linear subspace of $L^1(R^{k+\ell})$. Therefore it suffices to prove it for non-negative functions. If f is a non-negative function in L^1, let

$$f_n = (f \wedge n)k_{K_n}$$

where $K_1 \subset K_2 \subset K_3 \subset \cdots$ are compact sets which exhaust $R^{k+\ell}$. Lemma

3 tells us that Fubini's theorem is valid for each f_n, and since

$$f_1 \leq f_2 \leq f_3 \leq \cdots$$

$$\lim_n f_n(X) = f(X) \qquad \text{everywhere that } f(X) \text{ is defined}$$

Lemma 2 says the theorem is valid for f.

Corollary. *Let* S *be a measurable subset of* $R^{k+\ell}$. *Then, for almost every* X *in* R^k, *the section*

$$S_X = \{Y \in R^\ell ; (X, Y) \in S\}$$

is a measurable subset of R^ℓ, *and*

$$m(S) = \int_{R^k} m(S_X) \, dX.$$

Proof. This is just Fubini's theorem applied to the characteristic function of S.

Corollary. *If* S *is a measurable subset of* R^k *and* T *is a measurable subset of* R^ℓ, *then the Cartesian product* S $\times$ T *is a measurable subset of* $R^{k+\ell}$, *and* $m(S \times T) = m(S)m(T)$.

Proof. Once one has shown that $S \times T$ is measurable, the formula $m(S \times T) = m(S)m(T)$ is immediate from the previous corollary. We have left the proof that $S \times T$ is measurable to the exercises.

In practice we will often use the familiar notation

$$\int f(X) \, dX = \int_{R^k} f$$

to denote the Lebesgue-integral of f over R^k. In fact, we did this in the first corollary above. This notation is especially helpful in expressing the Fubini result:

$$(7.32) \qquad \int \left\{ \int f(X, Y) \, dY \right\} dX = \int f(X, Y) \, dX \, dY.$$

In the last integral we wrote $dX \, dY$ in place of $d(X, Y)$. This notation was avoided in the proof of Fubini's theorem, so that we would be forced to think carefully about what type of object each symbol represented.

Two comments are in order. First, obviously the order in (7.32) is not important, that is, we may just as well integrate first with respect to X and then with respect to Y. Second, it is extremely important to bear in mind that the validity of Fubini's theorem depends on the *a priori* knowledge that f is integrable over $R^{k+\ell}$. If f is merely measurable, both of the iterated integrals may exist yet be different. Fortunately this cannot happen if f is a non-negative measurable function; and the result for non-negative functions is the most useful tool for testing the integrability of functions on $R^{k+\ell}$ which are made up out of functions on R^k and functions on R^ℓ.

Theorem 21 (Fubini). *If* f *is a non-negative measurable function on* $R^{k+\ell}$, *then*

(i) *for almost every* X *in* R^k, *the function* f(X, ·) *is measurable;*

(ii) *the function*

$$g(X) = \int_{R^\ell} f(X, \cdot)$$

is a measurable function on R^k *if we allow* $+\infty$ *as a value;*

(iii) $$\int_{R^k} g = \int_{R^{k+\ell}} f.$$

Proof. What do we mean in part (ii) where we say "if $+\infty$ is allowed as a value"? At the simplest level we mean, of course, that since f is not assumed to be integrable we may have

$$\int_{R^\ell} f(X, \cdot) = \infty$$

for quite a few values of X. But we mean something more. We mean that the function g, which is defined almost everywhere on R^k and maps into the extended non-negative real number system, is measurable in the sense that there is a sequence of non-negative continuous functions of compact support which converges to g pointwise almost everywhere. For such a function g the truncated functions $g \wedge n$ are measurable in the usual sense and converge monotonely upward to g. Since the monotone convergence theorem works just as well for infinite limits as finite ones:

$$\left. \begin{array}{l} f_1 \leq f_2 \leq f_3 \leq \cdots \\ f = \lim_n f_n \\ \lim \int f_n = \infty \end{array} \right\} \quad \Longrightarrow \quad \int f = \infty$$

we have no trouble in defining the integrals of non-negative functions which are measurable in this extended sense. The proof of Fubini's theorem for these extended non-negative measurable functions is just (the proof of) Lemma 2 plus the fact that we know Fubini's theorem for the truncated functions,

$$(f \wedge n)k_{K_n}.$$

Corollary. *Let* f *be a measurable function on* $R^{k+\ell}$. *If either of the iterated integrals*

$$\int \left\{ \int |f(X, Y)| \, dY \right\} dX$$

$$\int \left\{ \int |f(X, Y)| \, dX \right\} dY$$

is finite, then (*they both are and*) f *is an integrable function.*

Proof. Since Fubini's theorem is valid for $|f|$, the finiteness of either integral will imply the integrability of f because they are both equal to

$$\int |f(X, Y)|\, dX\, dY.$$

EXAMPLE 5. Let f and g be integrable functions on R^k. Associated with the pair (f, g) is an important "product" known as the **convolution** of f and g:

$$(f * g)(X) = \int f(X - Y)g(Y)\, dY.$$

It is far from clear that the integral on the right makes sense. After all, the product of two integrable functions (although measurable) need not be integrable. But we can use the second Fubini theorem to show that, given f and g, it is true that for almost every X the product $f(X - Y)g(Y)$ is an integrable function of Y.

First note that, for any given f, the function

$$\bar{f}(X, Y) = f(X - Y)$$

is a measurable function of X and Y, i.e., a measurable function on $R^k \times R^k$. This can be verified as follows. If $\{f_n\}$ is a sequence of continuous functions which converges pointwise almost everywhere to f, we want to show that $\bar{f}_n$ converges pointwise to $\bar{f}$, almost everywhere on $R^k \times R^k$. If E is the set of points X in R^k such that $f_n(X)$ does not converge to $f(X)$, then $\bar{f}_n(X, Y)$ fails to converge to $\bar{f}(X, Y)$ on the set

(7.33) $\{(X, Y); (X - Y) \in E\}.$

We need to know that, if E is any set of measure zero in R^k, then (7.33) is a set of measure zero in $R^{2k} = R^k \times R^k$. We have left the proof of this to the exercises. Once it is verified, we have shown that $\bar{f}$ is a measurable function on R^{2k}, i.e., that $f(X - Y)$ is a measurable function of X and Y.

Now, given f and g in $L^1(R^k)$ we know that

$$h(X, Y) = f(X - Y)g(Y)$$

is a measurable function on R^{2k}. We test its integrability using Theorem 21, by integrating $|h(X, Y)|$, first with respect to X, then with respect to Y:

$$\int |h(X, Y)|\, dX = \int |f(X - Y)|\,|g(Y)|\, dX$$

$$= |g(Y)| \int |f(X - Y)|\, dX.$$

Evidently,

$$\int |f(X - Y)|\, dX = \int |f(T)|\, dT;$$

i.e., translation by $(-Y)$ does not effect the L^1-norm. (Why? Because this is obviously so for continuous functions of compact support.) Thus

$$\int |h(X, Y)| \, dX = \| f \|_1 \int |g(Y)| \, dY$$

$$= \| f \|_1 \| g \|_1.$$

Thus, $h \in L^1(R^{2k})$ and $\| h \|_1 = \| f \|_1 \| g \|_1$.

Now we may apply the Fubini theorem (Theorem 20) to conclude that for almost every X, the function $f(X - Y)g(Y)$ is integrable with respect to Y, that the function

$$(f * g)(X) = \int f(X - Y)g(Y) \, dY$$

is an integrable function in R^k, and that

$$\int (f * g)(X) \, dX = \int f(X - Y)g(Y) \, dX \, dY$$

$$= \left(\int f \right)\left(\int g \right).$$

We also know that

$$\| f * g \|_1 \leq \| f \|_1 \| g \|_1.$$

The operation of convolution on L^1:

$$(f, g) \longrightarrow f * g$$

plays a central role in the theory of Fourier series and Fourier integrals. At this stage, the point we wish to emphasize is how the use of the second Fubini theorem, integrating first with respect to X and then with respect to Y, makes it clear that $f(X - Y)g(Y)$ is integrable on R^{2k}. Thus, whereas it was not clear a priori that there need exist a single X such that $f(X - Y)g(Y)$ is integrable with respect to Y, it follows that this is so for almost every X.

Exercises

1. Let S be a set of measure zero in R^k and let T be a bounded subset of R^l. Prove that the Cartesian product $S \times T$ is a set of measure zero in R^{k+l}.

2. Use the result of Exercise 1 to prove that, if S and T are subsets of R^k and R^l, respectively, and if either S or T is a set of measure zero, then $S \times T$ is a set of measure zero.

3. Let T be a fixed subset of R^l. Let $\mathcal{S}_T$ be the family of all subsets S of R^k such that $S \times T$ is a measurable subset of R^{k+l}. Prove that

    (a) $\mathcal{S}_T$ contains every set of measure zero in R^k;
    (b) if $S \in \mathcal{S}_T$, then the complement $(R^k - S)$ is in $\mathcal{S}_T$;
    (c) if $\{S_n\}$ is a sequence of sets in $\mathcal{S}_T$, the union $\bigcup_n S_n$ is in $\mathcal{S}_T$.

Note that (b) and (c) imply that $\mathcal{S}_T$ is also closed under countable intersections.

4. Show that, if S is any measurable set in R^k, there is a set $\tilde{S}$ with these properties:

(a) $S \subset \tilde{S}$

(b) $\tilde{S} = \bigcap_n U_n$, where each U_n is open (and therefore the union of a sequence of boxes).

(c) $m(\tilde{S} - S) = 0$.

5. Use the results of Exercises 3 and 4 to prove that, if B is a box in R^ℓ, then $S \times B$ is a measurable subset of $R^{k+\ell}$, for every set S in R^k.

6. Use the results of Exercises 3, 4, and 5 to prove that, if S is a measurable subset of R^k and T is a measurable subset of R^ℓ, then $S \times T$ is a measurable subset of $R^{k+\ell}$. (Its measure is, of course, $m(S \times T) = m(S)m(T)$.)

7. Prove that, if f is a measurable function on R^k and g is a measurable function on R^l, then $h(X, Y) = f(X)g(Y)$ is a measurable function on R^{2k}. (*Hint:* Use Exercise 2.)

8. From Exercise 7, do you see a simple proof that the Cartesian product of two measurable sets is measurable?

9. If f is a real-valued function on R^k, the graph of f is a set of measure zero in R^{k+1}.

10. If f is a non-negative measurable function on R^k, the set

$$S = \{(X, y) \in R^{k+1}; 0 \leq y \leq f(X)\}$$

is a measurable subset of R^{k+1} and

$$m(S) = \int f.$$

11. Let

$$f(x, y) = \frac{1}{x + y}.$$

Is f integrable on the square $[0, 1] \times [0, 1]$? What about

$$f(x, y) = \frac{1}{x^2 + y^2}?$$

12. Is

$$f(x, y, z) = \frac{1}{x^2 + y^2 + z^2}$$

integrable on the cube $[0, 1] \times [0, 1] \times [0, 1]$?

13. Let $p(x_1, \ldots, x_n)$ be a polynomial in n variables, i.e., a polynomial function on R^n. Let N be the null set (zero set) of p:

$$N = \{X \in R^n; p(x_1, \ldots, x_n) = 0\}.$$

Show that N is a set of measure zero, unless $p = 0$. (*Hint:* Use Fubini and induction on n.)

14. Prove that almost every $k \times k$ matrix is invertible.

15. Let E be a set of measure zero in R^k. Prove that the set

$$S = \{(X, Y) \in R^{2k}; (X - Y) \in E\}$$

is a set of measure zero in R^{2k}. (*Hint:* It suffices to prove that each bounded part of S has measure zero. Handle such a bounded part just as in Exercise 1, except that the sets $X - Y = $ constant are used in place of the sets $X = $ constant.)

16. Prove that the operation of convolution on $L^1(R^k)$ has these properties.

    (i) $(f * g) * h = f * (g * h)$;

    (ii) $f * g = g * f$;

    (iii) $f * (cg + h) = c(f * g) + f * h$;

    (iv) If f is of class C^n, then $f * g$ is of class C^n for all $g \in L^1(R^k)$.

17. If $f \in L^1(R^k)$ and g is a bounded measurable function on R^k, the convolution

$$(f * g)(X) = \int f(X - Y) g(Y) \, dY$$

is a uniformly continuous function on R^k.

18. Use the result of Exercise 17 to prove the following. If E is measurable set of positive measure in R^k, the algebraic difference set

$$E - E = \{ X - Y; \; X \in E, \; Y \in E \}$$

contains a non-empty open set.

*19. True or false? Almost every real $n \times n$ matrix has a real characteristic value.

7.9. *Orthogonal Expansions*

    Let E be a fixed measurable subset of R^k. The collection of all measurable functions f on E such that $|f|^2$ is integrable over E forms a complete inner product space (Hilbert space) in a natural way. This makes it possible to expand each function in this space—a space which we denote $L^2(E)$—in a series of pairwise orthogonal functions. This can be done in a variety of ways, depending on the orthogonal sequence of functions which is of particular interest in a given mathematical situation. This section will provide only a glimpse of this important set of ideas. The only orthogonal system we shall discuss in any detail is the system of functions e^{inx}, $n = 0, \pm 1, \pm 2, \ldots$ on the subset $E = [-\pi, \pi]$ of the real line. This will round out the general aspects of our discussion of Fourier series.

    Notation. Let p be a positive real number, and let E be a measurable subset of R^k. We denote by $L^p(E)$ the collection of all measurable functions f on the set E for which

$$\int_E |f|^p < \infty.$$

    Note that if g is a non-negative measurable function on E and $0 < p < \infty$, then g^p is a measurable function on E. (See Exercise 10, Section 7.7.)

Thus the definition of $L^p(E)$ makes sense. In case $p = 1$, it agrees with our previous definition of $L^1(E)$.

Convention. As with $L^1(E)$, we agree to call two functions in $L^p(E)$ equal whenever they differ by a null function.

Now it is a fact that for every p, $1 \leq p < \infty$, $L^p(E)$ is a Banach space relative to the L^p-norm:

$$(7.34) \qquad \|f\|_p = \left(\int_E |f|^p \right)^{1/p}.$$

We carry out the proof only for the case we are interested in.

Lemma. *If* f *and* g *are functions in* $L^2(E)$, *then the product* fg *is in* $L^1(E)$. *Furthermore,*

$$(7.35) \qquad \langle f, g \rangle = \int_E fg^*$$

is an inner product on $L^2(E)$.

Proof. The properties required of an inner product are:

(i) $\langle cf_1 + f_2, g \rangle = c\langle f_1, g \rangle + \langle f_2, g \rangle$
(ii) $\langle g, f \rangle = \langle f, g \rangle^*$
(iii) $\langle f, f \rangle \geq 0$; if $\langle f, f \rangle = 0$, then $f = 0$.

We saw in Chapter 6 that any inner product satisfies the Cauchy-Schwarz inequality:

$$|\langle f, g \rangle|^2 \leq \langle f, f \rangle \langle g, g \rangle.$$

The (standard) proof of this goes as follows. Note that

$$0 \leq \|f + cg\|_2^2 = \langle f + cg, f + cg \rangle$$
$$= \langle f, f \rangle + c\langle g, f \rangle + c^*\langle f, g \rangle + |c|^2\langle g, g \rangle$$

for every complex number c. Choose c so that the sum of the last two terms is 0:

$$\langle f, g \rangle + c\langle g, g \rangle = 0.$$

We can do that unless $g = 0$. (If $g = 0$, the inequality which we want to prove is trivial.) Put that value of c into the inequality

$$0 \leq \|f + cg\|_2^2$$

and out comes the Cauchy-Schwarz inequality.

Thus we know that

$$(7.36) \qquad \left| \int fg^* \right|^2 \leq \int |f|^2 \int |g|^2$$

on any subspace where (7.35) is known to make sense. For instance we know (7.36) on the subspace of $L^2(E)$ consisting of the bounded measurable functions of compact support. From the monotone convergence theorem and the fact that

$$\left(\int |f| \, |g| \right)^2 \leq \int |f|^2 \int |g|^2$$

on this subspace it is immediate that $(fg) \in L^1(E)$ for all $f, g \in L^2(E)$.

The last minor point which requires comment deals with condition (iii) for an inner product. In the case at hand, this says

$$\langle f, f \rangle = \int f f^*$$
$$= \int |f|^2$$
$$\geq 0$$

and that if $\langle f, f \rangle = 0$, then $f = 0$. Note that the last statement is legitimate because of our convention that null functions are equal to 0.

Theorem 22 (Riesz-Fischer). $L^2(E)$ *is a Hilbert space, i.e., a complete inner product space.*

Proof. The triangle inequality

$$\|f + g\|_2 \leq \|f\|_2 + \|g\|_2$$

is immediate from the Cauchy-Schwarz inequality and the relation

$$\|f\|_2 = (\langle f, f \rangle)^{1/2}.$$

Thus what we are proving is that $(L^2, \|\cdots\|_2)$ is a Banach space. Consider a fast Cauchy sequence in $L^2(E)$:

$$\sum_{n=1}^{\infty} \|f_n - f_{n+1}\|_2 = M < \infty.$$

We want to show that $\{f_n\}$ converges. Let $g_n = f_n - f_{n-1}$ (with $f_0 = 0$) so that

$$\sum_{n=1}^{\infty} \|g_n\|_2 = M < \infty$$

and we want to prove that the series

$$\sum_{n=1}^{\infty} g_n$$

converges in $L^2(E)$. Let

$$h_n = \sum_{j=1}^{n} |g_j|$$

so that $h_n \in L^2(E)$, $0 \leq h_1 \leq h_2 \leq h_3 \leq \cdots$ and (by the triangle inequality)

$$\|h_n\|_2 \leq \sum_{j=1}^{n} \|g_n\|_2 \leq M.$$

Thus

$$0 \le h_1 \le h_2 \le h_3 \le \cdots$$

$$\int_E h_n^2 \le M^2$$

from which it follows (monotone convergence) that

$$h(X) = \lim_n h_n(X) < \infty, \qquad \text{a.e.}$$

$$h \in L^2(E)$$

$$\int_E h^2 = \lim_n \int h_n^2.$$

At any point X where $h(X) < \infty$, the series $\sum g_n(X)$ is absolutely convergent. If we let

$$f(X) = \sum_{n=1}^{n} g_n(X), \qquad \text{a.e.}$$

then since

$$f_n = \sum_{j=1}^{n} g_j$$

$$|f_n| \le h_n \le h$$

we see that $f \in L^2(E)$ and $|f| \le h$. Since f and f_n are dominated by h, we have

$$|f - f_n|^2 \le 4h^2$$

$$\lim_n f_n(X) = f(X), \qquad \text{a.e.}$$

so, by the dominated convergence theorem,

$$\lim_n \int |f - f_n|^2 = 0.$$

The completeness of L^2 is very useful in discussing various types of series expansions. In a variety of problems important in mathematics and physics, one encounters expansions into series of orthogonal functions.

Definition. *Let $\{f_n\}$ be a sequence in $L^2(E)$. We call the sequence* **orthogonal** *if*

$$\langle f_n, f_k \rangle = 0, \qquad n \ne k$$

and we call it **orthonormal** *if*

$$\langle f_n, f_k \rangle = \delta_{nk} = \begin{cases} 1, & n = k \\ 0, & n \ne k. \end{cases}$$

If we have an orthogonal sequence $\{f_n\}$, and if each $f_n \ne 0$, we can transform it into an orthonormal sequence by dividing each f_n by its norm. Thus, the two concepts are only a whisker apart. We might put it this way: In working with orthogonal sequences, we may as well normalize so that each function has norm 1.

EXAMPLE 6. The most famous orthonormal sequence is that given by the trigonometric functions on the interval $E = [-\pi, \pi]$:

$$f_n(x) = e^{inx}, \qquad n = 0, \pm 1, \pm 2, \ldots.$$

It is customary here to normalize so that E has measure 1. This means that we absorb a factor of $(2\pi)^{-1}$ into the inner product:

$$\langle f, g \rangle = \frac{1}{2\pi} \int_{-\pi}^{\pi} f(x)g(x)^* \, dx.$$

EXAMPLE 7. The **Legendre polynomials** P_n have these properties:

(i) P_n is a polynomial of degree n ($n = 0, 1, 2, \ldots$);
(ii) $\{P_n\}$ is an orthogonal sequence on the interval $[-1, 1]$.

Those conditions determine P_n to within a constant factor and the polynomials are:

$$P_0(x) = 1$$
$$P_n(x) = (2^n n!)^{-1} \frac{d^n(x^2 - 1)^n}{dx^n}, \qquad n \geq 1.$$

Lemma. *If $\{g_n\}$ is a linearly independent sequence in $L^2(E)$, there is a unique orthonormal sequence $\{f_n\}$ such that, for each n, f_n is a linear combination of $g_1, \ldots, g_n$.*

Proof. The sequence $\{f_n\}$ can be constructed recursively as follows. Let

$$f_1 = \frac{1}{\|g_1\|_2} g_1.$$

After $f_1, \ldots, f_{n-1}$ have been found, let

$$f_n = \frac{1}{\|h_n\|_2} h_n$$

where

$$h_n = g_n - \sum_{k=1}^{n-1} \langle g_n, f_k \rangle f_k.$$

The process in this lemma enables one to construct a great many orthonormal sequences. For example, on any bounded set E one may apply the process to the sequence of polynomial functions $1, x, x^2, \ldots$. The Legendre polynomials are the result of doing this on $E = [-1, 1]$.

Theorem 23. *Let $\{f_n\}$ be an orthonormal sequence in $L^2(E)$. The following are equivalent (all true or all false).*

(i) *If $h \in L^2$ and $\langle h, f_n \rangle = 0$ for all n, then $h = 0$.*
(ii) *For every f in L^2:*

$$f = \sum_n \langle f, f_n \rangle f_n \qquad \text{(series converging in L^2)}.$$

(iii) *For every f in* L^2:

$$\|f\|_2^2 = \sum_n |\langle f, f_n \rangle|^2.$$

Proof. Let $\{f_n\}$ be any orthonormal sequence. Let $f \in L^2(E)$. Let

$$c_k = \langle f, f_k \rangle$$

$$g_n = \sum_{k=1}^{n} c_k f_k.$$

Notice that

$$\langle g_n, f \rangle = \langle \sum_k c_k f_k, f \rangle$$

$$= \sum_{k=1}^{n} c_k \langle f_k, f \rangle.$$

Since

$$\langle f_k, f \rangle = \langle f, f_k \rangle^* = c_k^*$$

we have

$$\langle g_n, f \rangle = \sum_{k=1}^{n} |c_k|^2.$$

From the orthonormality relations $\langle f_n, f_k \rangle = \delta_{nk}$, it is easy to compute that

$$\|g_n\|_2^2 = \sum_{k=1}^{n} |c_k|^2.$$

Therefore,

$$\|f - g_n\|_2^2 = \langle f - g_n, f - g_n \rangle$$

$$= \langle f, f \rangle - \langle g_n, f \rangle - \langle f, g_n \rangle + \langle g_n, g_n \rangle$$

$$= \langle f, f \rangle - \sum_{k=1}^{n} |c_k|^2.$$

Thus we have **Bessel's inequality**:

$$\sum_{k=1}^{n} |c_k|^2 \leq \|f\|_2^2$$

$$c_k = \langle f, f_k \rangle.$$

Since that holds for each n

$$\sum_{k=1}^{\infty} |c_k|^2 \leq \|f\|_2^2.$$

Now, we can verify that $\{g_n\}$ is a Cauchy sequence in $L^2(E)$:

$$\|g_n - g_N\|_2^2 = \sum_{k=N+1}^{n} |c_k|^2, \qquad n > N.$$

$$\lim_N \sum_{k=N+1}^{\infty} |c_k|^2 = 0.$$

Since L^2 is complete, there exists $g \in L^2$ with

$$\lim \|g - g_n\|_2 = 0,$$

i.e., the series

$$g = \sum_{n=1}^{\infty} c_n f_n$$

converges in L^2. Let

$$h = f - g.$$

It is a simple matter to verify

$$\begin{aligned}
\|h\|_2^2 &= \|f - g\|_2^2 \\
&= \lim_n \|f - g_n\|_2^2 \\
&= \|f\|_2^2 - \sum_{k=1}^{\infty} |c_k|^2 \\
&= \|f\|_2^2 - \|g\|_2^2.
\end{aligned}$$

Similarly,

$$\langle h, f_n \rangle = 0, \qquad \text{all } n.$$

All of this is valid for any orthonormal sequence. Now, look at the properties (i), (ii), (iii) in the statement of the theorem. Property (i) says that, for every $f \in L^2$, the corresponding h is 0. Property (ii) says that, for every $f \in L^2$, the corresponding g is equal to f. Property (iii) says that, for every $f \in L^2$, $\|f\|_2 = \|g\|_2$. It should now be clear that those are equivalent properties of the sequence $\{f_n\}$.

What does Theorem 23 tell us? Suppose we call the orthonormal sequence $\{f_n\}$ **complete** provided 0 is the only element of L^2 which is orthonal to every f_n. We then know this. If $\{f_n\}$ is a complete orthonormal sequence in L^2, every f in L^2 can be expanded in a series

$$f = \sum_{n=1}^{\infty} c_n f_n$$

which converges to f in L^2-norm. The coefficients in the expansion are:

$$\begin{aligned}
c_n &= \langle f, f_n \rangle \\
&= \int_E f f_n^*.
\end{aligned}$$

The functions f_n serve as a basis of mutually perpendicular unit vectors. The coefficients c_n are the coordinates, much as in Euclidean space. Norms can be computed using these coordinates:

$$\|f\|_2^2 = \sum_{n=1}^{\infty} |c_n|^2.$$

In fact, if

$$g = \sum_{n=1}^{\infty} b_n f_n$$

then

$$\langle f, g \rangle = \sum_{n=1}^{\infty} c_n b_n^*.$$

The reader may have seen a proof that the trigonometric sequence

$$f_n(x) = e^{inx}, \qquad n = 0, \pm 1, \pm 2, \ldots$$

is complete in $L^2(-\pi, \pi)$. Suppose you know Weierstrass' approximation theorem, in this form (third corollary, Theorem 24, Chapter 5): If g is a continuous function on $[-\pi, \pi]$ and if $g(-\pi) = g(\pi)$, then g can be uniformly approximated by trigonometric polynomials

$$P(x) = \sum_{k=-n}^{n} a_k e^{ikx}.$$

Suppose one knows that. Suppose $h \in L^2$ and all of the Fourier coefficients of h are 0:

$$\int_E h f_n^* = 0, \qquad n = 0, \pm 1, \pm 2, \ldots.$$

By the Weierstrass theorem,

$$\int_E hg = 0$$

for every continuous g with $g(-\pi) = g(\pi)$. Use pointwise bounded approximation and the dominated convergence theorem, to see that

$$\int_E hg = 0$$

for every bounded measurable function g. Now choose a sequence of bounded measurable g_n's such that

$$|g_n| \leq |h|$$
$$g_n \longrightarrow h^*, \qquad \text{a.e.}$$

We obtain

$$\int_E h h^* = 0;$$

so $h = 0$.

Theorem 24. *Let* $f \in L^2(-\pi, \pi)$ *and define the Fourier coefficients of* f *by*

$$c_n = \frac{1}{2\pi} \int_{-\pi}^{\pi} f(x) e^{-inx} \, dx.$$

Then the partial sums of the Fourier series

$$\sum_{n=-\infty}^{\infty} c_n e^{inx}$$

converge to f *in* L^2 *norm, and*

$$\sum_n |c_n|^2 = \frac{1}{2\pi} \int_{-\pi}^{\pi} |f(x)|^2 \, dx.$$

If

$$\{c_n\}, \qquad n = 0, \pm 1, \pm 2, \ldots$$

is any sequence of complex numbers which is square-summable:

$$\sum_n |c_n|^2 < \infty$$

there is a (unique) function f in L^2 such that $\{c_n\}$ is the sequence of Fourier coefficients of f.

We can also expand each function in $L^2(-1, 1)$ in a series using the Legendre polynomials of Example 7. The completeness property of those polynomials can be proved just as we did it for the trigonometric functions, by using the Weierstrass theorem on approximation by ordinary polynomials. There are many other complete orthonormal sequences of polynomials which are important: Laguerre polynomials, Hermite polynomials, etc. The associated series expansions can be used to solve a variety of problems. Let's describe one such problem—to solve an integral equation.

EXAMPLE 8. Let g be a continuous function on the interval $[a, b]$. Let K be a continuous function on the square $[a, b] \times [a, b]$. We want to solve the equation

$$g(x) = f(x) + \int_a^b f(t)K(x, t)\, dt$$

for the function f. There is associated with the kernel K a complete orthonormal sequence in $L^2(a, b)$. The kernel K defines a mapping T from $L^2(a, b)$ into $L^2(a, b)$:

$$(Tf)(x) = \int_a^b f(t)K(x, t)\, dt, \qquad f \in L^2.$$

Actually, Tf is a continuous function so that

$$L^2(a, b) \xrightarrow{\;T\;} C[a, b].$$

Note that T is a linear transformation:

$$T(cf_1 + f_2) = cTf_1 + Tf_2.$$

Suppose K is real-valued and symmetric:

$$K(t, x) = K(x, t).$$

It can be proved that there is a complete orthonormal sequence $\{f_n\}$ in $L^2(a, b)$ such that each f_n is a characteristic vector of the transformation T:

$$Tf_n = \lambda_n f_n \qquad (\lambda_n \in R).$$

The various λ_n's are not all distinct.

Suppose one has such a sequence $\{f_n\}$. One description of how to solve the integral equation:

$$g(x) = f(x) + (Tf)(x)$$

is this: Expand

$$g = \sum_n a_n f_n$$

and the solution for f is

$$f = \sum_n c_n f_n$$

where

$$a_n = c_n + \lambda_n c_n,$$

i.e.,

$$c_n = a_n (1 + \lambda_n)^{-1}$$

That won't work if $\lambda_n = -1$ for some n. In the most important case, which is $K \geq 0$, one cannot have the difficulty because it can be shown that

$$\lambda_n \geq 0$$

$$\lim_n \lambda_n = 0.$$

The method described then solves the equation for every $g \in L^2(a, b)$.

Exercises

1. Carry out the proof that the Legendre polynomials (Example 7) form a complete orthonormal sequence in $L^2(-1, 1)$.

2. If f and g are in $L^2(-\pi, \pi)$, extend them periodically to the real line with period 2π. Show that the convolution

$$(f*g)(x) = \frac{1}{2\pi} \int_{-\pi}^{\pi} f(t)g(x - t)\, dt$$

is in $L^1(-\pi, \pi)$ and that its Fourier coefficients c_n are given by

$$c_n = a_n b_n$$

where a_n and b_n are the nth Fourier coefficients of f and g, respectively.

3. Assume from Chapter 6 the fact

$$\| X \|_p = (|x_1|^p + \cdots + |x_n|^p)^{1/p}, \qquad 1 \leq p < \infty$$

is a norm (satisfies the triangle inequality) on R^n (or C^n). Use this to prove that if f and g are simple functions,

$$\| f + g \|_p \leq \| f \|_p + \| g \|_p.$$

Then show that this inequality holds for all non-negative measurable functions and, hence, that $\| \cdots \|_p$ is a norm on L^p.

4. Prove that $L^p(E)$ is complete, $1 \leq p < \infty$. (*Hint*: Show that (basically) the same proof we gave for L^2 works for L^p.)

5. If $f \in L^2(-\pi, \pi)$, show that

$$f(re^{i\theta}) = \frac{1}{2\pi} \int_{-\pi}^{\pi} f(t) P_r(\theta - t)\, dt$$

defines a harmonic function on the open unit disk $\{z \in C; |z| < 1\}$ and that this function is complex-analytic if and only if the Fourier coefficients

$$c_n = \frac{1}{2\pi} \int_{-\pi}^{\pi} f(x)e^{-inx}\,dx$$

are 0 for $n < 0$. (P_r is the Poisson kernel from Chapter 5.)

6. If f is a bounded measurable function, define

$$\|f\|_\infty = \inf_g \sup |g|$$

where g ranges over all functions which agree with f almost everywhere. Prove that

$$\|f\|_\infty = \lim_{p \to \infty} \|f\|_p.$$

8. Differentiable Mappings

We are going to discuss the differentiation of maps from one Euclidean space into another. We begin by reviewing the basic facts which we shall need concerning linear transformations and partial derivatives.

8.1. Linear Transformations

The simplest class of (differentiable) maps from one Euclidean space into another consists of the transformations T which are linear:

$$T(cX + Y) = cT(X) + T(Y).$$

We have referred to such maps frequently in the examples and exercises. In this chapter they will play a central role in the development. Let us summarize their basic properties.

Let n and k be positive integers. We shall denote by $L(R^n, R^k)$ the vector space consisting of all linear transformations from R^n into R^k. The most basic fact about a linear transformation

$$R^n \xrightarrow{\ T\ } R^k$$

is that T is uniquely determined by its values on any n linearly independent vectors $X_1, \ldots, X_n$:

$$X = c_1 X_1 + \cdots + c_n X_n$$
$$T(X) = c_1 T(X_1) + \cdots + c_n T(X_n).$$

In particular T is determined by its values on the standard basis vectors $E_1 = (1, 0, \ldots, 0), \ldots, E_n = (0, 0, \ldots, 0, 1)$. Suppose

(8.1) $T(E_j) = (a_{1j}, a_{2j}, \ldots, a_{kj}), \qquad j = 1, \ldots, n.$

The $k \times n$ matrix

$$A = [a_{ij}], \qquad 1 \le i \le k, \quad 1 \le j \le n$$

is called the **standard representing matrix** for T. The equations (8.1) combine with the linearity of T to yield this explicit description of the action of T:

(8.2) $Y = T(X)$

$$AX^t = Y^t.$$

In (8.2), X^t is the transpose of X, that is, the $n \times 1$ matrix obtained by standing the n-tuple X up on its right end. Thus the second equation of (8.2) means:

$$\begin{bmatrix} a_{11} & a_{12} & \cdots & a_{1n} \\ a_{21} & a_{22} & \cdots & a_{2n} \\ \cdot & \cdot & & \cdot \\ \cdot & \cdot & & \cdot \\ \cdot & \cdot & & \cdot \\ a_{k1} & a_{k2} & \cdots & a_{kn} \end{bmatrix} \begin{bmatrix} x_1 \\ x_2 \\ \cdot \\ \cdot \\ \cdot \\ x_n \end{bmatrix} = \begin{bmatrix} y_1 \\ y_2 \\ \cdot \\ \cdot \\ \cdot \\ y_k \end{bmatrix}$$

$$y_i = \sum_{j=1}^{n} a_{ij} x_j.$$

There is no need to keep writing the "t" for transpose, if you can remember that $Y = T(X)$ corresponds to $Y = AX$, where the Y and X in the second equation are column vectors.

Any map from R^n into R^k is (or, can be) described by a k-tuple of real-valued functions on R^n

$$R^n \xrightarrow{\;T\;} R^k$$

$$Y = T(X)$$

(8.3) $y_1 = f_1(x_1, \ldots, x_n)$

$$\vdots \qquad \qquad \vdots$$

$$y_k = f_k(x_1, \ldots, x_n).$$

The function f_i is defined by "$f_i(X)$ is the ith coordinate of $T(X)$". In other words, f_i is the composition $f_i = \pi_i \circ T$, where π_i is the ith standard coordinate function on R^k:

$$\pi_i(Y) = y_i.$$

The map T in (8.3) is linear if and only if each f_i is linear (a linear functional).

If f is a linear functional on R^n, then f has the form

(8.4) $f(X) = a_1 x_1 + \cdots + a_n x_n$

where $a_1, \ldots, a_n$ are some fixed real numbers. The matrix $[a_1 \; \cdots \; a_n]$ is the standard representing matrix for the linear functional f. It is usually identified with the vector $A = (a_1, \ldots, a_n)$ so that (8.4) reads

(8.5) $$f(X) = \langle A, X \rangle$$

or

$$f(X) = A \cdot X$$

depending on your notation for inner (or dot) product.

Just remember this about the two standard ways of describing a linear transformation T. It is uniquely determined by the images of the standard basis vectors in R^n. The image of E_j under T is found by reading down the jth column of the representing matrix A. The transformation T is also determined by a k-tuple of linear functionals on R^n: $T = (f_1, \ldots, f_k)$. The form of f_i is

$$f_i(X) = \langle A_i, X \rangle$$

where A_i is the ith row of the standard matrix A.

From time to time, we shall have occasion to talk about affine maps on R^n. An **affine transformation** from R^n into R^k is a map

$$R^n \xrightarrow{\;F\;} R^k$$

which has the form

$$F(X) = Y_0 + T(X)$$

where Y_0 is a fixed vector in R^k and $T \in L(R^n, R^k)$. Thus, an affine transformation is a linear transformation followed by a translation. The linear transformations are the affine transformations which carry 0 into 0. Think of a real-valued function on the real line:

$$R \xrightarrow{\;f\;} R.$$

To say that f is affine means that the graph of f is a straight line. To say that f is linear means that the graph of f is a straight line through the origin.

8.2. Partial Derivatives

There are certain definitions and facts about differentiation which you should be familiar with from Chapter 4. We review them here for convenience. Let f be a real-valued function defined on an open set in R^n. The **partial derivatives** of f (if they exist) are

(8.6) $$(D_j f)(X) = \lim_{t \to 0} \frac{f(X + tE_j) - f(X)}{t}.$$

We will also denote $D_1 f, \ldots, D_n f$ by $\partial f/\partial x_1, \ldots, \partial f/\partial x_n$ when it seems useful to do so. The function f is **of class** C^1 on the open set U if $D_1 f, \ldots,$ $D_n f$ exist and are continuous functions on U.

If f has partial derivatives, we can ask whether they are differentiable. The second order derivative (if it exists)

$$D_i(D_j f) = \frac{\partial^2 f}{\partial x_i \, \partial x_j}$$

is denoted by $D_{i,j} f$. One of the most basic theorems here is that, if $D_{i,j} f$ and $D_{j,i} f$ exist and are continuous, then

(8.7)
$$D_{i,j} f = D_{j,i} f$$
$$\frac{\partial^2 f}{\partial x_i \, \partial x_j} = \frac{\partial^2 f}{\partial x_j \, \partial x_i}.$$

The function f is **of class** C^k on the open set U if all kth order derivatives

(8.8)
$$D_{i_1, \ldots, i_k} f = D_{i_1} D_{i_2} \cdots D_{i_k} f$$
$$= \frac{\partial^k f}{\partial x_{i_1} \cdots \partial x_{i_k}}$$

exist and are continuous on U. **Of class** C^∞ means of class C^k for all k. In dealing with higher order derivatives, it is convenient to use vector indices. If $i_1, \ldots, i_k$ are positive integers between 1 and n, then we let $i = (i_1, \ldots, i_k)$ and

(8.9)
$$D_i f = D_{i_1, \ldots, i_k} f.$$

Note that each of the differential operators $D_1, \ldots, D_n$ is a linear transformation

$$C^1(U) \xrightarrow{D_j} C(U)$$

from functions of class C^1 into continuous functions. Of course, we also have

$$C^k(U) \xrightarrow{D_j} C^{k-1}(U).$$

The second order differential operator

$$C^k(U) \xrightarrow{D_{i,j}} C^{k-2}(U)$$

is actually the composition $D_{i,j} = D_i \circ D_j$:

$$
\begin{array}{ccc}
C^k & \xrightarrow{D_j} & C^{k-1} \\
& \searrow{\scriptstyle D_{i,j}} & \downarrow{\scriptstyle D_i} \\
& & C^{k-2}
\end{array}
$$

More generally, (8.8) really means that if $i = (i_1, \ldots, i_k)$, then D_i is the composition of the linear transformations $D_{i_1}, \ldots, D_{i_k}$.

The concept of partial derivative (8.6) is a special case of directional derivative. The derivative $(D_j f)(X)$ measures the rate of change of f at X along the line through X which is parallel to the jth axis. Rates of change along other lines are equally important. If V is any vector in R^n, then V determines a line through the origin—the 1-dimensional subspace spanned by V. If we translate that line so that it passes through X, we have the line

$$L = \{(X + tV); t \in R\}$$

and we can try to differentiate f at X along L:

$$(8.10) \qquad (D_V f)(X) = \lim_{t \to 0} \frac{f(X + tV) - f(X)}{t}.$$

The derivative $(D_V f)(X)$ depends not only on the line L but on the vector V as well, because in (8.10) we used a scale and direction along L which depended on V as a unit. Sometimes, we may sloppily refer to $(D_V f)(X)$ as the derivative of f at X in the direction of the vector V; but, we shall try not to because $D_V f$ depends upon the length of V as well as its direction. If we let

$$W = \frac{1}{|V|} V$$

then W is the vector of unit length which has the same direction as V and it is $(D_W f)(X)$ which is properly called the **derivative of f at X in the direction of the vector** V. Of course,

$$D_W f = \frac{1}{|V|} D_V f.$$

That is a special case of the fact that $D_V f$ is a linear function of V (if f is, say, a fixed function of class C^1). For a fixed V, then D_V is a linear transformation

$$C^1(U) \xrightarrow{D_V} C(U)$$

and

$$D_{aV_1 + V_2} = aD_{V_1} + D_{V_2}.$$

In particular, if $V = (v_1, \ldots, v_n)$, then

$$D_V = v_1 D_1 + \cdots + v_n D_n$$

$$(8.11) \qquad D_V f = v_1 \frac{\partial f}{\partial x_1} + \cdots + v_n \frac{\partial f}{\partial x_n}, \qquad f \in C^1.$$

If f is of class C^1 on the open set U, the **gradient** of f is

$$f' = (D_1 f, \ldots, D_n f)$$

$$(8.12) \qquad = \left(\frac{\partial f}{\partial x_1}, \ldots, \frac{\partial f}{\partial x_n} \right).$$

Note that f' is a (continuous) map from U into R^n. According to (8.11), the gradient satisfies

$$D_V f = \langle V, f' \rangle$$
$$= V \cdot f'.$$

At any point X, the vector $f'(X)$ gives the direction in which f is increasing most rapidly.

Each (real-valued) affine function on R^n is of class C^∞. More generally, each polynomial

$$(8.13) \qquad f(X) = \sum_{i_1,\ldots,i_n=0}^{N} a_{i_1,\ldots,i_n} x_1^{i_1} \cdots x_n^{i_n}$$

is of class C^∞. The affine functions are the polynomials of degree 1 (or zero). Recall the use of vector indices which enables us to rewrite (8.13) as

$$f(X) = \sum_{|i|=0}^{N} a_i x_1^{i_1} \cdots x_n^{i_n}$$
$$= \sum_{|i|=0}^{N} a_i X^i.$$

We have used $|i|$ for the weight or degree of the vector index i:

$$|i| = i_1 + \cdots + i_n.$$

It gives the degree of X^i and it could be confused with the Euclidean length of the vector i in R^n. Let's solve that problem by agreeing not to get confused by that. The coefficient a_i is given by

$$a_i = \frac{1}{i!}(D_i f)(0)$$

where $i! = i_1! \cdots i_n!$.

You are supposed to know Taylor's theorem, which is of interest in attempting to approximate a smooth function by polynomials. The form of that theorem which we shall use says this. Let f be a function of class C^{k+1} on a convex neighborhood U of the point $A = (a_1, \ldots, a_n)$. Then

$$f(X) = \sum_{|i|=0}^{k} \frac{1}{i!}(D_i f)(A)(X - A)^i + R_k(X)$$

where the remainder $R_k(X)$ is given by

$$R_k(X) = (k+1) \sum_{|i|=k+1} \frac{(X-A)^i}{i!} \int_0^1 (D^i f)(A + t(X-A))(1-t)^k \, dt.$$

A special set of functions of class C^∞ on U consists of the **real-analytic functions** on U, those such that for each point $A \in U$ they can be expanded in a convergent power series

$$(8.14) \qquad f(X) = \sum_{|i|=0}^{\infty} c_i(X - A)^i.$$

If f is real-analytic on U and if A is a point of U, then the coefficients c_i in (8.14) are uniquely determined. They are

$$c_i = \frac{1}{i!}(D_i f)(A)$$

and there is a largest symmetric box about A

$$B = \{X; |x_i - a_i| < \epsilon, i = 1, \ldots, n\}$$

in which the series (8.14) converges.

Bear in mind that the real-analytic functions are extremely special. We shall be working more with the classes C^k and C^∞.

8.3. Differentiable Functions

What is a differentiable function on R^n? We could define it as a function f for which the partial derivatives $D_1 f, \ldots, D_n f$ all exist. There are several difficulties with that as a definition of differentiable function. First, the mere existence of $(D_1 f)(X), \ldots, (D_n f)(X)$ does not imply that f is differentiable in the directions of other vectors. (If $D_1 f, \ldots, D_n f$ exist and are *continuous* throughout an open set U, then there is no difficulty with directional derivatives.) Second, most mathematical definitions which depend upon a particular choice of coordinates lead to conceptual difficulties at some point. Third, what is going to play the role of the derivative of f? It would be nice to have "differentiable" mean that the "derivative" exists.

One way to counter the first and second objections is to require for differentiability at X that $(D_V f)(X)$ should exist, for every vector V in R^n— to require differentiability in every direction. The astute reader may be able to guess that that's not quite enough to require. Suppose that all the directional derivatives existed but the formula

$$(D_V f)(X) = v_1(D_1 f)(X) + \cdots + v_n(D_n f)(X)$$

broke down. That would create havoc. We want $(D_V f)(X)$ to be a linear function of V. So, we could say: f is differentiable at X if every directional derivative $(D_V f)(X)$ exists and if $(D_V f)(X)$ is a linear function of V. What, then, would play the role of the derivative of f? *Answer:* The linear functional ℓ:

$$\ell(V) = (D_V f)(X)$$

will be the derivative of f at X.

Now, we are going to start afresh on the question: "What should differentiable mean?" and look at it from a geometrical point of view. We'll come to the same conclusions which we just reached, but in a somewhat neater form.

Suppose f is a real-valued function on an interval of the real line. If x is a point of that interval, the derivative

$$f'(x) = \lim_{h \to 0} \frac{f(x + h) - f(x)}{h}$$

(if it exists) is the slope of the tangent line to the graph of f at x. Indeed, the definition of tangent line involves the limit of chords—a limit which

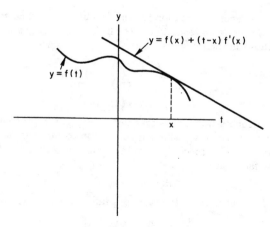

FIGURE 28

is the geometrical analogue of the definition of $f'(x)$. The tangent line has the equation

$$y(t) = f(x) + (t - x)f'(x)$$

and is the line which "best approximates" f near x. (See Figure 28.) What (precisely) does that mean? It means this. Suppose we search for an affine function

$$a(t) = b(t - x) + c$$

to approximate $f(t)$ near x. If we want

$$\lim_{t \to x} [f(t) - a(t)] = 0$$

the constant c must be chosen as $f(x)$. If we take $c = f(x)$ and $b = f'(x)$, then we get

$$\lim_{t \to x} \frac{f(t) - a(t)}{t - x} = 0.$$

i.e., $f(t) - a(t)$ goes to zero faster than $t - x$ as t approaches x.

Quite often, it is more convenient to phrase the last remarks this way:

$$f'(x) = \lim_{h \to 0} \frac{f(x + h) - f(x)}{h}$$

(8.15) $$f(x + h) = f(x) + hf'(x) + R(h; x)$$

$$\lim_{h \to 0} \frac{1}{h} R(h; x) = 0.$$

If x is fixed throughout some given discussion, analysts will often write (8.15) this way:

$$f(x + h) = f(x) + hf'(x) = o(h)$$

where $o(h)$ is a generic expression for any function of h which goes to zero faster than h does (as h tends to 0). The little "oh" may denote many different functions in the same discussion:

$$h^2 = o(h)$$
$$h^5 = o(h)$$
$$h \neq o(h).$$

The point of such a notation is that in approximation problems such as the one we have been discussing we obtain approximations such as (8.15) in which we don't care precisely what the remainder $R(h; x)$ is. All we care about is the fact that

$$R(x; h) = o(h).$$

Now, let's get back to the case of a function of several variables:

$$U \longrightarrow R, \qquad U \subset R^n.$$

Fix a point X in U. Differentiability of f at X will correspond to the existence of a tangent "plane" to the graph of f at the point $(X, f(X))$, as indicated in Figure 29. The tangent "plane" will actually be n-dimensional —a hyperplane in R^{n+1}—and it will be defined by an affine function

$$y = a(T)$$
$$= c_0 + a_1 t_1 + \cdots + a_n t_n.$$

Since we want the hyperplane to pass through the point $(X, f(X))$, it is

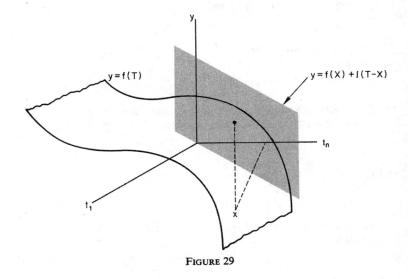

FIGURE 29

more convenient to write the affine function as

$$a(T) = c + b_1(t_1 - x_1) + \cdots + b_n(t_n - x_n)$$
$$= c + \ell(T - X)$$

where ℓ is a linear function. Clearly we want $c = f(X)$. What will it mean to say that we can find some hyperplane

$$y = f(X) + \ell(T - X)$$

which is tangent to the graph of f at $(X, f(X))$?

After a long hassle about the geometric meaning of tangency, we would eventually come around to agreeing that what the tangency means is that the hyperplane approximates the graph to the first order; i.e., that (as T approaches X) $f(T) - a(T)$ goes to zero faster than $T - X$ does. We can express that analytically in several ways

(i)
$$\lim_{T \to X} \frac{f(T) - a(T)}{|T - X|} = 0$$

(ii)
$$\lim_{T \to X} \frac{f(T) - f(X) - \ell(T - X)}{|T - X|} = 0$$

(iii)
$$f(X + H) = f(X) + \ell(H) + R(X; H)$$

where

$$\lim_{H \to 0} \frac{1}{|H|} R(X; H) = 0$$

(iv)
$$f(X + H) = f(X) + \ell(H) + o(H).$$

In (iv), $o(H)$ means $o(|H|)$.

Definition. *Let* f *be a real-valued function on an open set* U *in* R^n, *and let* X *be a point of* U. *The function* f *is* **differentiable** *at* X *if there exists a linear functional* ℓ *on* R^n *such that*

(8.16)
$$f(X + H) = f(X) + \ell(H) + o(H).$$

There is one simple thing we must verify, namely, if f is differentiable at X the linear functional ℓ in (8.16) is unique. We see that as follows. Suppose that we also have

$$f(X + H) = f(X) + \ell_1(H) + o(H).$$

Subtract and you conclude that

(8.17)
$$\ell(H) - \ell_1(H) = o(H).$$

Remember that $o(H)$ is not a specific function but a symbol for any function of H which goes to zero faster than H does as H goes to 0. If you forget that, you'll get confused by such statements as

$$o(H) - o(H) = o(H).$$

We want to deduce from (8.17) that $\ell = \ell_1$. What we need to show is that if λ:

$$\lambda(H) = a_1 h_1 + \cdots + a_n h_n$$

is a linear functional on R^n and $\lambda(H) = o(H)$, then $\lambda = 0$. You should be able to do that fairly easily.

Definition. *If* f *is differentiable at* X, *then the linear functional* ℓ *in* (8.16) *is called the* **derivative of** f **at** X *and is denoted by* $(Df)(X)$:

$$\ell = (Df)(X).$$

Theorem 1. *If* f *is differentiable at* X *and* $\ell = (Df)(X)$, *then all directional derivatives of* f *at* X *exist and*

$$(D_V f)(X) = \ell(V), \qquad V \in R^n.$$

Proof. Let V be a non-zero vector in R^n. We want to show that

$$(8.18) \qquad \lim_{t \to 0} \frac{f(X + tV) - f(X)}{t} = \ell(V).$$

Since f is differentiable at X,

$$(8.19) \qquad f(X + tV) = f(X) + \ell(tV) + o(tV).$$

Since V is a fixed non-zero vector and $|tV| = |t||V|$, $o(tV)$ may as well be replaced by $o(t)$. Since ℓ is linear, $\ell(tV) = t\ell(V)$. From (8.19) we have

$$\frac{f(X + tV) - f(X)}{t} = \ell(V) + \frac{1}{t} o(V) = \ell(V) + o(1).$$

What does that say? It is precisely (8.18).

A particular consequence of Theorem 1 is that the partial derivatives $(D_1 f)(X), \ldots, (D_n f)(X)$ exist and

$$(D_j f)(X) = \ell(E_j).$$

Thus we have

$$\ell(V) = v_1 (D_1 f)(X) + \cdots + v_n (D_n f)(X)$$

for every vector V in R^n. In other words, $(Df)(X)$ is the linear functional "inner product with the gradient $f'(X)$".

Theorem 2. *Let* f *be a real-valued function on an open set* U *in* R^n. *The two following statements are equivalent:*

(i) f *is a function of class* C^1.

(ii) f *is differentiable at each point of* U *and the derivative* Df *is continuous.*

Proof. Assume (i), that is, assume that the partial derivatives $D_1 f, \ldots, D_n f$ exist and are continuous functions on U. Fix a vector X in U. We showed in Chapter 4 that the directional derivatives of f at X

exist and are given by

(8.20) $$(D_V f)(X) = v_1 (D_1 f)(X) + \cdots + v_n (D_n f)(X).$$

Define

$$\ell(V) = (D_V f)(X), \qquad V \in R^n.$$

From (8.20) it is clear that ℓ is a linear function on R^n. The question is: Does $f(X + H) - f(X) - \ell(H)$ go to 0 faster than $|H|$ does as H tends to 0?

In the one-variable case, it is easy to see that it does. In fact, we already discussed that in motivating our definition of differentiable function. In the one-variable case, we do not need to know that f' is continuous; but, here is the proof for the one-variable case that parallels the proof for the n-variable case. The mean value theorem gives

$$f(x + h) - f(x) = f'(c)h$$

where c is a point between x and $x + h$. The point c depends on h; but, all that we need to know about it is that as h tends to 0 the point c approaches x (because it is trapped between x and $x + h$). Since f' is continuous

$$\lim_{h \to 0} f'(c) = f'(x).$$

Therefore,

$$f(x + h) - f(x) - hf'(x) = h[f'(x) - f'(c)] = o(h).$$

Now, let's look at the two-variable case. The n-variable case is a simple extension of this one. We look at $f(X + H) - f(X)$ as a function of h_1 and then as a function of h_2. We assume throughout that H is small enough so that $X + H$ stays in a square centered at X and contained in U. Write

(8.21) $$\begin{aligned} f(X + H) - f(X) &= f(x_1 + h_1, x_2 + h_2) - f(x_1, x_2 + h_2) \\ &\quad + f(x_1, x_2 + h_2) - f(x_1, x_2). \end{aligned}$$

Fix h_2 and let

$$g_1(t) = f(t, x_2 + h_2).$$

Then g_1 is a function of class C^1 on the closed interval between x_1 and $x_1 + h_1$. Furthermore, $g_1'(t) = (D_1 f)(t, x_2 + h_2)$. Thus

(8.22) $$g_1(x_1 + h_1) - g_1(x_1) = h_1 (D_1 f)(c_1, x_2 + h_2)$$

where c_1 depends upon h_1 but lies between x_1 and $x_1 + h_1$. If we define

$$g_2(t) = f(x_1, t)$$

then we also have

(8.23) $$g_2(x_2 + h_2) - g_2(x_2) = h_2 (D_2 f)(x_1, c_2)$$

where c_2 is between x_2 and $x_2 + h_2$. Let

$$P_1 = (c_1, x_2 + h_2)$$
$$P_2 = (x_1, c_2).$$

Then (8.21), (8.22), and (8.23) yield this:

$$f(X + H) - f(X) = h_1(D_1 f)(P_1) + h_2(D_2 f)(P_2).$$

We want to know about the behavior of

$$f(X + H) - f(X) - \ell(H)$$

where

$$\ell(H) = h_1(D_1 f)(X) + h_2(D_2 f)(X).$$

We have

$$(8.24) \qquad f(X + H) - f(X) - \ell(H) = \sum_{j=1}^{2} h_j[(D_j f)(P_j) - (D_j f)(X)]$$

The points P_1 and P_2 depend upon H. What is important is that each of them has its ith coordinate between x_i and $x_i + h_i$ so that

$$\lim_{H \to 0} P_j = X.$$

Since each $D_j f$ is continuous, it should be reasonably apparent from (8.24) that

$$f(X + H) - f(X) - \ell(H) = o(H).$$

We have proved that f is differentiable at each point of U. We have also asserted in (ii) that the derivative Df is continuous. What does that mean? For each X, $(Df)(X)$ is a linear functional on R^n. Thus Df is a map

$$U \xrightarrow{\ Df\ } L(R^n, R^1).$$

What we have asserted is that $(Df)(X)$ depends continuously on X. The continuity is defined relative to some norm on the space of linear functionals $L(R^n, R^1)$. Since the space is finite-dimensional, it doesn't matter which norm we use, because all norms on $L(R^n, R^1)$ will define the same convergent sequences. For instance, we could use the "natural" norm

$$\| \ell \| = \sup_{|X| \leq 1} |\ell(X)|.$$

This turns out to be identical with

$$\| \ell \| = |A|$$

where A is the vector in R^n which determines ℓ:

$$\ell(H) = \langle H, A \rangle = a_1 h_1 + a_2 h_2 + \cdots + a_n h_n.$$

We know that the linear functional $\ell = (Df)(X)$ is determined by the vector $A = f'(X)$. Since the partial derivatives $D_1 f, \ldots, D_n f$ are continuous, it is clear that $f'(X)$ depends continuously on S.

There is a second statement in Theorem 2, that (i) follows from (ii). We leave that as an exercise.

Exercises

1. If ℓ is a linear functional on R^n, the **norm** of ℓ is

$$\|\ell\| = \sup_{X \neq 0} \frac{|\ell(X)|}{|X|}.$$

Prove that

  (a) $\|\ell\| < \infty$;

  (b) $\|\ell\|$ is the smallest non-negative number M such that

$$|\ell(X)| \leq M|X|, \qquad \text{for all } X;$$

  (c) $\|\ell\| = |A|$, where A is the unique vector in R^n such that

$$\ell(X) = \langle X, A \rangle, \qquad X \in R^n.$$

2. Prove that, if ℓ is a linear functional on R^n and

$$\ell(H) = o(H)$$

then $\ell = 0$.

3. What does it mean to say

$$f(H) = o(1)$$

for small H?

4. If f is differentiable at X, then f is continuous at X.

5. If $f(x, y) = xy^2 + x^3y^4$ and $\ell = (Df)(5, 6)$, what is $\ell(7, 8)$?

6. Let $\langle \cdot, \cdot \rangle$ be an inner product on R^n and define $N(X) = \langle X, X \rangle^{1/2}$. Prove that N is differentiable at every point except $X = 0$ and

$$(DN)(X) = \frac{1}{N(X)}\langle \cdot, X \rangle.$$

7. If f is a differentiable (real-valued) function on a connected open set $U \subset R^n$ and if $Df = 0$, then f is constant.

8. Let f be a differentiable function on the open set U. If f has either a local maximum or local minimum at the point $X \in U$, then $f'(X) = 0$.

8.4. Differentiable Maps

Now we consider a map

$$U \xrightarrow{F} R^k, \qquad U \subset R^n$$

that is, a vector-valued function on an open set U in R^n. What shall it mean for F to be differentiable at a point X? Here, we do not try to give a geometrical motivation. Differentiability will mean that it is possible to approximate F sufficiently well by a linear function from R^n into R^k.

Definition. *Let* F

$$U \xrightarrow{F} R^k, \qquad U \subset R^n$$

be a map defined on an open set U *in* R^n, *mapping* U *into* R^k. *Let* X *be a point in* U. *We say that* F *is* **differentiable at** X *if there exists a linear transformation* L

$$R^n \xrightarrow{L} R^k$$

such that

(8.25) $$F(X + H) = F(X) + L(H) + o(H).$$

If F *is differentiable at each point of* U, *we call* F *a* **differentiable map.**

This definition subsumes the case of a real-valued map. Another elementary remark is that differentiability implies continuity.

If F is differentiable at X, then the linear transformation L in (8.25) is unique; it is called the **derivative of** F **at** X and denoted by $L = (DF)(X)$; the standard representing matrix for $(DF)(X)$ is denoted by $F'(X)$ and is called the **Jacobian matrix of** F **at** X. Note that this is consistent with our previous notation f', for the gradient of a function.

Theorem 3. *Let* $F = (f_1, \ldots, f_k)$ *be a map from an open set* U *into* R^k, *and let* X *be a point of* U. *The map* F *is differentiable at* X *if and only if each* f_j *is differentiable at* X. *If* F *is differentiable at* X, *then*

$$(DF)(X) = ((Df_1)(X), \ldots, (Df_k)(X)).$$

Proof. The map F is given by a k-tuple of real-valued functions: $f_i(X)$ is the ith coordinate of $f(X)$. In other words,

$$Y = F(X)$$
$$y_1 = f_1(x_1, \ldots, x_n)$$
$$\begin{matrix} \cdot & & \cdot \\ \cdot & & \cdot \\ \cdot & & \cdot \end{matrix}$$
$$y_k = f_k(x_1, \ldots, x_n).$$

If each f_i is differentiable at X, then for each i we have

(8.26) $$f_i(X + H) - f_i(X) - \ell_i(H) = o(H)$$

where ℓ_i is a linear functional on R^n. The map $L = (\ell_1, \ldots, \ell_k)$ is then a linear transformation from R^n into R^k. We assert that F is differentiable at X with derivative L. Let

$$G(H) = F(X + H) - F(X) - L(H).$$

According to (8.26), for each i, the ith coordinate of $G(H)$ goes to 0 faster than $|H|$ does as H tends to 0. Therefore,

$$G(H) = o(H).$$

Conversely, suppose that F is differentiable at X with $(DF)(X) = L$. Now L is given by a k-tuple of real-valued functions $L = (\ell_1, \ldots, \ell_k)$. Since L is linear, each ℓ_i is linear. The norm of $f_i(X + H) - F(X) - \ell_i(H)$ does not exceed the norm of $F(X + H) - F(X) - L(H)$. Therefore, we have (8.26), which is what we want.

With Theorem 3, we can build many differentiable maps by combining k-tuples of differentiable functions. From the relation $L = (\ell_1, \ldots, \ell_k)$ which we have just been discussing, we see the following. The linear functional ℓ_i is determined by the gradient of f_i at X. Therefore, the Jacobian matrix of F is given by

$$F'(X) = \begin{bmatrix} \dfrac{\partial f_1}{\partial x_1} & \cdots & \dfrac{\partial f_1}{\partial x_n} \\ \cdot & & \cdot \\ \cdot & & \cdot \\ \cdot & & \cdot \\ \dfrac{\partial f_k}{\partial x_1} & \cdots & \dfrac{\partial f_k}{\partial x_n} \end{bmatrix}.$$

For that reason, other standard notations for the Jacobian matrix of the map $F = (f_1, \ldots, f_k)$ are:

$$(8.27) \qquad\qquad \frac{dF}{dX} \quad \text{or} \quad \frac{dY}{dX}$$

$$(8.28) \qquad F' = \frac{\partial(f_1, \ldots, f_k)}{\partial(x_1, \ldots, x_n)} \quad \text{or} \quad F' = \frac{\partial(y_1, \ldots, y_k)}{\partial(x_1, \ldots, x_n)}.$$

This notation is frequently handy.

Now, let's take a look at some examples of maps which arise naturally, without any reference to coordinates.

EXAMPLE 1. Let L be a linear transformation

$$R^n \xrightarrow{\ L\ } R^k.$$

Then L is a differentiable map and $(DL)(X) = L$ for every X, because

$$L(X + H) = L(X) + L(H).$$

The Jacobian matrix for L is (constantly equal to) the standard representing matrix for L.

EXAMPLE 2. Suppose that we identify R^2 with the field of complex numbers by identifying the point $X = (x_1, x_2)$ with the complex number $z = (x_1 + ix_2)$. Then we can define a map from R^2 into R^2 by $F(z) = z^2$. More precisely, $F(x_1, x_2)$ is the point in the plane which corresponds to the point $(x_1 + ix_2)^2$. That means that

$$F(x_1, x_2) = (x_1^2 - x_2^2, 2x_1 x_2).$$

or

$$Y = F(X)$$

where

$$y_1 = f_1(x_1, x_2) = x_1^2 - x_2^2$$
$$y_2 = f_2(x_1, x_2) = 2x_1x_2.$$

Since f_1 and f_2 are polynomials, clearly F is a differentiable map. The Jacobian matrix for F is

$$F'(X) = 2\begin{bmatrix} x_1 & -x_2 \\ x_2 & x_1 \end{bmatrix}$$

so that the derivative of F at X is the linear transformation

(8.29)
$$\begin{aligned} L(H) &= 2\begin{bmatrix} x_1 & -x_2 \\ x_2 & x_1 \end{bmatrix}\begin{bmatrix} h_1 \\ h_2 \end{bmatrix} \\ &= 2(x_1h_1 - x_2h_2, \, x_2h_1 + x_1h_2). \end{aligned}$$

Notice what (8.29) corresponds to when phrased in terms of complex multiplication. If X corresponds to the complex number z and H corresponds to the complex number w, then (8.29) can be rewritten

(8.30) $$L(w) = 2zw.$$

That is, the derivative of the map F at the point z is the linear transformation "complex multiplication by $2z$".

That is quite easy to see without reference to coordinates, matrices, etc. We have

$$\begin{aligned} F(z + w) &= (z + w)^2 \\ &= z^2 + 2zw + w^2 \\ &= F(z) + 2zw + w^2. \end{aligned}$$

With z fixed, $2zw$ is a linear function of w. Also $w^2 = o(w)$. Therefore, it is immediate from the definition of derivative that $(DF)(z) = L$, where L is defined by (8.30).

EXAMPLE 3. In Example 2, it is tempting to say, "Of course z^2 is differentiable and of course its derivative is $2z$ because the function $F(z) = z^2$ is complex-differentiable and its complex derivative is $2z$:

$$\lim_{w \to 0} \frac{F(z + w) - F(z)}{w} = 2z".$$

But that is not the same derivative as the one which we are discussing in this chapter. The function $F(z) = z^*$ is not complex-differentiable; however, it is differentiable as a map from R^2 into R^2 because it is a linear transformation on R^2. Remember that complex-differentiability is a very special property.

Still, the occurrence of $2z$ in (8.30) could hardly be an accident. Let's see why. Let U be an open set in the plane and let

$$U \xrightarrow{\;F\;} C.$$

Suppose that F is complex-differentiable. Let F' be its complex derivative. Then

$$F(z + w) = F(z) + F'(z)w + o(w)$$

because that is the definition of complex derivative. Now $F'(z)w$ is a linear function of w. Therefore, the map which F defines from R^2 to R^2 is differentiable and the derivative of F at z is "multiplication by the complex number $F'(z)$". Notice that the F' here is not the Jacobian matrix. We hope that this example clears up more confusion that it causes.

EXAMPLE 4. Fix a positive integer n. Let $R^{n \times n}$ be the space of $n \times n$ (real) matrices. Then $R^{n \times n}$ is isomorphic to R^{n^2}, so that we may discuss the differentiability of maps on $R^{n \times n}$. Here is an interesting one. Let U be the set of invertible $n \times n$ matrices. Recall that U is an open subset of $R^{n \times n}$. The proof of that runs as follows. Suppose A is an invertible $n \times n$ matrix. Let's show that $A + H$ is invertible for all matrices H of small norm. We rewrite

$$A + H = A(I + A^{-1}H).$$

Now $I + A^{-1}H$ will be invertible provided $|A^{-1}H|$ is small, because we can use the power series for $1/(1 + x)$. Choose $\delta > 0$ so that $\delta |A^{-1}| < 1$. Then

$$(8.31) \quad \begin{aligned} (A + H)^{-1} &= (I + A^{-1}H)^{-1}A^{-1} \\ &= A^{-1} - A^{-1}HA^{-1} + A^{-1}HA^{-1}HA^{-1} - \cdots, \quad |H| < \delta. \end{aligned}$$

Now let F be the inversion map on U:

$$F(A) = A^{-1}.$$

Is F differentiable? If there is any justice, F should be differentiable with derivative $-A^{-2}$. That can't be correct, because $(DF)(A)$ must be a linear transformation from $R^{n \times n}$ into $R^{n \times n}$. Aha! It must be the linear transformation "multiplication by $-A^{-2}$". Multiplication on the left or on the right? There is no reason to prefer one side to the other, so we'll put one A^{-1} on each side. We are led to guess that $(DF)(A)$ is the linear transformation on the space of matrices defined by

$$L(H) = -A^{-1}HA^{-1}.$$

We can change this rapidly into more than a guess. Look at (8.31). It tells us that

$$F(A + H) = F(A) - A^{-1}HA^{-1} + A^{-1}HA^{-1}HA^{-1} - \cdots$$

so that

$$\begin{aligned} |F(A + H) - F(A) - A^{-1}HA^{-1}| &\leq \sum_{k=2}^{\infty} |H|^k |A^{-1}|^{k+1} \\ &= |H|^2 \sum_{k=2}^{\infty} |H|^{k-2} |A^{-1}|^{k+1} \\ &= o(|H|). \end{aligned}$$

There are a few basic facts about combining differentiable maps which we will now take up. If F and G are differentiable maps into R^k and if they are both differentiable at X, then $F + G$ is differentiable at X and $D(F + G) = DF + DG$. That is easy to verify. Let's turn to the product of differentiable maps.

Theorem 4. *Let* U *be an open set in* R^n. *Let* F *be a map from* U *into* R^k *and let* f *be a real-valued function on* U. *If both* f *and* F *are differentiable at* X, *with respective derivatives* ℓ *and* L, *then the product* fF *is differentiable at* X *and its derivative* $M = (Df)(X)$ *is given by:*

$$M(H) = \ell(H)F(X) + f(X)L(H)$$

that is,

$$(fF')(X) = f'(X)F(X) + f(X)F'(X).$$

Proof. Let $\ell = (Df)(X)$, $L = (DF)(X)$, and

$$g(H) = f(X + H) - f(X) - \ell(H)$$
$$G(H) = F(X + H) - F(X) - L(H).$$

Now

$$
\begin{aligned}
(fF)(X + H) - (fF)(X) &= [f(X + H) - f(X)]F(X + H) \\
&\quad + f(X)[F(X + H) - F(X)] \\
&= \ell(H)F(X + H) + f(X)L(H) \\
&\quad + g(H)F(X + H) + f(X)G(H) \\
&= \ell(H)F(X) + f(X)L(H) \\
&\quad + \ell(H)[L(H) + G(H)] \\
&\quad + g(H)F(X + H) + f(X)G(H).
\end{aligned}
$$

Now, all we have to do is to show that the sum of the last three terms is $o(H)$. Since $G(H) = o(H)$, $g(H) = o(H)$ and since $F(X + H)$ tends to $F(X)$, each of the last two terms tends to zero faster than $|H|$. We can dispose of $\ell(H)G(H)$ in a similar way, leaving us with $\ell(H)L(H)$ to worry about. Now

$$|\ell(H)L(H)| \leq ||\ell|| \, ||L|| \, |H|^2$$

because ℓ and L are linear, and that shows that $\ell(H)L(H) = o(H)$.

We remark that Theorem 4 is valid for the product of two matrix-valued functions, with essentially the same proof. Here is a very fundamental way of combining differentiable maps.

Theorem 5 (Chain Rule). *Let* U *be an open set in* R^n *and let* V *be an open set in* R^k. *Let* F *and* G *be maps*

$$U \xrightarrow{\ F\ } R^k$$
$$V \xrightarrow{\ G\ } R^d$$

such that $F(U) \subset V$. *Let* X *be a point in* U. *Suppose that* F *is differentiable at* X *and that* G *is differentiable at* $F(X)$. *Then the composition* $G \circ F$ *is differentiable at* X *and*

(8.32) $$D(G \circ F)(X) = (DG)(F(X)) \circ (DF)(X)$$

Proof. Let $L = (DF)(X)$ and $M = (DG)(F(X))$. What the theorem asserts is that $G \circ F$ is differentiable at X and its derivative is the linear transformation $M \circ L$, from R^n into R^d:

$$R^n \xrightarrow{\ L\ } R^k \xrightarrow{\ M\ } R^d.$$

Let $Y = F(X)$. By the definition of derivative

$$F(X + H) = F(X) + L(H) + |H| R(H)$$
$$G(Y + K) = G(Y) + M(K) + |K| S(K)$$

where

$$\lim_{H \to 0} R(H) = 0$$
$$\lim_{K \to 0} S(K) = 0.$$

Now

$$
\begin{aligned}
(G \circ F)(X + H) - (G \circ F)(X) &= G(F(X + H)) - G(F(X)) \\
&= G(Y + L(H) + |H| R(H)) - G(Y) \\
&= G(Y + K) - G(Y) \\
&= M(K) + |K| S(K)
\end{aligned}
$$

where $K = L(H) + |H| R(H)$. Since M is linear,

$$
\begin{aligned}
M(K) &= M(L(H)) + M(|H| R(H)) \\
&= M(L(H)) + |H| M(R(H)).
\end{aligned}
$$

The result will be proved if we can show that

$$|H| M(R(H)) + |K| S(K) = o(H).$$

The left-hand side (above) is not larger than

$$|H| \|M\| |R(H)| + [|L(H)| + |H| |R(H)|] |S(K)|$$
$$\leq |H| [\|M\| |R(H)| + \|L\| |S(K)| + |R(H)| |S(K)|].$$

Now

$$\lim_{H \to 0} K = 0$$

so that

$$\lim_{H \to 0} S(K) = 0.$$

Since $R(H)$ tends to 0 as H does, the result follows.

In terms of the Jacobian matrices, Theorem 5 says

(8.33) $$(G \circ F)'(X) = G'(F(X)) F'(X).$$

When calculating, it is often most convenient to phrase the chain rule this way

$$Y = F(X)$$
$$Z = G(Y)$$
$$\frac{dZ}{dX} = \frac{dZ}{dY}\frac{dY}{dX}$$

or

$$\frac{\partial(z_1, \ldots, z_d)}{\partial(x_1, \ldots, x_n)} = \frac{\partial(z_1, \ldots, z_d)}{\partial(y_1, \ldots, y_k)}\frac{\partial(y_1, \ldots, y_k)}{\partial(x_1, \ldots, x_n)}.$$

In other words,

(8.34) $$\frac{\partial z_i}{\partial x_j} = \sum_{r=1}^{k} \frac{\partial z_i}{\partial y_r}\frac{\partial y_r}{\partial x_j}.$$

It is important to remember exactly what (8.34) means. It means this. Suppose we are given $f_1, \ldots, f_k$—differentiable functions on an open set in R^n. They define a map F, variously described as

$$Y = F(X)$$

or

$$y_1 = f_1(x_1, \ldots, x_n)$$
(8.35)
$$y_k = f_k(x_1, \ldots, x_n).$$

Then suppose that we have $g_1, \ldots, g_d$—differentiable functions on (an open set containing) the range of F. They define a map G, variously described as

$$Z = G(Y)$$

or

$$z_1 = g_1(y_1, \ldots, y_k)$$
(8.36)
$$z_d = g_d(y_1, \ldots, y_k).$$

Now we "express the z's as functions of the x's" by substituting $f_r(x_1, \ldots, x_n)$ for y_r in (8.36). That describes the composed map $H = G \circ F$

$$Z = H(X)$$
$$z_1 = h_1(x_1, \ldots, x_n)$$
(8.37)
$$z_d = h_d(x_1, \ldots, x_n).$$

Remember that

$$h_i(x_1, \ldots, x_n) = g_1(f_1(x_1, \ldots, x_n), \ldots, f_k(x_1, \ldots, x_n)).$$

The meaning of (8.34) and the content of the chain rule is that

(8.38) $$\frac{\partial h_i}{\partial x_j}(X) = \sum_{r=1}^{k} \frac{\partial g_i}{\partial y_r}(F(X)) \frac{\partial f_r}{\partial x_j}(X).$$

Now obviously, in practice it is cumbersome to keep writing x_j, f_j, y_r, g_r, z_i, and h_i if you're trying to calculate something. Either (8.38) or (8.32) should be used as the chain rule when you're trying to understand carefully what's going on. But in calculations it is usually simpler to say that the map F is specified by giving the y's as certain functions of the x's:

$$y_r = y_r(x_1, \ldots, x_n)$$

and that the map G is specified by giving the z's as certain functions of the y's:

$$z_i = z_i(y_1, \ldots, y_k).$$

Under those constraints, the rate of change of z_i with respect to x_j is

$$\frac{\partial z_i}{\partial x_j} = \sum_{r=1}^{k} \frac{\partial z_i}{\partial y_r} \frac{\partial y_r}{\partial x_j}.$$

All you have to remember is what's a function of which and where to evaluate the derivatives.

Theorem 6 (Mean Value Theorem). *Let* g *be a differentiable function on* U, *a convex open set in* R^n. *Let* A *and* B *be points in* U. *There exists a point* P, *on the line segment from* A *to* B, *such that*

$$g(B) - g(A) = \langle B - A, g'(P) \rangle.$$

Proof. Let

$$F(t) = tB + (1 - t)A, \qquad 0 \leq t \leq 1.$$

The image of F is the line segment from A to B. Let $f = g \circ F$, i.e.,

$$f(t) = g(tB + (1 - t)A).$$

Then f is continuous on the closed interval $[0, 1]$ and differentiable (at least) on the open interval $(0, 1)$. Furthermore,

$$f'(t) = \sum_{j=1}^{n} \frac{\partial g}{\partial x_j} \frac{\partial x_j}{\partial t}$$

or, more precisely,

$$f'(t) = \sum_{j=1}^{n} \frac{\partial g}{\partial x_j} \frac{\partial f_j}{\partial t}$$

$$= \sum_{j=1}^{n} (b_j - a_j) \frac{\partial g}{\partial x_j}(F(t))$$

$$= \langle B - A, f'(F(t)) \rangle.$$

The mean value theorem in one variable tells us that there is a number c, $0 < c < 1$, such that $f'(c) = f(1) - f(0)$. Let $P = F(c)$.

Exercises

1. Repeat Exercises 1 and 2 of Section 8.3 for linear transformations rather than linear functionals. (Of course, the vector A in Exercise 1 will be replaced by the standard representing matrix A.) Prove the uniqueness of the derivative of a differentiable map.

2. A differentiable map is continuous.

3. Let U be an open set in R^n and let F and G be differentiable mappings from U into R^k. Define the real function $\langle F, G \rangle$ by $\langle F, G \rangle(X) = \langle F(X), G(X) \rangle$. Prove that $\langle F, G \rangle$ is differentiable on U and that (in a suitable sense)

$$(D\langle F, G \rangle)(X) = \langle (DF)(X), G(X) \rangle + \langle F(X), (DG)(X) \rangle.$$

4. State and prove an analogue of Theorem 4 for matrix-valued functions.

5. Fix a positive integer k, and let

$$R^{n \times n} \xrightarrow{F} R^{n \times n}$$

be the map on $n \times n$ matrices defined by $F(A) = A^k$. Show that F is differentiable at every point of $R^{n \times n}$, and find $(DF)(A)$.

6. Use Exercise 4 and the chain rule to find the derivative of the map $G(A) = A^{-k}$ defined on the set of invertible $n \times n$ matrices.

7. Let U be an open set in the plane and let f be a complex-valued function on U. Suppose that f is complex-differentiable. Let F be the map from U into R^2 which f defines

$$F = (u, v)$$
$$f = u + iv.$$

(We distinguish between f and F so as not to confuse the complex derivative f' with the gradient F'.) Show that if $L = (DF)(x, y)$, then (with proper interpretations),

$$L(a, b) = f'(x + iy)(a + ib).$$

8. Let F be a continuously differentiable map on a neighborhood of the origin in R^2, $F = (u, v)$. Let f be the associated complex-valued function $f = u + iv$. Show that f cannot be a square root of the identity function z, i.e., that

$$[f(x, y)]^2 = x + iy$$

cannot hold on a neighborhood of the origin.

9. If F is a differentiable map from an open set U into R^k, then the derivative is a map

$$U \xrightarrow{DF} L(R^n, R^k).$$

The space $L(R^n, R^k)$ is finite-dimensional—it is essentially R^{nk}. So, doesn't it make sense to talk about DF being differentiable? What kind of an object would $(D^2F)(X)$ be? By D^2F we mean $D(DF)$. What does D^2F have to do with second order partial derivatives? What is a map of class C^∞?

10. The mean value theorem gives the first step in Taylor's theorem in n variables:

$$f(X + H) = f(X) + \sum_{j=1}^{n} h_j \frac{\partial f}{\partial x_j}(P).$$

Look at $g(t) = f(X + tH)$, $0 \le t \le 1$, and use it to show that (if f is of class C^2)

$$f(X + H) = f(X) + \sum_{j=1}^{n} h_j(D_j f)(X) + \tfrac{1}{2}\Big(\Big[\sum_{j=1}^{n} h_j D_j\Big]^2 f\Big)(P)$$

where P is on the line segment from X to $X + H$: Generalize to functions of class C^{k+1}. Compare with the integral form of the remainder (Theorem 10 of Chapter 5).

8.5. Inverse Mappings

Suppose that we have a differentiable map

$$U \overset{F}{\longrightarrow} R^k, \qquad U \subset R^n$$

from an open subset of R^n into R^k. We would like to know when F has an inverse map

(8.39) $$R^n \overset{F^{-1}}{\longleftarrow} F(U)$$

which is differentiable. The inverse function (8.39) exists if and only if F is $1:1$. If F is $1:1$, then, in order to discuss the differentiability of F^{-1}, we will need to know that $F(U)$ is an open set. Suppose for the moment that F is $1:1$, $F(U)$ is open and F^{-1} is differentiable. The chain rule then tells us that (at each point)

$$(DF^{-1}) \circ (DF) = I_n$$
$$(DF) \circ (DF^{-1}) = I_k$$

where I_j is the identity transformation on R^j. Therefore, for any X in U, the Jacobian matrix $F'(X)$ has both a right and a left inverse. So, it must be a square matrix ($k = n$).

We conclude two things from that brief discussion. First, we need concern ourselves only with the case of a differentiable map

$$U \overset{F}{\longrightarrow} R^n, \qquad U \subset R^n.$$

Second, if F is to have a differentiable inverse, then it is necessary that the Jacobian matrix $F'(X)$ be invertible for each X in U. Now, we are going to prove that if the continuously differentiable map F is $1:1$ and $F'(X)$ is invertible at each point, then $F(U)$ is an open set and F^{-1} is a differentiable map. This should be known to the reader in the special case of a linear transformation: A linear transformation L which is invertible maps R^n onto R^n and the inverse L^{-1} is a linear transformation. We shall handle more general maps F through approximation of them by linear transformations.

Lemma. *Let* F *be a continuously differentiable map*

$$U \xrightarrow{\ F\ } R^n$$

where U *is an open set in* R^n. *Let* X *be a point of* U *such that* $(DF)(X)$ *is invertible, i.e., such that* $F'(X)$ *is a non-singular matrix. Then* X *has a neighborhood* W *for which there exist constants* $m > 0$ *and* $M > 0$ *such that*

$$m|B - A| \le |F(B) - F(A)| \le M|B - A|, \qquad A, B \in W.$$

Proof. It is the left-hand inequality which interests us. Let $L = (DF)(X)$, so that L is an invertible linear transformation on R^n, i.e., $\det F'(X) \ne 0$. To get an idea of how the proof of the lemma will proceed, let's observe that L has the property asserted in the lemma. We know that

$$|L(B) - L(A)| \le \|L\|\, |B - A|$$

for all A, B in R^n. The inverse transformation L^{-1} is also linear; hence,

$$\begin{aligned}
|B - A| &= |L^{-1}(L(B)) - L^{-1}(L(A))| \\
&= |L^{-1}(L(B) - L(A))| \\
&\le \|L^{-1}\|\, |L(B) - L(A)|.
\end{aligned}$$

Thus,

$$(8.40) \qquad \|L^{-1}\|^{-1}|B - A| \le |L(B) - L(A)| \le \|L\|\,|B - A|$$

for all A, B in R^n.

Let $\epsilon = \frac{1}{2}\|L^{-1}\|^{-1}$. We know that $F = (f_1, \ldots, f_n)$ where each f_i is a function of class C^1 on U, that is, f_i' is a continuous function on U. Choose a convex neighborhood W of X so that

$$(8.41) \qquad |f_i'(P) - f_i'(X)| < \frac{\epsilon}{n}, \qquad i = 1, \ldots, n;\ P \in W.$$

Let $A, B \in W$. Apply the mean value theorem to each f_i:

$$(8.42) \qquad f_i(B) - f_i(A) = \langle B - A, f_i'(P_i)\rangle$$

where P_i is a point on the line segment from A to B. Combine (8.41) and (8.42) and one has

$$(8.43) \qquad |f_i(B) - f_i(A) - \langle B - A, f_i'(X)\rangle| < \frac{\epsilon}{n}|B - A|.$$

The number

$$\langle f_i(B) - f_i(A)\rangle - \langle B - A, f_i'(X)\rangle$$

is the ith coordinate of the vector

$$F(B) - F(A) - L(B - A).$$

The norm of a vector does not exceed the sum of the absolute values of its coordinates. Therefore, (8.43) tells us that

$$(8.44) \qquad |F(B) - F(A) - L(B - A)| < \epsilon|B - A|, \qquad A, B \in W.$$

Thus

$$|F(B) - F(A)| \geq |L(B - A)| - \epsilon |B - A|$$
$$\geq \|L^{-1}\|^{-1} |B - A| - \epsilon |B - A|$$
$$= \tfrac{1}{2}\|L^{-1}\|^{-1} |B - A|, \qquad A, B \in W.$$

It should be clear that we can also arrange that

$$|F(B) - F(A)| \leq M |B - A|$$

by a similar argument; however, as we said, that is not of primary concern to us.

Theorem 7 (*Inverse Mapping Theorem*). *Let* F *be a continuously differentiable map from an open subset of* R^n *into* R^n. *Let* X *be a point of the domain of* F *such that the derivative* (DF)(X) *is invertible, i.e., such that* det $F'(X) \neq 0$. *Then* X *has an open neighborhood* W *such that*

(a) F *is* 1 : 1 *on* W;

(b) F(W) *is an open set;*

(c) *the inverse map*

$$W \xleftarrow{\ F^{-1}\ } F(W)$$

is continuously differentiable and its derivative is

$$(DF^{-1})(F(A)) = (DF)(A)^{-1}.$$

Proof. Let N be an open ball centered at X such that the closed ball $\bar{N}$ lies in the domain of F and on which we have an inequality

(8.45) $m|B - A| \leq |F(B) - F(A)|, \qquad A, B \in N.$

The last lemma guarantees the existence of such a ball N and constant $m > 0$. Since F is continuous, the inequality will hold on the closed ball $\bar{N}$ as well. That shows us that F is 1 : 1 on $\bar{N}$ because if $F(A) = F(B)$ with $A, B \in \bar{N}$, then (8.45) implies that $B = A$. Therefore, we could take $W = N$ and satisfy condition (a) of the theorem.

We will have to shrink N a bit in order to get a neighborhood which satisfies condition (b) as well as condition (a). First, let us shrink N a little so that $(DF)(A)$ is non-singular at each point of N. We can do that because F is of class C^1 so that det $F'(A)$ is a continuous function (which does not vanish at $A = X$). Let $Y = F(X)$. We want to show that Y is in the interior of $F(N)$, that is, we want to find a neighborhood of Y which lies entirely in the range of F. To that end, let S be the boundary of the ball N:

$$N = \{A; |A - X| < \delta\}$$
$$S = \{A; |A - X| = \delta\}.$$

The set S is compact and F is continuous; hence $F(S)$ is a compact subset of R^n. The point Y is not in $F(S)$ because $Y = F(X)$, F is 1 : 1 on $\bar{N}$ and $X \notin S$. The situation is represented by Figure 30. Let r be the distance

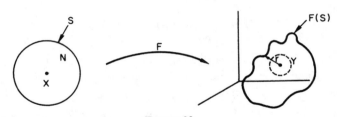

Figure 30

from Y to the compact set $F(S)$:

$$r = d(Y, F(S))$$
$$= \inf_{A \in S} |Y - F(A)|.$$

We shall show that $F(N)$ contains the open ball of radius $r/2$ about the point Y. Let Z be a point of R^n with

$$|Z - Y| < \frac{r}{2}.$$

Let's look at the distance from Z to the set $F(\bar{N})$

$$d(Z; F(\bar{N})) = \inf \{|Z - F(A)|; A \in \bar{N}\}.$$

Since $|Z - Y| < r/2$, we know two things:

(8.46)
$$d(Z, F(\bar{N})) < \frac{r}{2}$$
$$d(Z, F(S)) > \frac{r}{2}.$$

The function $g(A) = |Z - F(A)|^2$ is continuous on $\bar{N}$ and so there exists a point $B \in \bar{N}$ such that

$$|Z - F(B)|^2 = d(Z, F(\bar{N}))^2.$$

Since F is $1:1$ on $\bar{N}$, (8.46) tells us that $B \notin S$, that is, B is in the open ball N. Thus, the function

$$g(A) = \sum_{i=1}^{n} (z_i - f_i(A))^2$$

has a minimum on N at the point B. That implies that the partial derivatives of g all vanish at B

(8.47) $$0 = \frac{\partial g}{\partial x_j}(B) = -2 \sum_{i=1}^{n} (z_i - f_i(B)) \frac{\partial f_i}{\partial x_j}(B).$$

The sum on the right side of (8.47) is the jth coordinate of the vector (matrix)

$$[Z - f(B)]F'(B).$$

We chose N so that $F'(A)$ was invertible for every A in N. Thus the simultaneous equations (8.47) imply that $Z - F(B) = 0$. We have proved that $B \in F(N)$.

Let $W = \{A \in N; |Y - F(A)| < r/2\}$. Then W is an open neighborhood of X which satisfies conditions (a) and (b) in the conclusion of the theorem. Condition (c) can be verified for our set W as follows. Since F is $1:1$ on W and $F(W)$ is open, it makes sense to ask whether F^{-1}

$$W \xleftarrow{F^{-1}} F(W)$$

is differentiable. We do know right away that F^{-1} is continuous, since F is continuous and $1:1$ on $\bar{W}$ and $\bar{W}$ is compact. The continuity of F^{-1} is also apparent from the inequalities in the last lemma. Let's show that F^{-1} is differentiable at the point $Y = F(X)$ and that

$$(DF^{-1})(Y) = L^{-1}$$

where $L = (DF)(X)$. The proof for any other point of W is the same. For points Z in $F(W)$, define

(8.48) $R(Z) = F^{-1}(Z) - F^{-1}(Y) - L^{-1}(Z - Y)$.

We must show that

(8.49) $$\lim_{Z \to Y} \frac{R(Z)}{|Z - Y|} = 0.$$

Apply the linear transformation L to both sides of (8.48). We obtain

(8.50) $L(R(Z)) = L(A - X) - [F(A) - F(X)]$

where A is the unique point in W such that $F(A) = Z$. Since F is differentiable at X and $(DF)(X) = L$, (8.50) guarantees that

(8.51) $$\lim_{A \to X} \frac{L(R(Z))}{|A - X|} = 0.$$

Since

$$|L(R(Z))| \geq \|L^{-1}\|^{-1} |R(Z)|$$
$$|A - X| \leq 2\|L^{-1}\| |F(A) - F(X)|$$

(8.51) implies (8.49).

Definition. *Let* U *be an open subset of* R^n *and let* F *be a differentiable map from* U *into* R^n. *The* **Jacobian** *of* F *is the function* det F', *from* U *into* $R^{n \times n}$.

Theorem 8 (Inverse Mapping Theorem). *Let* U *be an open subset of* R^n *and let* F *be a continuously differentiable map from* U *into* R^n *such that the Jacobian of* F *is nowhere* 0 *on* U. *Then* F *is an open mapping and* F *is locally* $1:1$. *If* F *is* $1:1$ *on* U, *then the inverse map*

$$U \xleftarrow{F^{-1}} F(U)$$

is continuously differentiable and $(F^{-1})'(F(X)) = [F'(X)]^{-1}$.

Proof. Let V be an open subset U. Is $F(V)$ an open set? Yes, because (according to Theorem 7) each point $X \in V$ has a neighborhood W which F maps onto an open set, and we may take W small enough so that $W \subset V$. The neighborhood W also can be chosen so that F is $1 : 1$ on W—that is what we mean by saying that F is locally $1 : 1$. The statement about F^{-1} is immediate from Theorem 7.

EXAMPLE 5. **Polar coordinates** in the plane arise from a differentiable map. Let

$$U = \{(r, \theta) \in R^2; r > 0\}$$

and define

$$U \xrightarrow{\ F\ } R^2$$

by

$$F(r, \theta) \doteq (x, y)$$

where

$$x = r \cos \theta$$
$$y = r \sin \theta.$$

Obviously F is a map of class C^∞. The Jacobian matrix of F is

$$F'(r, \theta) = \begin{bmatrix} \dfrac{\partial x}{\partial r} & \dfrac{\partial x}{\partial \theta} \\ \dfrac{\partial y}{\partial r} & \dfrac{\partial y}{\partial \theta} \end{bmatrix}$$

$$= \begin{bmatrix} \cos \theta & -r \sin \theta \\ \sin \theta & r \cos \theta \end{bmatrix}.$$

The Jacobian of F is

$$\det F'(r, \theta) = r(\cos^2 \theta + \sin^2 \theta)$$
$$= r.$$

Thus, $\det F'$ is nowhere 0 on U so that F is locally $1 : 1$ and an open mapping. The range $F(U)$ is the punctured plane $R^2 - \{0\}$.

Notice that F is $1 : 1$ on any slice

$$a < \theta < b$$

for which $b - a < 2\pi$. The image under F of that slice is an open sector in the plane. (See Figure 31.) On such a sector, we may use r and θ as "coordinates," meaning that for each point (x, y) in the sector there is one and only one number pair (r, θ) such that $r > 0$, $a < \theta < b$, $x = r \cos \theta$, $y = r \sin \theta$. On that sector, r and θ may be expressed as "smooth" functions of x and y.

Without the explicit formulas

$$r = (x^2 + y^2)^{1/2}$$

$$\theta = \tan^{-1} \frac{y}{x} \quad \text{(suitably interpreted)}$$

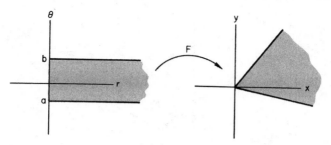

FIGURE 31

we can see something about derivatives. We know that

$$(DF^{-1})(x, y) = [(DF)(r, \theta)]^{-1}$$

and thus

$$(DF^{-1})(x, y) = \begin{bmatrix} \cos \theta & -r \sin \theta \\ \sin \theta & r \cos \theta \end{bmatrix}^{-1}$$

$$= \frac{1}{r} \begin{bmatrix} r \cos \theta & r \sin \theta \\ -\sin \theta & \cos \theta \end{bmatrix}.$$

Therefore,

$$\frac{\partial r}{\partial x} = \cos \theta, \qquad \frac{\partial r}{\partial y} = \sin \theta$$

$$\frac{\partial \theta}{\partial x} = -\frac{1}{r} \sin \theta, \qquad \frac{\partial \theta}{\partial y} = \frac{1}{r} \cos \theta.$$

Are you sure you understand what these equations mean?

EXAMPLE 6. The **spherical coordinates** for a point (x, y, z) in R^3 are are the numbers r, θ, ϕ defined by

$$x = r \cos \theta \sin \phi$$
$$y = r \sin \theta \sin \phi$$
$$z = r \cos \phi.$$

If $(x, y, z) \neq (0, 0, 0)$, there is a unique such triple with $r > 0$, $0 \leq \theta < 2\pi$, $0 \leq \phi \leq \pi/2$ (Figure 32). The map $F(r, \theta, \phi) = (x, y, z)$ behaves a great deal like the polar coordinate map in the last example. The Jacobian of F is $-r^2 \sin \phi$, which underscores the fact that spherical coordinates do not serve well along the z-axis where $\phi = 0$.

EXAMPLE 7. Suppose that we have any continuously differentiable map

$$U \xrightarrow{F} R^n$$

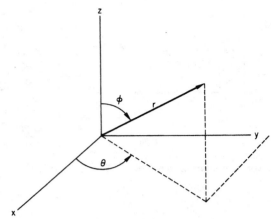

FIGURE 32

which is 1:1 and has a non-vanishing Jacobian:

$$\det F'(X) \neq 0, \qquad X \in U.$$

Then F defines a new "coordinate system" on U. With each point $X = (x_1, \ldots, x_n)$ in U is associated a unique n-tuple $Y = F(X) = (y_1, \ldots, y_n)$, where $y_i = f_i(x_1, \ldots, x_n)$. In other words, if it is convenient we can describe each point in U by giving its y coordinates. As X ranges over U, the possible n-tuples of y coordinates $(y_1, \ldots, y_n)$ range over the open set $V = F(U)$. The inverse map

$$U \overset{F^{-1}}{\longleftarrow} V$$

simply expresses the x coordinates as functions of the y coordinates: $x_i = g_i(y_1, \ldots, y_n)$. As in the previous example, it is quite common to introduce new coordinates on U by starting with the map F^{-1}, i.e., by starting with $x_i = g_i(y_1, \ldots, y_n)$.

The uses of different coordinate systems are not restricted to special ones of the type given in the two preceding examples. The proofs of certain general theorems are facilitated by the use of coordinate systems suited to the context. Here is a very simple illustration.

Suppose that g is a function of class C^1 on a neighborhood of the origin in R^3 with $g(0) = 0$. Roughly speaking, the zero set of g should look like a surface which passes through the origin. That is a valid idea if we do not allow certain degenerate cases. Suppose that $(Dg)(0) \neq 0$, that is, suppose that the gradient of g does not vanish at the origin. Then, one can show that the set $S = \{X; g(X) = 0\}$ is a genuine surface near the origin. There are other functions f which define the same surface. For example, let h be any function of class C^1 such that $h(0) \neq 0$. Let $f = gh$. Then, in a small neighborhood of the origin, f and g have the same zero set. Now ask this. Suppose we are given f, of class C^1 on a neighborhood of the origin, and suppose that f vanishes on the part of S which lies in that neighborhood. Is f divisible by g; that is, can we write $f = gh$ where h is a smooth function near the origin?

What does the inverse mapping theorem tell us about that question? It tells us that it suffices to worry about it in the case in which g is one of the coordinate functions x_i. Why? Since $(Dg)(0) \neq 0$, one of the partial derivatives of g is different from 0 at the origin. For convenience, suppose that $(D_1g)(0) \neq 0$. Then we may use g, x_2, and x_3 as coordinates for R^3 near the origin. The map

$$F(X) = (g(X), x_2, x_3)$$

is of class C^1 and has Jacobian matrix

$$F'(X) = \begin{bmatrix} \dfrac{\partial g}{\partial x_1} & \dfrac{\partial g}{\partial x_2} & \dfrac{\partial g}{\partial x_3} \\ 0 & 1 & 0 \\ 0 & 0 & 1 \end{bmatrix}.$$

Thus $\det F' = D_1g$ and so $\det F'(0) \neq 0$. Choose U, a neighborhood of the origin which F maps $1:1$ onto an open neighborhood of the origin. If f is of class C^1 on U and f vanishes on $S \cap U = \{X \in U; g(X) = 0\}$, then $f \circ F^{-1}$ is of class C^1 on V and vanishes on $F(S) = \{Y \in V; y_1 = 0\}$. It should be clear that proving that g divides f is the same as proving that y_1 divides $f \circ F^{-1}$. In short, we need consider only the problem: f is of class C^1 near the origin and f vanishes on the plane $x_1 = 0$; show that $f = x_1h$ where h is (at least) continuous. That is easy to show.

In the last example, we mentioned a surface S in R^3 which was defined by an equation $g(x_1, x_2, x_3) = 0$. Near the origin where $(D_1g)(0) \neq 0$, we were able to use g, x_2, and x_3 as coordinates. In particular, we could express x_1 as a function of g, x_2 and x_3 near the origin: $x_1 = p(y_1, x_2, x_3) = p(g(x_1, x_2, x_3), x_2, x_3)$. The part of the surface S which passes through the origin can then be described by $x_1 = p(0, x_2, x_3)$. In short, the relation $g(x_1, x_2, x_3) = 0$ implicitly defines x_1 as a function of x_2 and x_3; and that dependence can be made explicit near a point where $(D_1g)(X) \neq 0$. This is a special case of the following basic result.

Theorem 9 (Implicit Function Theorem). *Let* W *be an open subset of* R^{k+m}, *let* F *be a continuously differentiable map from* W *into* R^m *and let* (A; B) *be a point of* W *such that* F(A; B) = 0. *Suppose that the* m × m *matrix*

(8.52) $$M = [(D_{k+j}f_i)(A; B)]$$

is non-singular. There exists an open neighborhood U *of the point* A *and*

an open neighborhood V *of* B *such that if* $X \in U$, *there exists a unique
point* $G(X) \in V$ *such that* $F(X; G(X)) = 0$. *The map* G *which is thereby
defined is continuously differentiable.*

   Proof. Define a map

$$W \overset{\tilde{F}}{\longrightarrow} R^{k+m}$$

by $\tilde{F}(X; Y) = (X; F(X, Y); F(X; Y))$. Clearly $\tilde{F}$ is continuously differentiable on W. The Jacobian matrix of $\tilde{F}$ has the block form

$$\tilde{F}' = \begin{bmatrix} I & 0 \\ [D_j f_i] & [D_{k+j} f_i] \end{bmatrix}.$$

In particular,

$$\det \tilde{F}'(A; B) = \det M$$

where M is the matrix of (8.52). By hypothesis, $\det M \neq 0$. We apply the
inverse mapping theorem to the map $\tilde{F}$ at the point $(A; B)$. We obtain an
open neighborhood of $(A; B)$ on which $\tilde{F}$ is 1 : 1 with continuously differentiable inverse. We may assume that the neighborhood is of the form
$U \times V$, where U is a neighborhood of A and V is a neighborhood of B.

   Let's look at the inverse map for $\tilde{F}$ which is defined on $\tilde{F}(U \times V)$,
an open neighborhood of $\tilde{F}(A; B) = (A; 0)$. That map expresses $(X; Y)$
as a function of $(X; F(X, Y))$. Let $X \in U$. The point $\tilde{F}^{-1}(X; 0)$ has the
form $(X; Y)$ for some $Y \in V$. Define $G(X) = Y$.

   The implicit function theorem can be formulated as a result about
scalar-valued functions, and it is worthwhile to see what the result says
in that form. We are given a point $A = (a_1, \ldots, a_k)$ in R^k and a point
$B = (b_1, \ldots, b_m)$ in R^m. Suppose that we have m functions $f_1, \ldots, f_m$
of class C^1 on a neighborhood of the point $(A; B)$ in R^{k+m}. We are interested in the set of points $(X; Y)$ in R^{k+m} which satisfy simultaneously the
m relations

$$f_1(x_1, \ldots, x_k, y_1, \ldots, y_m) = 0$$

(8.53)

$$f_m(x_1, \ldots, x_k, y_1, \ldots, y_m) = 0.$$

Suppose that $(A; B)$ is such a point. The implicit function theorem states
the following. If

$$\left| \frac{\partial(f_1, \ldots, f_m)}{\partial(y_1, \ldots, y_m)} \right| (A; B) \neq 0$$

(vertical bars denote determinant), then there is an open set U about A
and an open set V about B such that, for each $X \in U$ there is precisely
one point $Y \in V$ for which the m relations (8.53) hold—and Y depends
upon X in a continuously differentiable way. In other words, there are
continuously differentiable functions $g_1, \ldots, g_m$ on U such that on the

set $U \times V$ the equations (8.53) are equivalent to

$$y_1 = g_1(x_1, \ldots, x_k)$$

(8.54)

$$\cdot \qquad \cdot$$
$$\cdot \qquad \cdot$$
$$\cdot \qquad \cdot$$

$$y_m = g_m(x_1, \ldots, x_k).$$

From the relations

(8.55) $\qquad f_i(x_1, \ldots, x_k, g_1, \ldots, g_m) = 0, \qquad 1 \leq i \leq m$

which hold identically on U we can compute the partial derivatives of $g_1, \ldots, g_m$. Apply the chain rule, to differentiate with respect to x_j:

(8.56) $\qquad \dfrac{\partial f_i}{\partial x_j}(X; G(X)) + \displaystyle\sum_{r=1}^{m} \dfrac{\partial f_i}{\partial y_r}(X; G(X)) \dfrac{\partial g_r}{\partial x_j}(X) = 0, \qquad 1 \leq i \leq m.$

This is a system of linear equations for the derivatives

$$\frac{\partial g_1}{\partial x_j}, \ldots, \frac{\partial g_m}{\partial x_j}.$$

The solution can be obtained from Cramer's rule because the coefficient matrix

$$\begin{bmatrix} \dfrac{\partial f_1}{\partial y_1} & \cdots & \dfrac{\partial f_m}{\partial y_m} \\ \cdot & & \cdot \\ \cdot & & \cdot \\ \cdot & & \cdot \\ \dfrac{\partial f_m}{\partial y_1} & \cdots & \dfrac{\partial f_m}{\partial y_m} \end{bmatrix}$$

is non-singular at $(X; G(X))$—its determinant is

$$\left| \frac{\partial(f_1, \ldots, f_m)}{\partial(y_1, \ldots, y_m)} \right| (X; G(X)).$$

The solution by Cramer's rule is

$$\frac{\partial g_r}{\partial x_j} = - \frac{\left| \dfrac{\partial(f_1, \ldots, f_m)}{\partial(y_1, \ldots, x_j, \ldots, y_m)} \right|}{\left| \dfrac{\partial(f_1, \ldots, f_m)}{\partial(y_1, \ldots, y_m)} \right|}.$$

The Jacobians on the right are evaluated at $(X; G(X))$, and $\partial(y_1, \ldots, x_j, \ldots, y_m)$ means replace y_r by x_j.

Of course, all that the last equations say is this. If $F = (f_1, \ldots, f_m)$ and $G = (g_1, \ldots, g_m)$, then

$$F(X; G(X)) = 0.$$

This says that $F \circ \tilde{G} = 0$, where

$$U \xrightarrow{\ \tilde{G}\ } U \times V$$
$$\tilde{G}(X) = (X; G(X)).$$

Thus

$$F'(X; G(X))\tilde{G}'(X) = 0.$$

Now the $(k + m) \times k$ Jacobian matrix $\tilde{G}'(X)$ has the block form

$$\tilde{G}'(X) = \begin{bmatrix} I \\ G'(X) \end{bmatrix}$$

where I is the $k \times k$ identity matrix. If we employ the shorthand

$$\frac{dF}{dX} = \begin{bmatrix} \dfrac{\partial f_1}{\partial x_1} & \cdots & \dfrac{\partial f_m}{\partial x_k} \\ \cdot & & \cdot \\ \cdot & & \cdot \\ \cdot & & \cdot \\ \dfrac{\partial f_m}{\partial x_1} & \cdots & \dfrac{\partial f_m}{\partial x_k} \end{bmatrix}$$

$$\frac{dF}{dY} = \begin{bmatrix} \dfrac{\partial f_1}{\partial y_1} & \cdots & \dfrac{\partial f_m}{\partial y_m} \\ \cdot & & \cdot \\ \cdot & & \cdot \\ \cdot & & \cdot \\ \dfrac{\partial f_m}{\partial y_1} & \cdots & \dfrac{\partial f_m}{\partial y_m} \end{bmatrix}$$

then the equation $F'(X; G(X))\tilde{G}'(X) = 0$ becomes

$$\frac{dF}{dX}(X; G(X)) + \frac{dF}{dY}(X; G(X))G'(X) = 0.$$

Thus

$$G'(X) = -\left(\frac{dF}{dY}\right)^{-1} \frac{dF}{dX}.$$

Sloppily speaking, along the set where $F(X, Y) = 0$, one has

$$\frac{\partial Y}{\partial X} = \frac{-\dfrac{dF}{dX}}{\dfrac{dF}{dY}} \qquad \text{(remember this is nonsense).}$$

The situations in which one uses the implicit function theorem do not usually come packaged quite as neatly as what we have been describing. So, let us state the theorem another way. Suppose that P is a point in R^n and that F is a continuously differentiable map from a neighborhood of P into R^m, where $m \leq n$. The Jacobian matrix $F'(P)$ is an $m \times n$ matrix. Suppose that the rank of $F'(P)$ is as large as it can be: rank $F'(P) = m$. Then there are some m indices $i_1 < \cdots < i_m$ such that

$$\left| \frac{\partial(f_1, \ldots, f_m)}{\partial(x_{i_1}, \ldots, x_{i_m})} \right|(P) \neq 0.$$

The implicit function theorem says the following. Look at S, the level set of F through P

$$S = \{X; F(X) = F(P)\}.$$

Then that part of S which is near P can be described by expressing the coordinates $x_{i_1}, \ldots, x_{i_m}$ as certain continuously differentiable functions of the remaining $n - m$ coordinates.

Exercises

1. Let $f(x) = x^2$, $g(x) = x^3$. Answer the following for each function:

    (a) At which points x is the derivative invertible?

    (b) Is the function $1:1$ on a neighborhood of 0?

    (c) Is the image an open subset of R^1?

    (d) What can you say about the differentiability of the inverse function (wherever it is defined)?

2. Let $f(x) = x^4 \sin(1/x)$, $x \neq 0$, and $f(0) = 0$.

    (a) Show that f is of class C^1 on the real line.

    (b) Show that the range of f is open but f is not $1:1$ on any neighborhood of 0.

3. Prove that, if f is a real-valued function of class C^1 on the real line and f is locally $1:1$, then the image $f(R)$ is open and f is $1:1$.

4. Let f be a function of class C^1 on a neighborhood of the origin in R^n and suppose that $f(X) = 0$ on the set $\{X; x_1 = 0\}$. Prove that $f(X) = x_1 g(X)$, where g is continuous on a neighborhood of the origin.

5. Under the hypotheses of the inverse mapping theorem, prove that if F is of class C^k then F^{-1} is of class C^k.

6. Let F be the map from R^2 into R^2:

$$F(x, y) = (x, x^2 + y^2).$$

    (a) What is the image of F?

    (b) What is the matrix $F'(x, y)$?

    (c) At which points (x, y) is F locally $1:1$?

7. Let F be the map from R^2 into R^2:

$$F(x, y) = (e^x \cos y, e^x \sin y).$$

    (a) Show that $(DF)(x, y)$ is invertible at every point (x, y).

    (b) Show that F is an open mapping, i.e., that $F(U)$ is open for every open set U.

    (c) Show that F is not a $1:1$ map.

    (d) What is the image $F(R^2)$? Is there an open set U which F maps $1:1$ onto $F(R^2)$?

8. Refer to Exercise 7 of Section 8.4. Use the inverse mapping theorem to prove that, if

$$U \xrightarrow{\ f\ } C$$

is a complex-analytic function, then f^{-1} (where defined) is a complex-analytic function.

9. Consider the real-valued function

$$f(x, y) = x^2 + y^2 - 5$$

on R^2. What does the implicit function theorem say in a neighborhood of $(2, 1)$? In a neighborhood of $(\sqrt{5}, 0)$?

10. Suppose everything is of class C^1 and the equations

$$f(x, y, z) = 0 \quad \text{and} \quad h(x, y, z) = 0$$

can be solved for y and z as functions of x. What is dy/dx?

11. Assume the implicit function theorem and use it to prove the inverse mapping theorem. (*Hint:* Consider $G(X, Y) = X - F(Y)$.)

8.6. *Change of Variable*

We now turn to "change of variable" theorems, sometimes called substitution theorems in calculus. The 1-dimensional theorem says this. Suppose h is an integrable function on the interval $[a, b]$ and we wish to compute

$$\int_a^b h(x) \, dx.$$

Suppose $x = g(t)$, in this sense: We are given a smooth function g such that g maps $[c, d]$ onto $[a, b]$ with $g' > 0$. Then

$$\int_a^b h(x) \, dx = \int_c^d h(g(t))g'(t) \, dt.$$

In higher dimensions the substitution will be given by a map G: $R^n \longrightarrow R^n$

$$x_1 = g_1(t_1, \ldots, t_n)$$
$$\vdots \qquad \vdots$$
$$x_n = g_n(t_1, \ldots, t_n).$$

To be precise, suppose we have

$$U \overset{G}{\longrightarrow} V$$

where

(i) U, V are open sets in R^n;
(ii) G is of class C^1;
(iii) G is $1 : 1$ from U onto V;
(iv) $\det G'(X) \neq 0$ throughout V.

What we shall prove is this. If $h \in L^1(V)$, then $(h \circ G) \in L^1(U)$ and

$$\int_V h(y) \, dy = \int_U h(G(X)) |\det G'(X)| \, dX.$$

In case h is the characteristic function of a measurable set E, the result will say

$$m(E) = \int_{G^{-1}(E)} |\det G'(X)| \, dX.$$

Or, one can read it this way. If A is a measurable subset of U, then

$$m(G(A)) = \int_A |\det G'(X)| \, dX.$$

Lemma 1. *If* E *is a box in* V, *then*

(8.57) $$m(E) = \int_{G^{-1}(E)} |\det G'(X)| \, dX.$$

Proof. Later.

Lemma 2. *Suppose Lemma 1 is valid for a particular map* G. *Then* G^{-1} *maps sets of measure* 0 *into sets of measure* 0.

Proof. Exercise.

Lemma 3. *Suppose Lemma 1 is valid for a particular map* G. *Then, for each* $h \in L^1(V)$, *the function* $(h \circ G) |\det G'|$ *is in* $L^1(U)$ *and*

(8.58) $$\int_V h(Y) \, dY = \int_U (h \circ G)(X) |\det G'(X)| \, dX.$$

Proof. From Lemma 1 it is clear that (8.58) holds for step functions—finite linear combinations of characteristic functions of boxes. Given $h \in L^1(V)$ we can find a sequence $\{h_k\}$ such that

(i) each h_n is a step function;

(ii) $\int_V |h - h_k| \longrightarrow 0$

(iii) $\lim_{k \to \infty} h_k(Y) = h(Y)$, a.e.

From (ii)

$$\lim_{j, k \to \infty} \int_V |h_j - h_k| = 0.$$

Now $|h_j - h_k|$ is also a step function. Apply (8.58) to those functions. We get

$$\lim_{j, k \to \infty} \int_U |h_j(G(X)) - h_2(G(X))| |\det G'(X)| \, dX = 0.$$

If we let

$$U_k(X) = h_k(G(X)) |\det G'(X)|$$

then $U_k \in L^1(U)$ and

$$\lim_{j, k \to \infty} \int_U |U_j(X) - U_k(X)| \, dX = 0.$$

Also

$$\lim_{k \to \infty} U_k(X) = h(G(X)) |\det G'(X)|, \qquad \text{a.e.}$$

because of Lemma 2. Thus, $(h \circ G) |\det G'|$ is in $L^1(U)$ and

$$\int_U h(G(X)) |\det G'(X)| \, dX = \lim_{k \to \infty} \int_U U_k(X) \, dX$$

$$= \lim_{k \to \infty} \int_V h_k(Y) \, dY$$

$$= \int_V h(Y) \, dY.$$

We are trying to prove this theorem.

Theorem 10. *Let* U, V *be open sets in* $\mathbf{R}^n$. *Suppose* G *is a map*

$$U \xrightarrow{\ \mathbf{G}\ } V$$

such that

   (i) G *is of class* C^1;
  (ii) G *is* $1:1$;
 (iii) $\det G'(X) \neq 0, X \in U$.

If $h \in L^1(V)$, *then* $(h \circ G) \cdot \det G'$ *is in* $L^1(U)$ *and*

$$\int_V h(Y) \, dY = \int_U h(G(X)) |\det G'(X)| \, dX.$$

We have noted several things about the proof.

(a) It is true when $n = 1$.

(b) For any given G, the theorem follows once we can show that

(8.59) $$m(E) = \int_{G^{-1}(E)} |\det G'(X)| \, dx$$

for every box $E \subset V$.

We also wish to observe that:

(c) It suffices to verify (8.59) for boxes E and maps G such that

$$\det G' > 0 \quad \text{on} \quad G^{-1}(E)$$

$$D_1 g_1 \quad \text{has constant sign on} \quad G^{-1}(E).$$

To see this, subdivide the open set V into boxes, using gridworks. Make sure that each box E is small enough that $\det G'$ has constant sign on $G^{-1}(E)$. One of the derivatives $D_i g_j$ must be different from zero at any point where $\det G' \neq 0$; hence we can choose the boxes E small enough that $\det G'$ has constant sign on $G^{-1}(E)$ and that some one of the partial derivatives $D_i g_j$ also has constant sign on that set. By reordering (some of the) coordinates, we may arrange that (c) is satisfied; hence it is the only case we need to consider.

Now, we proceed to prove the last assertion when $n > 1$, by induction. We have the theorem for $n = 1$. We shall now show that, if the theorem is true for $n - 1$, then Lemma 1 is true for n. So take a box E and a map G, as in (c).

We define a map F

$$G^{-1}(E) \overset{F}{\longrightarrow} R^n$$

by

$$F(x_1, \ldots, x_n) = (g_1(x_1, \ldots, x_n), x_2, \ldots, x_n).$$

(See Figure 33.) We shall apply the theorem in dimension $n - 1$,

once to the map $G \circ F^{-1}$

once to the map F^{-1}.

Of course, those are maps from R^n to R^n; however, each is independent of one of the variables.

Let's look at F^{-1}:

$$S \overset{F^{-1}}{\longrightarrow} G^{-1}(E).$$

Why is F a $1:1$ map, and how is F^{-1} defined? Let's see. If $z \in S$, then

$$F^{-1}(z_1, \ldots, z_n) = (x_1, z_2, \ldots, z_n)$$

where

$$z_1 = g_1(x_1, z_2, \ldots, z_n)$$

and that makes sense, because, if we fix $z_2, \ldots, z_n$, the condition "$D_1 g_1$ of constant sign" tells us that g_1 is $1:1$ on the segment $\{x; x_j = z_j, j = 2, \ldots, n\}$.

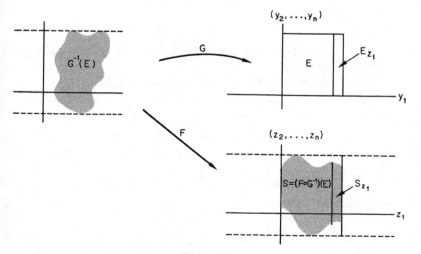

FIGURE 33

Now f is independent of $x_2, \ldots, x_n$; so the 1-dimensional theorem tells us something about F, or F^{-1}. It's F^{-1} that interests us. Suppose we fix $z_n, \ldots, z_n$. Look at the 1-dimensional map g:

$$g(z_1) = \text{the first coordinate of } f^1(z_1, \ldots, z_n)$$

$$= t, \qquad \text{where } g_1(t, z_2, \ldots, z_n) = z_1.$$

We know that

$$\int h(t)\, dt = \int h(g(t))\, |g'(t)|\, dt.$$

Now

$$(F^{-1})'(z) = \begin{bmatrix} D_1(F^{-1})_1 & D_2(F^{-1})_1 & \cdots & D_n(F^{-1})_1 \\ 0 & 1 & 0 & \cdots & 0 \\ 0 & 0 & 1 & \cdots & 0 \\ \cdot & \cdot & \cdot & & \cdot \\ \cdot & \cdot & \cdot & & \cdot \\ \cdot & \cdot & \cdot & & \cdot \\ 0 & 0 & 0 & \cdots & 1 \end{bmatrix}.$$

So that

$$\det (F^{-1})'(z) = g'(z_1).$$

Enough of that, for the moment. Let's look at the map $G \circ F^{-1}$

$$(G \circ F^{-1})(z_1, \ldots, z_n) = (y_1, \ldots, y_n)$$

where

$$y_1 = z_1$$
$$y_j = g_j(x_1, z_2, \ldots, z_n)$$

and x_1 is determined by

$$g_1(x_1, z_2, \ldots, z_n) = z_1.$$

So $G \circ F^{-1}$ leaves the first coordinate fixed. That is the point of defining F, namely, that $G \circ F^{-1}$ is like a map on R^{n-1}.

Specifically, suppose we fix z_1. The set

$$S_{z_1} = \{z \in S;\, \pi_1(z) = z_1\}$$

is mapped by $G \circ F^{-1}$ onto the set

$$E_{z_1} = \{y \in E;\, y_1 = z_1\}.$$

That means that, by the $(n - 1)$-dimensional theorem, we can compute

$$m(E_{z_1}) = \int_{S_{z_1}} |\det H'_{z_1}(z_2, \ldots, z_n)|\, dz_2 \cdots dz_n$$

where H_{z_1} is the map

$$S_{z_1} \overset{H_{z_1}}{\longrightarrow} E_{z_1}$$

$$H_{z_1}(z_2, \ldots, z_n) = H(z_1, \ldots, z_n) = (G \circ G^{-1})(z_1, \ldots, z_n)$$

that is, $H = G \circ F^{-1}$. But, how do we compute H'_{z_1}? That's easy.

$$H = (h_1, \ldots, h_n)$$

$$h_1(z_1, \ldots, z_n) = z_1$$

$$H'(Z) = \begin{bmatrix} 1 & 0 & \cdots & 0 \\ 0 & D_2 h_2 & \cdots & D_2 h_2 \\ \cdot & \cdot & & \cdot \\ \cdot & \cdot & & \cdot \\ \cdot & \cdot & & \cdot \\ 0 & D_2 h_n & \cdots & D_n h_n \end{bmatrix}$$

$$H'(Z) = \begin{bmatrix} 1 & 0 & \cdots & 0 \\ 0 & & & \\ \cdot & & \\ \cdot & & H'_{z_1}(z_2, \ldots, z_n) \\ \cdot & & \\ 0 & & \end{bmatrix}.$$

So det $H'_z(z_2, \ldots, z_n) = \det H'(Z)$. Thus,

$$m(E_{z_1}) = \int_{S_{z_1}} \det H'(z) \, dz_2 \cdots dz_n.$$

Now

$$m(E) = \int m(E_{z_1}) \, dz_1$$

$$= \int dz_1 \int_{S_{z_1}} \det H'(z) \, dz_2 \cdots dz_n$$

$$= \int_S \det H' \, dz_1 \cdots dz_n.$$

Since

$$H = G \circ F^{-1}$$

$$H'(Z) = G'(F^{-1}(z))(F^{-1})'(Z).$$

Thus,

$$m(E) = \int_S \det G'(F^{-1}(z)) \det (F^{-1})'(z) \, dz.$$

Apply the 1-dimensional theorem to F^{-1} in a manner analogous to the application just made to $G \circ F^{-1}$. We obtain

$$m(E) = \int_{G^{-1}(E)} \det G'(X) \, dX.$$

Corollary. *If* L *is a linear transformation from* R^n *into* R^n *then, for every measurable set* S *in* R^n, *the image set* L(S) *is measurable and its measure is* m(L(S)) = |det L| m(S).

Proof. If L is non-singular, apply the theorem with $G = L$. Then det G' has the constant value det L. If L is singular, the image $L(R^n)$ is a proper subspace of R^n and hence has measure zero. So the result is trivially true.

The corollary is much more elementary than the change of variables theorem. It can be proved directly in a variety of ways. Some proofs of Theorem 10 proceed by verifying the corollary first and then using it to (help) prove the more general result.

EXAMPLE 7. An elementary example with which the reader is probably familiar from calculus deals with polar coordinates. Let

$$U = \{(r, \theta); r > 0, 0 < \theta < 2\pi\}.$$
$$G(r, \theta) = (r \cos \theta, r \sin \theta).$$

Then G is $1 : 1$ and maps U onto

$$V = R^2 - \{(x, 0); x \geq 0\}.$$

We have

$$G'(r, \theta) = \begin{bmatrix} \cos \theta & -r \sin \theta \\ \sin \theta & r \cos \theta \end{bmatrix}$$
$$\det G'(r, \theta) = r.$$

Since $R^2 - V$ is a set of measure zero in the plane, $L^1(V) = L^1(R^2)$. Theorem 10 states that if f is an integrable function on R^2, then $f(r \cos \theta, r \sin \theta)$ is integrable over U and

$$\int_{R^2} f = \int_U f(r \cos \theta, r \sin \theta) r \, dr \, d\theta$$
$$= \int_0^{2\pi} \int_0^\infty f(r \cos \theta, r \sin \theta) r \, dr \, d\theta.$$

(a) Let

$$f(x, y) = \frac{1}{x + iy}, \qquad x^2 + y^2 \neq 0.$$

Then f is a measurable function on R^2, and since

$$| f(r \cos \theta, r \sin \theta)| = \frac{1}{r}$$

we see that f is integrable on each bounded subset of the plane:

$$\int_{\{|z| \leq r_0\}} |f| = \int_0^{2\pi} \int_0^{r_0} dr \, d\theta$$
$$= 2\pi r_0 < \infty.$$

(b) The reader should be (or become) familiar with the calculation of

$$\int_{-\infty}^\infty e^{-x^2} \, dx.$$

Obviously $f(x) = e^{-x^2}$ is integrable on R^1 because $e^{-x^2} < e^{-|x|}$ for large x. Note that (since $f \geq 0$)

$$\left(\int f\right)^2 = \int_{-\infty}^{\infty} e^{-x^2}\, dx \int_{-\infty}^{\infty} e^{-y^2}\, dy$$

$$= \int_{R^2} e^{-(x^2+y^2)}\, dx\, dy$$

$$= \int_0^{2\pi} \int_0^{\infty} e^{-r^2} r\, dr\, d\theta$$

$$= \int_0^{2\pi} \left\{ \int_0^{\infty} re^{-r^2}\, dr \right\} d\theta$$

$$= \int_0^{2\pi} \left\{ -\tfrac{1}{2} e^{-r^2} \Big|_0^{\infty} \right\} d\theta$$

$$= \tfrac{1}{2} \int_0^{2\pi} d\theta$$

$$= \pi.$$

Thus $\int f = \sqrt{\pi}$.

Exercises

1. Evaluate

$$\int_0^1 \int_0^1 \frac{1}{x+y}\, dx\, dy$$

by a suitable change of variables. Justify each step, using the Fubini theorem or the change of variable theorem.

2. Prove Lemma 2 of this section.

3. From spherical coordinates (r, ϕ, θ) in R^3 we obtain a formula

$$m(E) = \int_E f(r, \phi, \theta)\, dr\, d\phi\, d\theta$$

for the measure (volume) of a set E. Find $f(r, \phi, \theta)$ explicitly.

4. What is the measure of the ball

$$B(0; r) = \{X \in R^4; |X| \le r\}?$$

5. Find

$$\int_0^1 \int_0^1 \int_0^1 \frac{1}{x^2 + y^2 + z^2}\, dx\, dy\, dz.$$

***6.** Let G be a map of class C^1, from an open set $U \subset R^n$ into R^n, and let $S = \{X \in U; \det G'(X) = 0\}$. Prove that the (image) set $G(S)$ is a set of measure zero.

Appendix.
Elementary
Set Theory

A.1. Sets and Functions

We describe some basic terminology. We shall talk frequently about sets. We make no attempt to define what a set is, except to say that a set is any collection of objects. A set will be called a collection, class, or family on various occasions. The objects in the set S are called the **members, elements,** or **points** of S. If x is a member of S, we write $x \in S$. Most often, a set is specified by giving a rule for membership in the set. A convenient piece of shorthand for presenting such a rule is the notation $\{x; \ldots\}$, to be read "the set of all x such that. . . ."

The concept of set is simple and basic. Experience indicates that large numbers of people comprehend readily what the abstract idea is. But, indiscriminate use of set terminology does have some problems and pitfalls which are not immediately apparent. We shall try to avoid such difficulties by not venturing into the mathematical contexts in which they arise. If the reader meets a set and feels that we have not given a well-defined rule by which we can determine for each x whether or not x is a member of the set, then he or she should feel free to inquire or complain about it.

The set S is a **subset** of the set T if each member of S is a member of T. If S is a subset of T, we write $S \subset T$. If $S \subset T$ and $S \neq T$, then S is called a **proper subset** of T. Note that every set is a subset of itself. The **empty set** is the set with no members. It will be denoted by $\varnothing$. It is a subset of every set.

The **union** of the sets S and T is the set $S \cup T$, which consists of the objects which are either in S or in T:

$$S \cup T = \{x; x \in S \text{ or } x \in T\}.$$

The **intersection** of S and T is the set $S \cap T$, which consists of the objects which are both in S and in T:

$$S \cap T = \{x; x \in S \text{ and } x \in T\}.$$

We always have

$$S \cap T \subset S \cup T$$

i.e., "or" does not preclude the possibility of "and". The sets S, T are called **disjoint** (from one another) if their intersection is the empty set:

$$S \cap T = \varnothing.$$

The **complement of T relative to S** is

$$S - T = \{x \in S; x \notin T\}.$$

Frequently, one works in a context in which all sets of interest are subsets of one "universal" set S. In such a case, the complement of T relative to S is referred to as (simply) the **complement** of T.

The **Cartesian product** of S and T is the set $S \times T$, which consists of all (ordered) pairs (s, t) with s in S and t in T:

$$S \times T = \{(s, t); s \in S, t \in T\}.$$

Probably the most important concept in mathematics is that of "function". A **function** consists of

  (i) a pair of sets D, Y;

  (ii) a rule f, which associates with each object x in D an object $f(x)$ in Y.

We usually describe the above in a slightly imprecise way by saying that f is a **function from D into Y** and by writing

$$D \xrightarrow{\ f\ } Y.$$

We call D the **domain (of definition)** of f and Y the **range** of f. A function is also called a transformation, map, mapping, or operator.

What we have given is not really a definition in the strict sense, because we used the word "rule" in the so-called definition; and what is a rule? So, we try to further clarify the concept of function by discussing graphs of functions. The **graph** of f is a subset of the product set $D \times Y$, namely the subset

$$G = \{(x, f(x)); x \in D\}.$$

The Cartesian product $D \times Y$ (usually) has very many subsets, very few of which are graphs of functions from D into Y. If G is a subset of $D \times Y$, when is G the graph of a function from D into Y? *Answer*: If and only if, for each x in D, there is precisely one y in Y such that $(x, y) \in G$. Some people prefer to give this subset characterization as the definition of a function.

Let f be a function

$$D \xrightarrow{f} Y.$$

If T is a subset of Y, the **inverse image of T under f** is

$$f^{-1}(T) = \{x \in D; f(x) \in T\}.$$

A **level set** of f is the inverse image of a point $y \in Y$:

$$\{x; f(x) = y\}.$$

(Actually, we only call this a level set if it is non-empty.) A function is **constant** if it has precisely one level set.

If A is a subset of D, the set of values which are assumed by f on A is called the **image of A under f**:

$$f(A) = \{y \in Y; y = f(x) \quad \text{for some} \quad x \in A\}.$$

The image set $f(D)$, i.e., the set of all values assumed by f, is called the **image** of f. If the image of f is (all of) Y, we say that f is a function from D **onto** Y (or sometimes, "f is onto").

The function f is called **1:1** (read "one-to-one") if

$$f(x_1) \neq f(x_2), \qquad x_1 \neq x_2.$$

In other words, f is 1:1 provided $f(x_1) = f(x_2)$ implies $x_1 = x_2$. Suppose that f is 1:1 and onto. There is then defined an **inverse function**

$$D \xleftarrow{f^{-1}} Y$$

as follows: $f^{-1}(y)$ is the (unique) point $x \in D$ such that $y = f(x)$. A 1:1 function from D onto Y is also called a **1:1 correspondence** between the members of D and the members of Y.

If

$$D \xrightarrow{f} Y \xrightarrow{g} Z$$

there is defined the **composition**

$$D \xrightarrow{g \circ f} Z$$

by

$$(g \circ f)(x) = g(f(x)).$$

For example, if f is 1:1 and onto we have

$$D \xrightarrow{f} Y \xrightarrow{f^{-1}} D$$

and the composition $f^{-1} \circ f$ is the **identity function I** on D:

$$I(x) = x.$$

Similarly, $f \circ f^{-1}$ is the identity function on Y.

If Y is a set, a **sequence** (of elements) in Y is a function from the set of positive integers

$$Z_+ = \{1, 2, 3, \ldots\}$$

into the set Y. Usually a sequence in Y will be indicated as $\{y_n\}$, meaning the function

$$Z_+ \xrightarrow{\ f\ } Y$$

for which $f(n) = y_n$.

We shall sometimes deal with sequences of sets. If $\{S_n\}$ is a sequence of sets, we have the union and intersection

$$\bigcup_n S_n = \{x; x \in S_n \text{ for some } n\}$$

$$\bigcap_n S_n = \{x; x \in S_n \text{ for every } n\}.$$

We will also deal with families of sets which are indexed by sets other than the positive integers. If we talk about the family of sets $\{S_\alpha; \alpha \in A\}$, that means that associated with each $\alpha \in A$ we have a set S_α. The notations

$$\bigcup_\alpha S_\alpha, \qquad \bigcap_\alpha S_\alpha$$

should have meanings which are clear.

At one point, we shall make use of the **axiom of choice** which says the following. If $\{S_\alpha; \alpha \in A\}$ is a collection of non-empty non-overlapping sets

$$S_\alpha \neq \varnothing$$

$$S_\alpha \cap S_\beta = \varnothing, \qquad \alpha \neq \beta$$

then there exists a set which consists of exactly one element from each S_α. (Loosely, there is a way to choose one element from each set.) This is called an axiom (assumption), for the following reason. What is asserted is that, given $\{S_\alpha; \alpha \in A\}$, there is a rule (function) which selects an element $x_\alpha \in S_\alpha$. But, what exactly is the rule? Since we can't say, we just assume that there is one.

A.2. Cardinality

The question of how many members a set has becomes much more complicated for infinite sets than it is for finite sets. One tends to think that the set of positive integers $\{1, 2, 3, \ldots\}$ has more members in it than does the set of even positive integers $\{2, 4, 6, \ldots\}$. On the other hand, if we list one set under the other thusly:

$$1 \quad 2 \quad 3 \quad 4 \quad \cdots$$
$$2 \quad 4 \quad 6 \quad 8 \quad \cdots$$

it is difficult to convince anyone that the first set has more members than the second. The lesson to be learned from this is the following: If we deal with infinite sets and wish to discuss whether or not one set has more members than another, at least one of our intuitive ideas about "more than" will have to fall by the wayside. It turns out that the thing to scrap is the intuitive idea which suggests that $\{1, 2, 3, \ldots\}$ has more members than does $\{2, 4, 6, \ldots\}$, namely, the idea that, if S is a proper subset of T, then T has more members. In particular, face the fact that there are just as many even integers as there are integers.

We are not going to give anything resembling a thorough discussion of the circle of ideas involved with the number of elements in a set. But we need to know just a bit about the question of when two sets have the same number of elements. We say that S and T **have the same cardinality** if there exists a 1 : 1 correspondence between the members of S and the members of T. Thus S and T have the same cardinality if there exists a function

$$S \xrightarrow{f} T$$

which is 1 : 1 with $f(S) = T$.

The most basic fact about cardinality is this. Suppose that S and T are sets such that there exists a 1 : 1 function from S into T, and also there exists a 1 : 1 function from T into S; then S and T have the same cardinality. Intuitively, if T has at least as many elements as does S and vice versa, then they have the same number of elements. We shall not stop to prove this fact. (It's a non-trivial mental exercise.) We shall content ourselves with some special cases.

A set S is **finite** if (it is empty or) there exists a positive integer n such that S has the same cardinality as $\{1, 2, 3, \ldots, n\}$. A set is **infinite** if it is not finite. Every infinite set exhibits the phenomenon we saw with the integers and even integers, that is, every infinite set can be put in 1 : 1 correspondence with a proper subset of itself. The set S is **countable** if it is finite or has the same cardinality as the set of positive integers. The term countable stems from the following.

Lemma. *The non-empty set* S *is countable if and only if there exists a function from the set of positive integers onto* S, *i.e., if and only if there is a sequence* $\{x_n\}$ *such that*

(i) *for each* n, $x_n \in S$;
(ii) *if* $x \in S$, *then* $x = x_n$ *for some* n.

Proof. If S is countable but not finite, we have a sequence $\{x_n\}$ such that, if $x \in S$, then $x = x_n$ for precisely one n. If S is finite then $S = \{x_1, x_2, \ldots, x_n\}$. Define $x_k = x_1$ for $k > n$ and $\{x_k\}$ is the sequence we want.

So, a non-empty countable set is one such that there is a listing $x_1, x_2, x_3, \ldots$ of its elements; i.e., a nonempty countable set is the image of a sequence.

Theorem A.1. *Every subset of a countable set is countable.*

Proof. Suppose S is countable and T is a subset of S. Suppose T is non-empty. Then S is non-empty, and we have a sequence $\{x_n\}$ with S as its image:

$$S = \{x_n; n \in Z_+\}.$$

Let n_1 be the smallest positive integer k such that $x_k \in T$. Let n_2 be the smallest positive integer k such that $k > n_1$ and $x_k \in T$. Continue by induction, to obtain the sequence of positive integers $n_1 < n_2 < n_3 < \cdots$ such that n_j is the least positive integer k such that $k > n_{j-1}$ and $x_k \in T$. Let

$$t_j = x_{n_j}$$

and it is obvious that T is the image of the sequence $\{t_n\}$.

Corollary. *The set* S *is countable if and only if* S *has the same cardinality as some subset of the positive integers.*

Theorem A.2. *Suppose that* $\{S_n\}$ *is a sequence of countable sets. Then the union*

$$\bigcup_n S_n$$

is a countable set.

Proof. Obviously we may throw out any S_n's which are empty. Thus we have sequences which list the members of the sets S_n:

$$S_n = \{x_{n_k}; k \in Z_+\}.$$

We then list the elements of the union by the following scheme:

Corollary. *The set of all rational numbers is countable.*

Proof. A rational number has the form m/n, where m is an integer and n is a positive integer. For each n, let S_n be the set of all numbers m/n, where n is an integer. The set of integers is countable by Theorem A.2, or by the scheme

$$1 \quad 2 \quad\quad 3 \quad 4 \quad\quad 5 \quad \ldots$$
$$0 \quad 1 \quad -1 \quad 2 \quad -2 \quad \ldots$$

So the union of the sets S_n is countable.

The reader should be very familiar (henceforth) with the essence of the last corollary, which is the fact that the set of positive rational numbers is countable via the scheme:

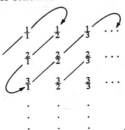

Theorem A.3. *The set of real numbers* (*points on a line*) *is uncountable* (*not countable*).

Proof. It will suffice to show that the set of real numbers x, $0 \le x \le 1$, is not countable. Each x in that interval has a decimal representation

$$x = .a_1 a_2 a_3 \ldots, \qquad 0 \le a_k \le 9.$$

The only ambiguity in the representation is caused by an infinite string of repeating 9's. The decimal

$$.a_1 a_2 \ldots a_n 999 \ldots$$

represents the same number as does

$$.a_1 a_2 \ldots (a_n + 1)000 \ldots.$$

That is a minor technicality.

Suppose that we have a sequence of points x_n in the interval $0 \le x \le 1$. We shall show that the image of the sequence $\{x_n\}$ cannot exhaust the interval. (Hence, the interval is not a countable set.) Given the sequence $\{x_n\}$, look at the decimal representations:

$$x_1 = .a_{11} a_{12} a_{13} \ldots$$
$$x_2 = .a_{21} a_{22} a_{23} \ldots$$
$$x_3 = .a_{31} a_{32} a_{33} \ldots, \text{ etc.}$$

We shall exhibit a number x, $0 \le x \le 1$, which is not in the list x_1, x_2, x_3, ... We describe x by its decimal representation. For each n, choose an integer a_n, $0 \leq a_n \leq 9$, such that $a_n \neq a_{nn}$. Let

$$x = .a_1 a_2 a_3. \ldots$$

Then $x \neq x_n$, because the nth digit of x is different from the nth digit of x_n. Of course, there is a little confusion caused by the fact that different digits do not imply different numbers (repeating 9's). Therefore, we should choose the digits a_n so that $0 \leq a_n \leq 9$ *and* $|a_n - a_{nn}| \geq 2$. Then $x \neq x_n$ for every n, for certain. We have produced an x, $0 \leq x \leq 1$, which is not in the list $x_1, x_2, x_3, \ldots$. *Conclusion*: No such list can exhaust the points in that interval.

Thus, although the set of rational numbers has the same cardinality as the set of positive integers, the set of real numbers does not (i.e., it has a larger cardinality). One may reasonably ask: Does there exist a subset S of the set of real numbers which is uncountable yet does not have the same cardinality as the set of all real numbers? That turns out to be a very sophisticated question which plagued mathematical logicians for many years. It was settled completely only a few years ago. The answer in imprecise form is: You will not prove that such a set exists nor will you prove that no such set exists—unless mathematics as we know it is logically inconsistent.

List of Symbols

Bibliography

Here are a few basic references, roughly divided into three groups of increasing levels of difficulty.

Group I

COURANT, R., *Differential and Integral Calculus*, Vols. I, II. New York: Interscience, 1937.

SEELEY, R., *An Introduction to Fourier Series and Integrals*. New York: Benjamin, 1966.

TOEPLITZ, O., *The Calculus, a Genetic Approach*. Chicago: University of Chicago Press, 1963.

Group II

BARTLE, R. G., *The Elements of Real Analysis*. New York: Wiley, 1964.

BUCK, R. C., *Advanced Calculus*. New York: McGraw-Hill, 1956.

KAPLAN, W., *Advanced Calculus*. Reading, Mass.: Addison-Wesley, 1953.

LANDAU, E. G. H., *Foundations of Analysis*. New York: Chelsea, 1951.

RUDIN, W., *Principles of Mathematical Analysis*. New York: McGraw-Hill, 1964.

Group III

COURANT, R., and D. HILBERT, *Methods of Mathematical Physics*, Vol. 1. New York: Interscience, 1953.

DYM, H., and H. P. McKEAN, *Fourier Series and Integrals*. New York: Academic Press, 1972.

RUDIN, W., *Real and Complex Analysis*. New York: McGraw-Hill, 1966.

TITCHMARSH, E. C., *The Theory of Functions*, 2nd ed. New York: Oxford University Press, 1939.

Index